A SEMIPROGRAMMED APPROACH

INTRODUCTION TO STATISTICAL ANALYSIS

A SEMIPROGRAMMED APPROACH

CELESTE McCOLLOUGH

McGRAW-HILL BOOK COMPANY

New York St. Louis San Francisco Düsseldorf Johannesburg Kuala Lumpur London Mexico Montreal New Delhi Panama
Paris São Paulo Singapore Sydney Tokyo Toronto

Library of Congress Cataloging in Publication Data

McCollough, Celeste.
Introduction to statistical analysis.

Published in 1963 under title: Statistical concepts: a program for self-instruction.
Includes bibliographical references.
1. Statistics. 2. Statistics—Problems, exercises, etc. I. Title. 519 73-9508
HA29.M168 1974
ISBN 0-07-044805-1
ISBN 0-07-044804-3 (pbk.)

**INTRODUCTION TO STATISTICAL ANALYSIS:
A SEMIPROGRAMMED APPROACH**

123456789 MAMM 9876543

This book was set in Helvetica and Caledonia by Progressive Typographers. The editors were Robert P. Rainier and James R. Belser; the designer was Anne Canevari Green; and the production supervisor was Joe Campanella. The drawings were done by Vantage Art, Inc. The Maple Press Company was printer and binder.

CONTENTS

PREFACE

This book developed through my coauthorship of *Statistical Concepts: A Program for Self-Instruction* (McGraw-Hill, 1963). Although some examples and illustrations have been adapted from that book, the text is entirely new in content as well as format.

Field tests have encouraged my belief that the semiprogrammed style, by its alternation of programmed material and straight text, provides more instruction with less monotony than a fully programmed style. A double-page layout has therefore been adopted as a convenient method of combining textual and programmed materials gracefully. The reader can use the right-hand page of programmed frames to confirm his understanding of materials just presented on the left-hand page. This pattern achieves greater economy of presentation without sacrificing the student's opportunity for active response to questions.

Greater economy permits a more complete coverage of elementary statistics than is possible in a fully programmed book of the same length. Users of *Statistical Concepts* will find some topics here which had to be omitted from the earlier book, in particular, one-way analysis of variance, product-moment correlation, and the most popular distribution-free tests. This book also contains enough detail on computational methods for a student to apply basic statistical tests without consulting additional sources.

In writing this book I have continued to seek ways of presenting statistical concepts without presuming more preparation than elementary algebra. Every student, however scant his mathematical background, is surely entitled to understand the ideas which shape modern statistical reasoning. It is not quite correct to say that the book presents these concepts nonmathematically; the ideas *are* mathematical, and they should be confidently understood as such. Rather, I have tried to make certain mathematical ideas clear enough to become part of a student's conceptual framework even when he usually thinks of himself as "not mathematically inclined."

This volume will fulfill the same needs served by the earlier book—as introduction, aid, or review to accompany a beginning course. With the additions mentioned above, it will also be able to serve as a primary textbook in introductory statistics courses whenever these courses are intended for students not majoring in mathematics. Accordingly, the student who selects this book for self-instruction will be introduced to all the topics normally covered in such beginning courses. He will even get some practice in computation, with emphasis on application through understanding rather than on a "cookbook" use of methods.

I am grateful to Professor Loche Van Atta, coauthor of *Statistical Concepts*, for supplying new examples for use in the correlation lessons. He also contributed illustrations to the earlier book which have been included here with his permission, in particular, those on frequency distributions, measures of central tendency and dispersion, z-score distributions, and sampling distributions of means. Lesson 19 on linear functions follows closely a lesson which he prepared for the earlier book. I also wish to thank him for arranging the field testing of this book.

My thanks are due to Professor Norman D. Henderson, of Oberlin College, who gave advice on the selection of topics to be included; to Professor William Tedford, of Southern Methodist University, who pointed out the advantages of treating the t test as a special case of analysis of variance; and to Professor C. M. Dayton, of the University of Maryland, whose field testing of the trial version in a summer-session course helped in adjusting the difficulty level of these lessons. I am grateful also to Mrs. Marian McConnell and Mrs. Eleanor Williamson for their patience and efficiency in preparing the trial-version copies, handling mailing problems, and tabulating the results.

Celeste McCollough

TO THE STUDENT

Each lesson in this book contains three parts: an introductory page; four or five separate sections, each with one or two double-page layouts; and a two-page review. The double-page layouts present new information as straight text on the left page, together with numbered questions and answers on the right page. Follow three rules to obtain the best learning and greatest enjoyment from this book:

1. Study the text on the left page carefully before you begin to work on the right page. Mark whatever seems most important in your own way.

2. As you work on the right page, conceal the answers to each numbered statement (called a FRAME) until you have written down your own answers. Refer to the left page whenever necessary.

3. Before beginning the Review, look back over the lesson sections. Then try to complete the review frames without referring to the text again.

What is meant by a semiprogrammed text A textbook is written to provide the reader with a good pathway from start to finish in his study of a new subject. A *programmed* text is simply more explicit about the steps in the pathway. Each step is a numbered frame, and great care is taken to make sure that each step follows logically from the steps that have gone before. Some part of each frame is purposely left incomplete so that the reader who is following the pathway closely can supply the missing words from the knowledge he has recently acquired.

A *semiprogrammed* text contains straight text material alternating with programmed frames. This alternation gives a change of pace; the basic new information can be studied in an integrated form, then examined and applied actively in the numbered frames which follow.

This semiprogrammed text has been constructed so that each lesson is a nested set of steps. There are four or five main steps (the titled sections), each containing a few subsidiary points. You may find it easy to achieve the aim of complete mastery: to be able to "think" the lesson in its entirety at once. After study, the four or five sections can be brought to mind as a group, and each section will carry with it the memory of its subsidiary points.

Why and how to use the mask Rule 2 says to conceal the answers. But since the answers are printed just to the right of each frame, why not read the answers along with the frames? Why bother to write them down for yourself?

Writing down your own answers is the only way to find out *at once* whether you are getting the essential points or not. How many times have you read placidly for an hour, only to find out later that you have missed something essential without realizing it? If you look at the answers too soon, the same sort of thing can happen. You may give yourself the *illusion* of having understood, as you have sometimes done in reading other textbooks, only to find out later that you have "got lost" because something was not clear to you.

To help you conceal the answers conveniently, you will need a mask properly shaped to cover the answers for a frame until you have written down your own answers. You can make a mask by cutting heavy paper or light cardboard to the following shape and dimensions:

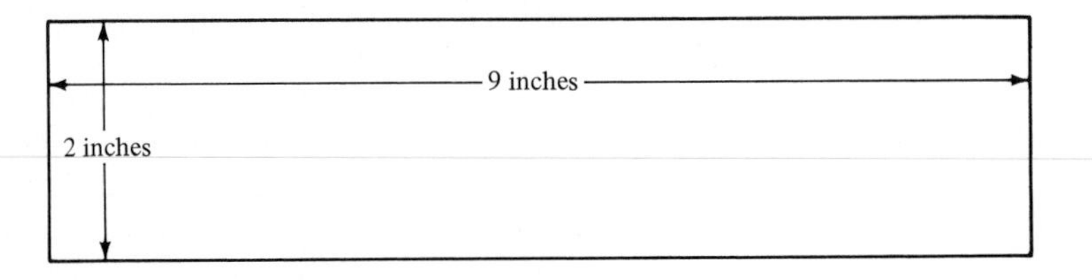

Place the mask lengthwise over the answer column at the right-hand side of the page. Slide the mask down the column frame by frame as you complete the items. Always write down your best idea of the missing answers for the *entire* frame before you slide the mask down to look at the printed answers.

If you follow this procedure carefully, you will find that the material you are studying continues to make sense to you all the time. You will not necessarily recall all the logical steps you have taken. But when you have completed a series of sections, a brief reexamination of those sections will bring to mind all the main points you have worked on. The review will then help you make sure those main points are still clear and ready for use in the later lessons.

How long should a study period be? A single section in one of these lessons can be completed in 10 or 15 minutes. This is therefore a text which can be picked up at odd moments—before mealtimes, between classes, or while you are waiting for someone. Plan at first to study only one or two sections at a time. Each time you return to the lesson after an interval, look at the frames at the end of the last section you studied. These frames will remind you of what you had just learned, and you can then pick up where you left off. You will be able to complete an entire lesson in about three short study periods.

The lessons were written with the idea that each lesson might be completed by an "average" student in about 1 hour. Of course, an average student is a purely statistical idea, and no individual is truly described by a statistical average. But if you find that a lesson holds your interest for longer periods of time, you may wish to complete a whole lesson at a single sitting.

Symbols used within frames A blank line within a frame indicates that a word, number, or symbol is to be supplied. A long blank indicates that a *word* is required, rather than a number or symbol.

Example Each section of a lesson has one or two double-page layouts. The left page presents new information, and the ________ page contains programmed items called frames. *right*

If the blank requires two or more words to be supplied, the long blank will be a double line. A double line is also used for compound words spelled with a hyphen.

Example To use the programmed material most effectively, a student always conceals the printed answers until he has ________. *written his own answers*

A short blank indicates that a number or symbol is required. A single line is used even when a short sequence of numbers (such as an equation) is to be supplied. Whenever you are expected to provide a *symbol* or a combination of symbols and numbers rather than a number, the short blank will be a double line.

Example 5 + ____ = 10; 5 × ____ = 10. The sign + indicates that two numbers are to be added together; the sign ____ indicates that two numbers are to be multiplied. *5; 2* ×

As these examples have illustrated, the answer to a blank appears on the same line of print with the blank to which it belongs. When two blanks appear on the same line, the two answers are separated by a semicolon. Occasionally, the answers which belong to two successive blanks may be interchangeable, i.e., correct in either order. In such cases, the semicolon is replaced by a double-headed arrow, and such an arrow is included even when the two blanks do not fall in the same line of print.

Example A long blank calls for a ________ to be supplied; a short blank calls for either a ________ or a ________. *word* *number ↔ symbol*

In some frames you are expected to choose one of two or more words which are given inside parentheses (in alphabetical order). These multiple choices are distinguished also by italics, and you may simply cross out the incorrect answer or answers.

Example When you are working on a double-page section, you should study the (*programmed; textual*) material first. *textual*

Other aids to study At the end of each lesson you will find a page of Problems which require the application of concepts or techniques taught in that lesson. Solving these problems will give you additional practice in statistical methods. Answers to all problems are given in a separate section starting on page 361.

You may wish to follow up some topics in greater detail, particularly if you are interested in the mathematical development of basic points. References are suggested on page 369 to guide you in further study.

A topical index is also provided at the end of this book. It can be used when you want to locate a concept or symbol you have previously studied. The index can be an aid in reviewing for an examination, and it will also enable you to use this book as a convenient reference when, at some later time, you need to review a concept or method which you do not frequently use.

Celeste McCollough

PART ONE
Basic Concepts

LESSON 1 INTRODUCTION TO STATISTICAL INFERENCE

Most decisions about practical matters have to be taken on the basis of incomplete evidence. It is rarely possible to know everything one would like to know before making up one's mind. The ability to draw conclusions which will continue to appear correct as further evidence accumulates may indeed be the key quality which distinguishes successful business and political administrators.

In scientific investigation, also, it is necessary to draw conclusions from a limited amount of evidence. The collection of evidence is a costly and time-consuming procedure. An investigator needs to know when he has obtained sufficient evidence to permit him to draw a conclusion in which he can have a certain degree of confidence.

The special branch of mathematics called STATISTICS provides a logical analysis of this problem of making decisions on the basis of incomplete evidence. From the study of statistics one learns how to examine the available evidence for the conclusions which can be drawn from it. One learns, further, how to estimate the degree of confidence one may have in these conclusions.

The bits of evidence available for making a decision are the DATA for statistical analysis. If they are to provide the maximum information, these data must be both collected and analyzed according to certain clear-cut rules. From the study of statistics one should learn how to collect data so that the subsequent analysis can lead to meaningful decisions.

The plan which guides data collection is called the DESIGN of the investigation. Since this design determines the kind of STATISTICAL ANALYSIS which can be applied to these data, the investigator must already know what kind of analysis he will use before he even begins to collect data. The main job of statistical analysis, of course, is to produce the DECISION which prompted the investigation. However, we shall consider that the job of statistical analysis is never complete without an estimation of the DEGREE OF CONFIDENCE one is justified in having about that decision. Your study of statistics should enable you to design a procedure for data collection that will enable you to make a decision on the basis of your data and to know the degree of confidence you can place in that decision. The methods you learn will apply equally well, whether the particular decisions you may need to make lie in education, research, engineering, business, or public administration.

The first three lessons in this book deal with ideas which are fundamental to statistical decision making. Lesson 1 introduces some of these ideas in connection with the question, "Why are statistical methods necessary?"

A. SAMPLES AND POPULATIONS

If you wished to make an investigation of the intelligence level of college students in the United States, you might undertake to obtain the IQ scores of every individual enrolled in a college in the nation. This set of scores would be a complete, or EXHAUSTIVE, set of observations relevant to your investigation. With such a complete set, you could make certain statements about United States college-student IQs with perfect confidence. For example, you could make an exact statement about the average of all United States college-student IQs in that particular year (or moment) of investigation.

Usually it is not convenient or even possible to collect more than a small fraction of the relevant observations. The entire exhaustive set of relevant observations is called a POPULATION; any part, or SUBSET, of these observations is called a SAMPLE. Each observation actually collected is a DATUM (singular), and it is the sample made up of these DATA (plural) which serves as evidence for decision making. As soon as you have to deal with a sample instead of with the entire population, you must begin to use the methods of statistics.

In ordinary usage, as in the phrase "population of the United States," the word "population" refers to a set of elements whose individual members are persons. In technical usage, however, a population is not a set of persons but a set of OBSERVATIONS. In some cases these observations refer to characteristics of persons, such as heights, incomes, number of offspring, and the like, but even in such cases the elements of the population being considered in statistics are not the actual persons themselves. The elements are the observations about some particular characteristic of these persons.

Any set of observations about which we would like to draw conclusions is a population. If we want to draw conclusions about the IQs of all United States college students, the population is the set of IQ scores of all United States college students. The IQ score of any individual college student is an element of this population. Notice that we can speak of such a population even though not all students have taken the particular IQ test we have in mind. The concept of a population of observations applies to all the observations of a particular sort which *might* be obtained.

The population to which a sample belongs is called its PARENT POPULATION. A limited number of college-student IQs would be a sample from the parent population of college-student IQs.

We shall distinguish between relatively small populations, called FINITE populations, and extremely large, or INFINITE, populations. A population is said to be finite when it is possible to *enumerate* all its elements in such a way that there is both a first element and a last element. Populations which cannot be so enumerated are said to be infinite. In practice any extremely large population is treated as an infinite population.

Now consider what happens when a given individual's intelligence is "measured." The total population of observations relevant to this measurement, i.e., the parent population, would include every imaginable behavior of this person which indicates his intelligence level. The data are a set of responses to questions on a test, and such data are only a sample from the total population. In this case, since it is impossible to enumerate all the elements of the population set, the population is infinite.

1-1 The term "population statistics" is used in government and economics to mean the description of a population of persons living in a particular geographical area. The elements of such a population are ________. *persons*

1-2 In statistics, as a branch of applied mathematics, a population is a set of observations. Even when these observations are characteristics of persons (such as heights), the elements of the population are not the persons but the ________. *observations*

1-3 There are 27 persons living in a particular apartment house, and a social psychologist wishes to draw conclusions about the attitudes of these 27 persons. He chooses every third person in the house for study and obtains a single score on an attitude questionnaire from each person chosen. In this example, the number of elements in the *population* is ______, even though not all these observations are actually obtained. *27*

1-4 The number of elements in the *sample* of questionnaire scores from this population is ______. The parent population would be called (*finite; infinite*) since it is possible to enumerate a first and last member. *9* *finite*

1-5 A student of French is assigned 1,000 vocabulary words during a term, and the examination tests him on 75 words. Conclusions are to be drawn about his knowledge of the assigned vocabulary. The number of elements in the population is ______; the number of elements actually sampled is ______. The population is (*finite; infinite*). *1,000; 75* *finite*

1-6 All the IQ scores of all the students enrolled in a particular college would make up a ________ with respect to questions about IQs in that college at that time. The same set of scores would make up a ________ with respect to questions about IQs of all college students. *population* *sample*

1-7 Any *one* of these scores is the result of an intelligence test taken by one student. The test score summarizes a series of observations which are the student's responses to the test questions. The population of all imaginable observations indicating his intelligence level is the ________ population for this investigation. *parent*

1-8 However, the actual test contains only a ________ from this parent population. Such a population is (*finite; infinite*). *sample* *infinite*

B. KINDS OF CONCLUSIONS DRAWN FROM SAMPLES

When a sample is used, the investigator's interest is not in the sample for its own sake but in the population from which it has been drawn. He intends to make INFERENCES about the population from the evidence contained in his sample.

However, it is customary to speak of two kinds of statistics, descriptive and inferential, even though the descriptive branch (concerned only with summarizing characteristics of the sample) is a very small part of the whole subject. Most statistical methods belong to the inferential branch, and these methods can be usefully divided into three sorts. The structure of the whole subject is indicated in the diagram below.

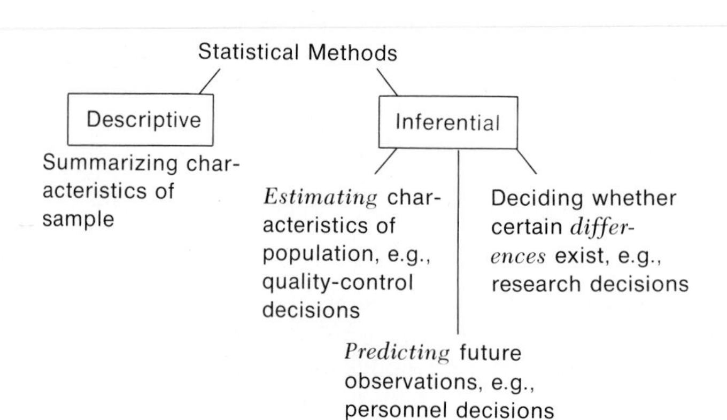

All the inferential methods are methods for statistical decision making. When we infer or estimate the characteristics of the sample's parent population, we make a decision about the probable characteristics of that population. QUALITY-CONTROL DECISIONS are a typical example. In order to decide whether a large batch of some product meets acceptable standards, the quality-control officer examines a sample drawn from that batch (population) and estimates the probable number of defective elements in the batch. If the estimated number is too high, he decides to reject the entire batch.

RESEARCH DECISIONS are usually of the second kind. A French teacher interested in determining which of two vocabulary-learning methods produces the better results may teach two classes by different methods and compare their vocabulary-test scores. He must decide whether a difference in average test score indicates that method 1 is really better than method 2.

The third kind of inferential method enables us to make informed guesses about observations we have not yet obtained. Since these are usually observations which could become available only at a future time, we call such inferences PREDICTIONS. Thus, an admissions officer may try to predict the future college performance of applicants for admission, using college aptitude-test scores as a basis for the inference. When his predictions lead to decisions about the admission or the classification of individual students, he has used statistical inference to make a PERSONNEL DECISION.

1-9 Statistical methods are classifiable into a lesser branch, called __________ methods, and a more extensive branch, called __________ methods.

descriptive
inferential

1-10 Descriptive statistics enable us to summarize the characteristics of the __________; inferential statistics permit us to draw conclusions about the __________.

sample
population

1-11 A person may decide whether to accept or reject as defective a large lot of some product, such as flashlight batteries. His decision is called a __________ decision.

quality-control

1-12 Quality-control decisions provide an example of decisions which require us to estimate the characteristics of a __________ from the observed characteristics of a __________.

population
sample

1-13 In this example the entire lot of flashlight batteries is the __________. Rather than examine all the batteries in the lot for defects, the quality-control officer examines a sample and __________ the number of defective batteries in the lot.

population

estimates

1-14 Research decisions usually involve the examination of samples from two or more populations which might be different. The investigator uses the samples to decide whether __________ really do exist among these populations.

differences

1-15 The French teacher who has used two different methods will obtain two samples of vocabulary-test scores, one from each class. He must decide whether the observed differences between the __________ can be taken as evidence that a difference exists between the two methods of vocabulary learning.

samples

1-16 A personnel officer must try to predict the future performance of some applicant for a job or for an educational program. The observations from which he makes his prediction are often scores on an aptitude test, such as a college entrance examination. When he uses his prediction to decide the applicant's suitability, he makes a __________ decision.

personnel

1-17 Inferential statistics may involve estimating __________ characteristics, deciding whether __________ exist, or predicting __________ observations. Descriptive statistics summarize a __________.

population
differences
future
sample

C. WHY STATISTICAL INFERENCE IS NECESSARY

Whenever the entire population is available for study, statistical inference is not necessary. A descriptive statement can be made about the population simply by applying the methods of descriptive statistics to the population itself. A comparison can be made between two populations without any need for statistical inference, provided that both populations are available for study. Evidently, then, the need for statistical methods of inference arises only when the available observations are limited to a sample (or samples) drawn from the population (or populations) about which inference is to be made.

On the other hand, a descriptive statement about a whole population cannot be made from a sample without special methods of inference. Different samples from the same population are not apt to be exactly alike; it would therefore not be wise to assume that the population is exactly like any one of the samples.

Since the characteristics of the parent population are likely to differ from those of its samples, two different words are used to refer to these characteristics. It is customary to talk about population characteristics as population PARAMETERS and to talk of sample characteristics as sample STATISTICS. Thus, when we infer population characteristics from a sample, we estimate population parameters from sample statistics.

Research decisions also require statistical inference. When we must use samples to decide whether differences exist between populations, we are again working from statistics in order to say something about parameters. The following example will help to show why special methods are necessary.

Case 1 The average height of a sample of students from Alpha College is found to be 1/4 inch greater than the average height of an Omega College sample. Should you conclude that Alpha students tend to be taller than Omega students?

You may well hesitate to draw this conclusion, on the grounds that such a small amount of difference might disappear if you were to draw a second sample from each college population. Your hesitation, if any, arises because all you have available is a pair of sample statistics (the average height for each sample). You know that these two statistics are different, but you hesitate to affirm that the population parameters also differ.

Before taking up another example, let us apply some useful labels to the main features of this one. We want to decide whether the *college* these students attend has anything to do with their average *height*. Two colleges are involved; a large number of heights are observed. We shall use the term VARIABLE for both "college" and "height." A variable, in mathematics, is something which can take on more than one VALUE, and this statement is true of these variables. The variable "college" can take on either of two values, "Alpha" and "Omega"; the variable "height" can take on any of a very large number of values. While the values of the variable "height" are all numbers, the values of the variable "college" are names. "College" is a CATEGORICAL variable because its values are categories into which observations can be sorted. (Some writers call these NOMINAL variables, since the values are category *names*.)

1-18 Characteristics of a sample are called sample ________. Characteristics of a population are called population ________.

statistics
parameters

1-19 The set of heights of *all* the students at Alpha College would be a (*population; sample*) in case 1. The *average* of all these heights would be a (*parameter; statistic*).

population
parameter

1-20 The average obtained from the Alpha *sample* is a (*parameter; statistic*). Since we have only a sample, can we assume that the average of the parent population is the same as the average of the sample? ________

statistic
No

1-21 If you had measured the heights of *all* the students at each of these two colleges and found that the average height at Alpha was 1/4 inch greater than the average at Omega, you (*would; would not*) hesitate to state, "The average height at Alpha is slightly greater than the average height at Omega."

would not

1-22 You would then have no need for methods of statistical inference because both ________ of height measurements would be completely known. Statistical inference is needed only when populations must be described or compared on the basis of ________ drawn from those populations.

populations
samples

1-23 In this example we speak of both "college" and "height" as ________.

variables

1-24 A "constant" in mathematics is something which always has the same value. A "variable" is something which can take on more than one ________.

value

1-25 In case 1, the variable "college" can take on only ________ different values. Since these values are categories into which observations can be sorted, we call "college" a ________ variable.

2
categorical

1-26 One variable, "height," indicates what kind of observations we are studying; the other variable, "college," tells us how they are sorted into distinct populations. We want to decide whether these two populations of height observations actually ________ from each other.

differ

1-27 Since statistical inference is always needed when a decision must be made from samples, we shall need ________ inference in order to decide whether a difference exists.

statistical

1-28 We examine the (*parameters; statistics*) from the sample to decide about a difference in the population ________.

statistics
parameters

This book is largely concerned with statistical inference in two-variable problems resembling case 1. Such two-variable problems arise in various forms, however, and it will be worthwhile here to survey another type.

Case 2 The French teacher already mentioned has directed three classes to use different vocabulary-learning methods and has obtained vocabulary-test scores from the members of each class. The class using method 1 scored 10 points higher, on the average, than the classes using methods 2 and 3. Does this mean that method 1 produces superior vocabulary learning?

Again, there is a variable indicating the kind of observations available (like "height" in case 1). Here the observations are vocabulary-test scores; we can call the variable "achievement scores," since this test is a particular kind of achievement test. The other variable is of course "learning method," and it is a categorical variable even though its categories happen to be designated by the numbers 1, 2, and 3. These numbers are merely category names. The population of imaginable achievement scores of students learning by method 1 is infinite; only a sample is available. The populations of scores from methods 2 and 3 are also infinite, and the teacher must decide on the basis of his three samples whether these three populations actually differ from each other.

Achievement-test scores are usually obtained by counting the number of correct responses on a many-item test. Case 2 therefore differs from case 1 in an important respect. The variable "achievement score" can take on any of a very large number of values, but these values will all be simple whole numbers. There is no possibility that any score will fall into the space between a score of 70 and a score of 71; there is a gap between each pair of successive whole numbers into which no value can possibly fall. Since such gaps exist, the values of this variable are separate (discrete), and achievement score is called a DISCRETE variable.

The variable "height," on the other hand, can take on all sorts of fractional values, and it is always possible to imagine a height which might fall in between any two other heights, no matter how close together they are. Even 62.5 and 62.6 inches are not too close together; we might observe a height of 62.55 inches, or 62.59 inches, or 62.599 inches, and so on. It is not possible to think of any gaps in the series of numbers into which a height value could not conceivably fall. Because of this property, we call height a CONTINUOUS variable.

All categorical variables are also discrete variables. Each of the categorical variables we have so far examined ("college" and "method") has served to separate the observations into two or more populations for comparison. When a variable serves this function, its values are often given a special name. Instead of calling them values, we commonly refer to them as LEVELS. This word suggests the discrete, clearly separated positions of the variable's two or more values.

1-29 In case 1, the variable indicating the kind of observations made is "height"; in case 2, the corresponding variable is __________. *achievement score*

1-30 A discrete variable has distinct gaps between its values into which no value can possibly fall. Is height a discrete variable? __________. Is achievement score? __________ *No; Yes*

1-31 A variable whose values form a smooth continuum has no such gaps. Such a variable is called a __________ variable. *continuous*

1-32 Height is a continuous variable because it is always possible to think of a height which might fall in between any two other heights, no matter how __________ the distance between them. *small*

1-33 Suppose the French teacher uses a test with 117 items and converts all the scores into percentages. A score of 78 correct will become $78/117 \times 100$ percent, and since $78/117 = 2/3$, the percentage in this case will be ______. (Express the percentage to one decimal place.) *66.7*

1-34 If the scores are expressed as percentages, the variable is now "percent correct." This variable is (*continuous; discrete*). *continuous*

1-35 "Time" is a variable which can be measured with any desired degree of accuracy, to tenths, hundredths, even thousandths of a second. No matter how close together two time-values are, it is always possible to think of a value which would fall between them. Time is a __________ variable. *continuous*

1-36 All categorical variables are discrete variables. The variable "college" is a categorical variable, and it is therefore also a __________ variable. *discrete*

1-37 "College" is a variable which serves to separate the height observations into ______ distinct populations for comparison. It is therefore a variable with ______ levels. *2* *2*

1-38 "Method" is a variable serving the same function in case 2. However, "method" has three __________. *levels (or values)*

1-39 For each of these three levels, we have a sample of achievement scores. We know that the method 1 sample differs from the other samples, but we must use __________ inference to decide whether its population differs from the other populations. *statistical*

D. SIGNIFICANT DIFFERENCES

Within the population of height measurements at Alpha College, there will be a great deal of variation among individual students. In fact, there is likely to be more variation in height *within* either one of these colleges than there is *between* the two colleges. The difference between the samples from Alpha and Omega might actually be no larger than the difference one might frequently obtain between two different samples, both drawn from Alpha. The one sample actually obtained from Alpha might contain an unusually large number of high values or an unusually large number of low ones.

The variation between samples drawn from the same population is called SAMPLING VARIABILITY. It is due to the accidental factors which operate during SAMPLING, the process of selecting observations to make up the samples.

The variation between samples which are drawn from different populations will generally be larger than that between samples from the same population, provided that the populations themselves have different parameters. Sampling variability will be present in both cases, but in the first case the actual difference between the populations will also operate to make the samples differ from each other. Two factors operate to produce differences between the samples, and these differences will therefore tend to be larger when the populations differ than when the populations are the same.

The size of a difference between samples therefore contains a clue to whether two samples come from the same population. If the difference is so large that it would rarely occur because of sampling variability alone, this difference signifies that the two samples are probably from different populations. Such a large difference is called a SIGNIFICANT DIFFERENCE.

For example, the average height of a group of men is likely to be greater than the average height of a group of women, and the difference is likely to be large enough to occur only rarely because of sampling variability alone. That is, it would be rare to get such a large difference between two samples drawn from the population of women's heights (and equally rare to get such a large difference between two samples drawn from the population of men's heights). The difference is a significant difference, signifying that the two samples did not come from the same population but from two populations which have different parameters.

There is no question about the significance of a difference in heights between two individuals, but such a question ought to be raised about an apparent difference in their IQs. An intelligence-test score for one individual is not like a height measurement because the IQ score summarizes a *sample* of the intelligence-indicating behavior of that person. If another comparable test were given to the same person, a different score might be obtained simply because of sampling variability. In order to conclude that there is a significant difference between two persons' IQs, it is necessary to show that the difference is so large that it would rarely occur because of sampling variability alone.

1-40 The process of selecting observations for inclusion in a sample is called sampling. Variation between samples drawn from the same population is called __________ variability. *sampling*

1-41 Sampling variability is due to accidental factors which operate during the process of __________. Any variation between two samples of heights drawn from Alpha College will be called __________. *sampling* *sampling variability*

1-42 Is there any sampling variability present in the variation between two samples, one from Alpha and the other from Omega? __________, because both samples were selected through a sampling process. *Yes*

1-43 In this comparison, however, there may be a second factor affecting the size of the difference between the two samples. If the average height for one population differs from the average height for the other, this population difference will tend to __________ the size of the difference between the samples. *increase*

1-44 When two samples have been drawn from populations which differ, the differences between the samples are likely to be (*larger; smaller*) than the differences between samples from the same population. *larger*

1-45 If a difference is so large that it would rarely occur because of sampling variability alone, it is called a __________ difference. *significant*

1-46 A significant difference between two samples signifies that the samples are probably from __________ populations. *different*

1-47 In comparing two persons for height and IQ, the question of significance of a difference does not arise with regard to __________, but it does arise with regard to __________. *height; IQ*

1-48 In comparing IQs, we are comparing the results of two *samples* of intelligence-indicating behavior, one from each of two individuals. Two samples might differ from each other because of __________, even when the populations from which they come do not differ. *sampling variability*

1-49 Whenever we compare populations on the basis of samples alone, we have to use methods of __________ inference. These methods will enable us to decide whether the difference between the samples is a __________ difference. *statistical* *significant*

1-50 Statistical inference is used to decide on the significance of differences whenever populations must be compared on the basis of __________ alone. *samples*

E. STRENGTH OF ASSOCIATION

You must have noticed that a difference may be *significant*, in the statistical sense, without being particularly *interesting*. Perhaps you reflected on case 1, "So what? Suppose there really is a 1/4-inch difference between the two colleges in average height. What is so significant about that?" Clearly it is one thing to decide that a true difference between populations exists and quite another to decide that the difference deserves further attention. When we conclude that a difference is statistically significant, we have simply ruled out the possibility that the populations are really just alike and that an observed difference between samples is merely due to sampling variability.

Whenever we are able to conclude that a significant difference exists in a two-variable case, we can also conclude that the two variables are associated with each other. But this conclusion will be interesting only if there is some reason for caring about the association itself. In case 1 there probably is no such reason; in case 2 there surely is. But unless the association in case 2 is a strong one, we may still be wasting our time to give it further attention.

In most cases (but not all), the data which permit a decision about the significance of a difference will also provide a measure of the strength of association between the two variables involved. This measure is usually in terms of *predictive value*. For example, certain kinds of color blindness are closely associated with the variable "sex." There is a highly significant difference between the number of males and the number of females showing such color defects, and we can use this association to predict whether children as yet unborn will be colorblind. Our prediction will not be infallible, but it will certainly be much more accurate than it could be without knowledge of the association.

Some associations are quite weak; and because they are weak, they are almost useless for prediction. In case 1 the small size of the observed height difference (1/4 inch) makes one suspect that the association, if it exists at all, will turn out to be a very weak one. Knowledge of a student's height will not help very much in guessing which college he attends, nor will information on his college give much help in guessing his height.

In case 2 we can measure the strength of association by determining how much our prediction of vocabulary score can be improved by knowing the learning method which was used. Of course, we are not interested in this prediction for its own sake; we are really interested in improving the quality of learning by choosing a superior method. But we reason thus: If the association helps prediction quite a bit, the association between method and learning is strong, and vocabulary learning can be markedly improved by using the best method of the three.

Wherever the data will allow it, we should not stop our analysis when we have found evidence for an association, i.e., evidence of a statistically significant difference. We should go on to determine the strength of that association in order to know how valuable our finding is likely to be.

1-51 To be meaningful or interesting, a difference between two samples must first be shown to be statistically __________. *significant*

1-52 A difference in school grades between nine children who attended nursery school and seven children who did not is potentially very interesting. However, until this difference has been shown to be significant, there is a possibility that the observed difference is due merely to __________ variability and that there is no true difference between the __________ from which these samples came. *sampling* *populations*

1-53 Once we know that the difference is significant, we know that the two variables "nursery school attendance" and "school grades" are associated with each other. We do not yet know how strong this __________ may be. *association*

1-54 We shall be less interested in this association if it is very weak than if it is very __________. *strong*

1-55 A strong association allows good prediction of one variable from the other. A weak association is practically useless for __________. *prediction*

1-56 In case 2 the French teacher may find a significant difference between method 1 and method 2 or 3. He will then measure the __________ of __________ between method and achievement score in order to determine how useful and informative his finding can be. *strength; association*

1-57 A strong association would enable him to predict future achievement scores from information about the __________ used in vocabulary learning. His prediction would not be perfect, of course, but it would be better than it could be without knowledge of the __________ between the two variables. *method* *association*

1-58 However, he is probably not strictly interested in making such predictions. He is interested in improving his teaching. Can he improve his teaching any better by finding a strong rather than a weak association? __________ *Yes*

1-59 He can, because a strong association means that method makes a large amount of difference in achievement score, while a weak association means that method makes only a __________ amount of difference. *small*

1-60 Statistical analysis should not stop at finding evidence of association; it should go on to determine the __________ of the association. *strength*

REVIEW

1-61 To answer many practical questions, it is necessary first to collect some evidence. The study of statistics should teach you how to __________ a procedure for data collection so that the data will be informative. *design*

1-62 Statistical analysis is then performed in order to arrive at a __________ and to determine the degree of __________ you can have in it. *decision; confidence*

1-63 The entire set of observations relevant to a decision is called a __________. A subset of these observations is called a __________. The subset actually collected constitutes the __________ for statistical analysis. *population* *sample* *data*

1-64 With reference to the population, each observation is called an __________ in the population. With respect to the data collected, each observation is a __________. *element* *datum*

1-65 A finite population is a population whose elements can be __________ in such a way that there is a __________ member and also a __________ member. *enumerated; first* *last*

1-66 The population to which a sample belongs is called its __________ population. An investigator is usually interested in the sample only because of what it can tell him about the __________. *parent* *parent population*

1-67 We distinguish two kinds of statistical methods: __________ methods and __________ methods. *descriptive ↔ inferential*

1-68 Descriptive statistics provides ways of __________ characteristics of a sample. These characteristics are called sample __________. *summarizing* *statistics*

1-69 If we wish to estimate characteristics of the parent population, we must use methods of __________ statistics. Population characteristics are called __________. *inferential* *parameters*

1-70 Some kinds of decisions are based mainly on the __________ of population parameters. An example of such a decision is the __________ decision. *estimation* *quality-control*

1-71 Certain methods of inferential statistics are used for making decisions about the existence of __________ between two or more populations. *differences*

1-72 A third kind of decision based on statistical inference is the decision which involves predicting ________ observations. *future*

1-73 When an admissions officer infers something about an applicant's probable grades from seeing his aptitude scores, he makes such a decision. In this case, the decision is called a ________ decision. *personnel*

1-74 Most research decisions are actually decisions whether ________ exist between populations. *differences*

1-75 In the vocabulary-learning example, there are two variables. The variable on which the observations are made is ________. *achievement score*

1-76 The other variable is ________ of learning. It serves to ________ the observations into three populations for comparison with each other. *method* *separate*

1-77 A variable is something which can take on more than one ________. *value*

1-78 When the values are categories into which observations are sorted, the variable is called a ________ variable. *categorical*

1-79 When its values are numbers, and when there are *no gaps* in the series of numbers which may become values, it is called a ________ variable. If there are such gaps, it is called a ________ variable. *continuous* *discrete*

1-80 All categorical variables are also (*continuous; discrete*) variables. When a variable serves, like "method," to distinguish different populations, its values are commonly called ________. *discrete* *levels*

1-81 Statistical inference is necessary to decide about the existence of differences whenever the data include only a ________ of the relevant observations instead of including the entire ________. *sample* *population*

1-82 The process of selecting observations to make up a sample is called ________. Accidental factors operating during this process give rise to ________, which is the variation between samples with the same ________ population. *sampling* *sampling variability* *parent*

1-83 A difference between samples which is too large to be attributed to sampling variability is called a ________ difference. Such a difference signifies that the two variables in the problem are ________. *significant* *associated*

PROBLEMS

1. Could sampling variability ever result in the selection of two samples from two *actually different* populations in which the samples are so much alike that we would mistakenly conclude that the populations are *not* different?

2. In some extrasensory perception (ESP) experiments a "percipient" who is at some distance from an "agent" tries to guess or tell correctly the shapes of patterns which are presented only to the agent. When such experiments are carefully performed with great attention to detail, they often show a significant but very weak association between the percipient's responses and the stimuli actually presented to the agent. Since the association is extremely weak, why do you think it is considered worthwhile to expend large amounts of time and energy in this kind of research?

3. Two tenth-grade math classes are given the same examination, and a sample of 10 out of the 35 scores is drawn from each class. One sample has an average of 87.5; the other's average is 90. Should we question the statistical significance of the difference between classes?

4. Two brothers in these classes took the math exam; one scored 85 and the other scored 95. Can we conclude without statistical inference that the brothers differ in ability to do the kind of math tested?

5. Two baseball players are being compared on the basis of their batting averages. Does the question of significance of difference arise in comparing the two batting averages? Does it arise in comparing the batting *ability* of the two players?

6. Can there ever be a question of significance of difference in the comparison of two physical measurements of length? Consider comparing the distance between New York and Boston with the distance between New York and Washington.

7. On page 11 we described "achievement score" as a discrete variable, but we concluded that when such scores are converted to "percent correct," they become values of a continuous variable. In our example, the maximum score possible is 117. Work out (to one decimal place) the percentage equivalent for a score of 79; then compare it with the percentage equivalent for a score of 78. Since it would never be possible, *in this test,* for a value of "percent correct" to fall between these two percentage values, do you think "percent correct" ought to be called a continuous variable?

8. When two variables are associated in the statistical sense, should we conclude that variations in one variable are actually brought about by changes in the other variable? (Read "brought about" in the sense of "caused to occur.")

9. Suppose that you are a physician. You receive publicity from a pharmaceutical company reporting that a new pain reliever has been tried on a sample of surgical patients. These patients made fewer requests for additional medication to relieve pain than another group who were given only aspirin. As a student of statistical inference, what question would you ask before deciding to prescribe the new drug instead of aspirin? In the investigation as reported, are the observations collected on a discrete or continuous variable? What are the two variables? Name the levels of the variable separating the populations.

LESSON 2
PRINCIPLES OF STATISTICAL DECISION MAKING

Lesson 1 has introduced the single most common type of statistical decision called for in scientific or technical research, a decision about the significance of an observed difference between samples. The other types of decision (prediction and parameter estimation) require methods which depend on those used to decide about differences. Therefore, if we concentrate on decisions about the existence of differences, we shall lay the groundwork for understanding all three kinds of statistical decisions.

Each investigation which seeks to demonstrate differences must be carefully designed. Lesson 2 shows the style of thinking which must be adopted if statistical methods are to be used effectively. We begin with a brief examination of the sampling process.

A. RANDOM AND BIASED SAMPLES

Statistical methods permit us to make inferences about the parent populations only when we have samples of the proper kind. These special samples are called RANDOM SAMPLES. Every element in a random sample must be selected at random, in the manner of drawing numbered slips from a bowl in a lottery. Only accidental, i.e., chance or *random*, factors can influence a selection made in this way. Thus only random factors will be present to produce sampling variability, the differences among different samples from the same parent population.

Sampling in a random way is seldom as straightforward as drawing lottery numbers. In every field of investigation the student must learn special techniques to use when he needs a random sample. We shall not try to discuss these techniques, since they differ from one subject to another. But there are general rules for telling whether a sample is random or not. Whatever the nature of the observations being sampled, the sample is random only if the following conditions are met:

1 EQUAL CHANCE. A sample meets the condition of equal chance if it is selected in such a way that every observation in the entire population sampled has an equal chance of being included in the sample.

2 INDEPENDENCE. A sample meets the condition of independence when the selection of any single observation does not affect the chances for selection of any other observation.

Every report of a scientific or technical investigation based on samples must state how the sampling process has been carried out. From this description it is possible to determine whether the conditions of equal chance and independence have been met. Samples which do not meet these conditions are called BIASED samples.

In a relatively small population, the chance that a particular element will be selected is *not independent* of the selections of other elements unless the same element can be selected more than once. Let us make this point clear by an example. Suppose you are sampling from a relatively small population, the population of 52 playing cards. In order to meet the condition of independence, you must be able to select the same card more than once. It will be necessary for you to put back (or *replace*) the card you have just selected at random as soon as you have noted its identity. This procedure is called SAMPLING WITH REPLACEMENT. If you sample WITHOUT REPLACEMENT, the selection of the second card will be dependent upon the first card drawn; that second card cannot be the same as the first card, and therefore there are only 51 possibilities (instead of 52) for the second card. The available population has shrunk by 1, and its membership depends upon the identity of the first card drawn. In the same way, your third selection will be dependent upon the first and second choices, the fourth upon the first three choices, and so on.

When it is put to you in this way, you may readily agree that sampling without replacement will violate the independence condition. However, you may still feel uneasy about sampling *with* replacement. The very idea that a certain element might appear more than once in a small sample goes against intuition. Intuitively, you may be thinking of a sample as a small-scale replica of its population; yet how could a sample of five playing cards which contains two queens of hearts be "representative" of the set of 52 unique cards? If this troubles you, you must revise your concept of a proper sample. Instead of thinking of it as the population writ small, you must think of it as the mathematician does: a collection of elements whose selection has been governed by two conditions—equal chance and independence—and by nothing else.

A sample, when it is random, gives quite a bit of information about its parent population. Its average "resembles" the population average, for one thing. But it does not resemble the population in all respects. The population of playing cards does not contain any two cards which are alike; a random sample from this population probably will not either, but it *may*, and this possibility must exist in order for it to be truly random. In this respect the sample does not resemble the population, and it would be quite wrong to conclude from a sample with two queens of hearts that the parent population contained more than one such card.

You need not worry about replacement when a population is large. As the size of the population approaches an infinitely large number, sampling without replacement becomes approximately the same as sampling with replacement, and sample elements will be independently chosen even when sampling is without replacement. When sampling from a small population *cannot* be done with replacement, it is sometimes possible to make decisions on the data, but quite different procedures will be required because the condition of independence is being violated. All the procedures in this book assume that the population is infinite or if it is not, that sampling is carried out with replacement.

Sample A The IQ scores of all 10,000 Omega College students are written on identical slips of paper and shaken up in a container. A blindfolded person draws out 100 of the slips to obtain a sample from which conclusions are to be drawn about the IQs at Omega.

Sample B A sample of voter opinions in a certain district is selected by choosing every hundredth man from the list of registered voters and questioning both him and his wife.

2-1 The methods in this book can only be used to analyze the kind of samples called __________ samples.

random

2-2 A sample is random if it meets two conditions. To meet the condition of equal chance, every observation in the population must have an __________ of being included in the sample.

equal chance

2-3 Sample *B*, above, will not meet this condition. Unmarried women are voters, but the chance of their being selected is not __________ to the chance of selection of married women.

equal

2-4 To meet the condition of independence, the selection of any single observation must not affect the chances for any other observation to be __________.

selected

2-5 Since, in sample *B*, a married woman will be selected if (and only if) her husband is selected, this sample will not meet the condition of __________.

independence

2-6 A sample which is not random is a biased sample. The procedure in *B* will give a sample which is __________.

biased

2-7 Is the sample in *A* a biased or a random sample? __________

Random

2-8 If some students' scores were written separately on two different slips of paper and both slips were included in the container, which condition would not be met? __________

Equal chance

2-9 If several students' scores were written on the same slip of paper, which condition would not be met? __________

Independence

2-10 Sample elements are sometimes drawn from a small population one at a time. If each element is replaced after it is drawn, so that the same element can be drawn more than one time, sampling is said to be done with __________.

replacement

2-11 A sample drawn without replacement will be random only if the population is very __________.

large

B. STATISTICAL HYPOTHESES

Decisions about differences always begin with certain hypotheses; the decision problem is cast in such a way that it can be seen as a decision between two opposing hypotheses.

How could the decision about a college-height difference be cast in this way? In discussing the concept of a significant difference, we have already said that any difference we might find between the two samples will either (1) be small enough to be attributed to sampling variability alone or (2) be large enough to indicate that the parent populations are also different. Thus, we can cast the problem as a decision between the hypothesis (1) that there is *no* true difference between the height populations and the hypothesis (2) that there *is* a true difference between the height populations.

In order to simplify our terms, we shall begin to refer to the true (but unknown) averages of the two populations by using the Greek letter μ (mu). The true average of the Alpha population will be μ_A; the true average of the Omega population, μ_O. Hypothesis 1 can now be put as an equation: $\mu_A - \mu_O = 0$.

A hypothesis that there is *no difference* between the populations which are being compared is commonly called a **NULL HYPOTHESIS**. Hypothesis 1 is therefore a null hypothesis.

Hypothesis 2 can also be stated in the form of an equation. We may write (2*A*): $\mu_A - \mu_O \neq 0$; but this hypothesis can also be written in many other ways. If, before collecting our data, we already have good reason to expect that Alpha students will turn out to be taller, we can frame it in the form (2*B*): $\mu_A - \mu_O > 0$, to indicate that we expect a difference in the *positive* direction. This form of hypothesis 2 is described as a **DIRECTIONAL HYPOTHESIS**. If, on the other hand, we expect Omega students to be taller, we will write the directional hypothesis (2*C*): $\mu_A - \mu_O < 0$.

These three forms of hypothesis 2 are all **INEXACT**, in that they allow the difference $\mu_A - \mu_O$ to take on any one of a range of values. Hypothesis 2*B* allows the difference to take any value which is not zero or negative; hypothesis 2*C* allows it to take any value which is not zero or positive; and hypothesis 2*A* allows it to take any value at all, so long as it is not zero. There are an infinitely large number of ways to state hypothesis 2 as an **EXACT HYPOTHESIS**, one specifying a particular value which the difference is expected to take. We could write $\mu_A - \mu_O = 1/4$ inch, or $\mu_A - \mu_O = -3$, or any other equation indicating a specific nonzero value for the difference.

Here are the hypotheses we have just discussed.

	Hypothesis 1	Hypothesis 2
Inexact		Nondirectional: $\mu_A - \mu_O \neq 0$ Directional: $\mu_A - \mu_O > 0$ $\mu_A - \mu_O < 0$
Exact	$\mu_A - \mu_O = 0$	$\mu_A - \mu_O = 1/4$ $\mu_A - \mu_O = -3$

2-12 Decision problems are always cast as a decision between two opposing ________.

hypotheses

2-13 The decision will be made on the basis of samples, but the hypotheses are about the populations. In many cases, like the college-height case, the hypotheses are about the true averages of the ________.

populations

2-14 The true average of a population is generally unknown. We represent this true average by the Greek letter ______.

μ

2-15 The hypothesis $\mu_A - \mu_O = 0$ states that the true difference between the population averages is expected to be ______.

0

2-16 The symbols $>$ and $<$ are called inequality signs. The point is placed toward that side of the equation which has the (*greater; smaller*) value.

smaller

2-17 The hypothesis $\mu_A = \mu_O > 0$ states that the true difference is expected to be (*greater; less*) than zero.

greater

2-18 The symbol $\neq$ (read "not equal") also indicates an inequality, but it does not specify the direction of the inequality. It is nondirectional, and $\mu_A - \mu_O \neq 0$ is therefore a ________ hypothesis.

nondirectional

2-19 A hypothesis which specifies an exact value for the difference is called an exact hypothesis. $\mu_A - \mu_O = 0$ is an ________ hypothesis; $\mu_A - \mu_O \neq 0$ is an ________ hypothesis.

exact; inexact

2-20 $\mu_A - \mu_O \neq 0$ and $\mu_A - \mu_O > 0$ are both inexact hypotheses, but only one of them is a directional hypothesis. It is ______.

$\mu_A - \mu_O > 0$

2-21 $\mu_A - \mu_O = 0$ is called a null hypothesis because it states that there is ________ between the true averages of the populations.

no difference

2-22 A null hypothesis is always exact, but not every exact hypothesis is a null hypothesis. For example, $\mu_A - \mu_O = 1/4$ is an ________ hypothesis which is not a null hypothesis.

exact

2-23 Suppose we chose to decide between $\mu_A - \mu_O = 3$ and $\mu_A - \mu_O > 3$. Is either of these hypotheses a null hypothesis? ________. Which one is exact? ______ Is the other one directional or nondirectional? ________

No; $\mu_A - \mu_O = 3$
Directional

2-24 $\mu_A - \mu_O > 3$ is inexact; it allows the difference to assume any value greater than ______.

3

It is a fundamental rule that an investigator must select his two hypotheses *before* he has seen the data which he intends to use in deciding between them. Of course, he may have seen other relevant data from previous studies. But he ought not to choose his hypotheses in the college-height case with the knowledge that a difference of 1/4 inch, favoring Alpha, has already been found, unless he intends to gather *new* data for use in making his decision.

Once the two hypotheses have been chosen, the procedure in decision making is to treat one of these hypotheses as if it were true and examine the consequences of that assumption. The hypothesis chosen is called the ASSUMED hypothesis. If the consequences are acceptable in the light of whatever evidence is available, the investigator continues to regard that hypothesis as true. If the consequences are not acceptable, he *rejects* the assumed hypothesis and *accepts* the other one. The other hypothesis is called the ALTERNATE hypothesis.

We shall always refer to the assumed hypothesis as H_1 (read "H one") and the alternate hypothesis as H_2. Because it is much easier to calculate the consequences of an exact hypothesis than of an inexact hypothesis, H_1 is invariably an exact hypothesis. H_2 may be any sort of hypothesis, exact or inexact, which the investigator is prepared to accept in the event that he must reject H_1. For he must now face his data and ask himself, "Could I have reasonably expected to get these data if the assumed hypothesis is true?" When the answer is yes, he continues to regard H_1 as true. When the answer is no, he rejects it in favor of an alternate hypothesis under which the data could more reasonably be expected to occur.

Suppose, now, that you are an investigator faced with the array of hypotheses at the bottom of page 22. You do not know anything yet about the sample averages. Which exact hypothesis would you choose to assume as H_1? Like most investigators, you probably feel that it would be more informative to examine the consequences of assuming $\mu_A - \mu_O = 0$ than those of the other exact hypotheses. Behind your feeling of preference lies, perhaps, a sense that a difference of *any* size is the matter which interests you, not some particular size of difference.

There are a few times when one might be interested only in a particular value of the difference (such as 1 inch), and one would then write H_1 with precisely that value. Nevertheless in most cases one wishes to show only that the true difference is not zero, and for this reason *the assumed hypothesis is ordinarily a null hypothesis.* In fact, this is so generally the case that we have no conventional name for the assumed hypothesis other than null hypothesis. An investigator will almost invariably refer to H_1 and H_2 as his null hypothesis and alternate hypothesis, even though these terms are not strictly parallel. ("Null" refers to the content of the hypothesis, while "alternate" refers to the role it plays in decision making.) Many writers designate the two hypotheses as H_0 and H_1 because H_0 directly suggests a hypothesis of zero difference. It is important to remember, however, that not every assumed hypothesis need be a null hypothesis, and we shall use the symbols H_1 and H_2 for this reason.

2-25 Statistical hypotheses are formulated (*after; before*) the data have been collected and examined. — *before*

2-26 If the investigator *already knows* that two samples have shown a 1/4-inch difference, he may decide to formulate a directional hypothesis, $\mu_A - \mu_O > 0$. However, in order to make his decision, he will now have to collect two new ________. — *samples*

2-27 After choosing his two hypotheses, the investigator assumes one of them to be true. It thus becomes his ________ hypothesis. — *assumed*

2-28 The other hypothesis is called the ________ hypothesis. It is the one he intends to accept if he has to reject his ________ hypothesis. — *alternate*, *assumed*

2-29 We are designating the assumed hypothesis as (H_1; H_2) and the alternate hypothesis as ______. — *H_1*, *H_2*

2-30 It is easier to calculate the consequences of a hypothesis when it is exact. Therefore, one of these hypotheses must always be exact; which one? ______ — *H_1*

2-31 A null hypothesis is an exact hypothesis. Most of the time, H_1 is a ________ hypothesis. — *null*

2-32 The assumed hypothesis H_1 is usually a null hypothesis because one is usually interested in showing simply that a true ________ exists, no matter what its size. — *difference*

2-33 The investigator looks at his data and asks, "Could I have reasonably expected to get these data if my ________ hypothesis is true?" — *assumed*

2-34 If the answer is yes, he decides in favor of (H_1; H_2). If the answer is no, he decides in favor of ______. — *H_1*, *H_2*

2-35 When we find that the difference between samples is small enough to be due to sampling variability alone, we decide in favor of (H_1; H_2). — *H_1*

2-36 A decision in favor of H_2 is a decision to ________ H_1 and to ________ H_2. — *reject*, *accept*

2-37 The most important fact to remember in choosing H_2 is that it should be a hypothesis which will make the data appear reasonable in the event that H_1 must be ________. — *rejected*

C. H_1 AS THE "CONSERVATIVE HYPOTHESIS"

The assumed hypothesis is more often than not a null hypothesis, not only because a null hypothesis is conveniently exact but also because it is a *conservative* hypothesis. Statistical decision rules are constructed to keep their users from climbing out on dangerous limbs without adequate evidence.

To justify the word "conservative," let us return to the case involving vocabulary learning. When we read that two methods have been compared, we are interested in hearing whether the population of achievement scores resulting from one method is or is not different from that obtained by the other method. If the French teacher is proposing a new method, the burden of proof is on him to show us that it is really better than a conventional method. He can expect us, as his readers, to remain skeptical until he has made a good case for the new method.

When he starts out to compare the average achievement scores of his two samples, his hypotheses are similar to those for the height case. Letting μ_1 stand for the true population average for the new method 1, he may write two hypotheses: $\mu_1 - \mu_2 = 0$ and $\mu_1 - \mu_2 > 0$. He intends to assume one of these hypotheses, and he will continue to regard it as true unless his data force him to reject it. Which one should be given this role?

He knows that his readers will assume $\mu_1 - \mu_2 = 0$ unless he is able to show them otherwise; the burden of proof is on him. Therefore *he places the same burden of proof on his data.* He chooses to assume $\mu_1 - \mu_2 = 0$ also, rejecting it only if the data seem to make that hypothesis unreasonable; $\mu_1 - \mu_2 = 0$ thus becomes his H_1.

Suppose, after examining his data, he decides to reject H_1 when in fact H_1 is true. He has then made an *error,* although of course he does not know it. This kind of error is known as a TYPE I ERROR; we can describe it as an *unwarranted claim.* It is a claim that a true difference exists when in fact it does not.

If, on the other hand, he decides to continue believing H_1 when in fact H_2 is true, he has made another kind of error called a TYPE II ERROR. We can describe it as a *missed discovery,* since he has failed to discover a difference which truly exists.

Scientists usually consider type I errors to be less excusable than type II errors. Once they can identify the hypothesis which if accepted falsely (mistakenly) would amount to an "unwarranted claim," they will ordinarily formulate that hypothesis as H_2. H_1 will then be a hypothesis which does not make this claim.

Thus, however confident he may be in his project, an investigator will avoid assuming the hypothesis which describes what he hopes to demonstrate. He prefers to assume that his hope is unfounded until proved otherwise. His strategy is a conservative one, and the hypothesis he assumes is a conservative hypothesis. The investigator does not assume H_1 because he hopes it is true; in most cases he hopes he can reject it. He assumes H_1 to protect himself against a type I error.

2-38 Before looking at his data, the investigator *assumes* that one hypothesis is true. This hypothesis is designated as _______. — *H_1*

2-39 After analyzing his data, he *accepts* one of the hypotheses as true. If the data do not look too unreasonable in the light of H_1, the hypothesis he accepts is _______. — *H_1*

2-40 However, if the data look very unreasonable in the light of H_1, he rejects H_1 and ___________ H_2. — *accepts*

2-41 The actual statistical decision is made when he decides which hypothesis to (*accept; assume*). — *accept*

2-42 The statistical decision is *wrong* when the investigator decides to (*accept; assume*) a hypothesis which is not true. — *accept*

2-43 Accepting one hypothesis means rejecting the other. Accepting H_1 is the same as ___________ H_2. — *rejecting*

2-44 Since there are two hypotheses which might be falsely rejected, there are logically two different kinds of errors which might be made. These are called type _______ and type _______ errors. — *I*, *II*

2-45 Memorize the types by the name of the *rejected* hypothesis. Rejecting H_1 falsely is a type I error; rejecting _______ falsely is a type II error. — *H_2*

2-46 A type I error is made when _______ is rejected and yet is actually true. — *H_1*

2-47 When H_2 is true, rejecting H_2 means making a type _______ error. — *II*

2-48 If H_2 is *accepted* falsely, an unwarranted claim has been made. Accepting H_2 means rejecting _______. Therefore, an unwarranted claim is a type _______ error. — *H_1*, *I*

2-49 If H_2 is accepted *correctly*, a discovery has been made. Rejecting H_2 when it should have been accepted means that a ___________ has been missed. — *discovery*

2-50 A missed discovery occurs when H_2 is rejected falsely. This error is type _______. — *II*

2-51 An unwarranted claim occurs when _______ is rejected mistakenly. This error is type _______. — *H_1*, *I*

2-52 Type _______ errors are considered less excusable than type _______ errors. — *I*, *II*

D. TYPE I VS. TYPE II ERRORS

As we have just seen, the statistical decision-making process is designed to avoid type I errors more vigorously than type II errors. Yet in trying to avoid an unwarranted claim, the scientist risks failing to make a *warranted* one and thus missing what may be an important discovery. Is this consequence desirable?

Actually, of course, some type I errors are worse than others. Compare a similar situation in a quite different field. Suppose you and a friend are each betting on a horse race. Your friend bets on Laconic, the horse generally expected to win; you play a hunch and put your money on Capricorn. Betting on Laconic is like accepting H_1, the conservative hypothesis; betting on Capricorn is like accepting H_2. If Laconic wins, you have made a type I error. If Capricorn wins, your friend has made a type II error.

In placing your bet, you might have thought, "Money on Laconic won't pay off much even if I win; money on Capricorn will perhaps be lost, but I needn't bet very much money, and if he should win, I stand to gain a great deal." If your hunch is based on a pretty reliable tip, you might even decide to risk a large bet.

When Laconic wins, your type I error may be laughed at as foolish. The laughs will be louder if Capricorn was known to be a worn-out nag. They will not be so loud, however, if he was merely an unknown, a possible dark horse. It is the same with statistical decisions. Where H_2 itself is not too implausible, the punishment for making a type I error will be less.

Like the payoffs on different horses, the advantages of being *right* about H_1 or H_2 also vary from one case to another. In general, one gets little comfort from accepting H_1 even when it is true. H_1 represents the state of affairs everyone supposed was true anyway, and one has invested time and energy without turning up any new information. It is always more rewarding to find that one can reject H_1 and accept H_2; when H_2 is initially considered very implausible, accepting H_2 correctly may amount to a major scientific advance.

Individual investigators may differ widely in the way they regard the two types of errors. A young investigator with little prestige to lose may be willing to risk more on an interesting hypothesis. He is less concerned in avoiding a type I error than an established investigator would be and more concerned in not missing the discovery which will make his reputation. Furthermore, the relative costliness of type I and type II errors varies with the stage of the research project. In an early stage experiments are primarily exploratory. Accepting H_2 incorrectly will mean only that the investigator continues to spend time on a project which ultimately fails to work out. At a later stage such an error is likely to result in a published mistake. Scientists will generally be less cautious about making type I errors in exploratory experiments than in experiments which they intend to publish.

In gambling, these differences are reflected in the amount of money wagered. Section E shows how they can be incorporated in statistical decision-making strategy.

2-53 One of the two hypotheses describes what the investigator hopes to demonstrate. This hypothesis is (H_1; H_2). — *H_2*

2-54 Other people might consider H_2 as very unlikely. A person who accepts such an H_2 without strong evidence would be regarded as making an ________ claim. — *unwarranted*

2-55 An unwarranted claim is a type ________ error. It is generally avoided, but it will be avoided (*less; more*) carefully when the claim is very unlikely than when it is quite plausible. — *I*; *more*

2-56 By deciding to accept H_1, one can avoid a type ________ error. But even if he is right in his decision, he does not gain any new information. New information is gained when one can accept ________. — *I*; *H_2*

2-57 At an early exploratory stage, an investigator avoids type I errors (*less; more*) carefully than later in his research, when they are likely to come out in print. — *less*

2-58 Ordinarily, investigators avoid type ________ errors more carefully than the other type. At an exploratory stage in his project, an investigator is casting about for a promising new lead. He is likely to avoid type ________ errors more carefully at this stage, even though doing so means that he may make a type ________ error. — *I*; *II*; *I*

2-59 Once a decision between H_1 and H_2 has been made, the investigator may publish his study or go on to a new stage, or both. If he is going to publish it, he wants to be sure that he has not made a type ________ error. — *I*

2-60 When H_2 is generally considered to be a "way-out" hypothesis, he knows that people will be skeptical of a decision to accept H_2. Compared to another case with a more plausible H_2, this case calls for (*less; more*) caution in accepting H_2 and rejecting H_1. — *more*

2-61 A type I error is made when ________ is rejected falsely. A type II error is made when ________ is rejected falsely. — *H_1*; *H_2*

2-62 An unwarranted claim is a type ________ error. A missed discovery is a type ________ error. — *I*; *II*

2-63 More information has been gained when one is able to accept ________ correctly than when one is able to accept ________ correctly. — *H_2; H_1*

E. CHOOSING A REJECTION PROBABILITY

The decision between H_1 and H_2 will depend upon the data the investigator is going to collect. With these data in hand, he will ask, "Could I reasonably expect to get such data if H_1 is really true?" At this point we should notice that the word "reasonably" is ambiguous. To the young investigator intent on making his reputation with a startling discovery, it may mean "easily." He may plan to reject H_1 and accept H_2 unless the data can easily be seen to fit H_1. Another person more fearful of a type I error will not reject H_1 as long as the data can conceivably be seen to fit it.

It will help to put the question in a less ambiguous way. Let the French teacher ask, "If the populations are actually alike, could such a large difference between the samples arise from sampling variability?" Answers to this kind of question are not simply yes or no; they are answers like "yes, quite commonly," or "yes, but not very often."

Equivocal answers like these can be enormously improved by putting them into some kind of number language; "99 times in 100" or "once in a million" would be easily understood. Therefore the answer to the teacher's question should be given as a number representing the PROBABILITY of such a large difference when the populations sampled do not themselves differ.

Thus the investigator, raising this question in statistical decision making, does not get a simple yes or no. *He gets a maybe with a number attached.* Let us call this number p_o (read "p sub oh," the probability of an observed result). No matter how outlandish his observed results may appear in the light of H_1, it is always at least remotely possible that H_1 is still true: p_o can be very small, but it can never be zero.

Realizing this fact, each investigator must know in advance, before he sees his data, just *how small* p_o will have to be in order for him to reject H_1 and accept H_2. The number he chooses will be called his REJECTION PROBABILITY, and it is in this choice that his personal attitudes of caution or boldness may enter. He may, if he is very cautious, decide to reject H_1 if the data could occur only once in a hundred times when H_1 is true; his rejection probability is then .01 (or 1 percent). If he is less cautious, he may prefer to reject H_1 if the results have a probability no greater than 1 in 20 when H_1 is true.

We shall call this rejection probability ALPHA, using the Greek letter α as its symbol. An investigator plans to reject H_1 when p_o is less than or equal to α; that is, when $p_o \leqslant \alpha$. A difference which leads him to reject H_1 will then be said to be "significant at the level α," and the value of the rejection probability thus becomes the α level or the significance level of his decision.

Since p_o is the probability of his results when H_1 is true, p_o is the probability of a type I error. But before he is able to calculate p_o, the investigator can already say that the probability of his making a type I error will not be greater than α, because he intends to reject H_1 only when $p_o \leqslant \alpha$.

2-64 The question, "If the populations are actually alike, could such a large difference between the samples arise from sampling variability?" is not answered yes or no. It is answered in terms of a ________, represented here by the symbol p_o.

probability

2-65 If we find that $p_o = .10$ (or 10 percent), we learn that such a large difference could arise from sampling variability alone about ______ times in 100 when (H_1; H_2) is actually true.

10; H_1

2-66 The Greek letter α also stands for a probability. When p_o is less than or equal to ______, H_1 will be rejected and H_2 accepted.

α

2-67 One of these probabilities is *chosen* in advance by the investigator; the other one emerges from statistical analysis of his data. The one he chooses in advance is ______.

α

2-68 He chooses α when he decides, before seeing his data, that he will reject H_1 if it turns out that ($p_o = \alpha$; $p_o < \alpha$; $p_o \leqslant \alpha$). We call α his ________ probability.

$p_o \leqslant \alpha$
rejection

2-69 If it turns out that p_o is equal to or less than α, he plans to accept ______. However, even in this case, there is still a small probability that he will make an error. If he does, it will be a type ______ error, and its probability will not be greater than ______.

H_2

I

α

2-70 The probability p_o can be vanishingly small, but it can never actually reach ______. Therefore, one can never actually be *certain* that ______ is untrue.

0
H_1

2-71 A significant difference is a difference between ________ which signifies a true difference between ________.

samples
populations

2-72 Since H_1 can never be disproved, it is never possible to be certain that a true population difference exists. Therefore, a difference is never significant in an absolute sense; it is significant at a certain level of probability called the significance ________.

level

2-73 The significance level is also called the ______ level. When a difference is "significant at the .05 level," we know that p_o is not greater than ______.

α

.05

2-74 Before the data are analyzed, the probability of a type I error is (α; p_o). After the data analysis, the probability of a type I error is ______.

α
p_o

REVIEW

2-75 This chapter has focused attention on decisions about the existence of ________.

differences

2-76 The procedures discussed in this book are based on the assumption that the samples being analyzed are ________ samples.

random

2-77 In order to be a random sample, a sample must meet two conditions. Give their names: ________.

equal chance and independence

2-78 A sample meets the equal-chance condition if every observation in the population has an equal chance of ________.

being included in the sample

2-79 It meets the independence condition if the selection of one ________ does not affect the ________ for selection of any other ________.

observation; chance(s)
observation

2-80 A sample which fails to meet one or both these conditions is a ________ sample.

biased

2-81 When the population is small, the sample will be random only if sampling is done ________.

with replacement

2-82 Therefore a sample of 9 attitude questionnaires taken from a population of 27 will not be a random sample from that population unless it is possible for a particular questionnaire to be included more than ________.

once

2-83 In planning his study, an investigator formulates two hypotheses. He intends to ________ one of these hypotheses at the outset and determine whether his data will fit in with it.

assume

2-84 The other hypothesis is called his ________ hypothesis. After he has examined his data, he will ________ one of these hypotheses and ________ the other.

alternate
accept ↔
reject

2-85 We designate the assumed hypothesis as ________ and the alternate hypothesis as ________. One of these hypotheses must be an exact hypothesis; it is ________.

H_1
H_2
H_1

2-86 A hypothesis which states that there is no true difference between the populations is called a ________ hypothesis.

null

2-87 $\mu_1 - \mu_2 = 0$ and $\mu_1 - \mu_2 = 10$ are both examples of hypotheses which are ________; which one is a null hypothesis? ________

exact
$\mu_1 - \mu_2 = 0$

2-88 An alternate hypothesis need not be exact. $\mu_1 - \mu_2 \neq 0$ and $\mu_1 - \mu_2 > 0$ are both (*exact; inexact*) hypotheses. One of them is directional; which one? ______

inexact
$\mu_1 - \mu_2 > 0$

2-89 These two hypotheses must always be formulated ______ the data have been examined.

before

2-90 The investigator hopes to be able to accept ______, but he plans to accept ______ unless his data force him to reject it. He places the burden of proof on his data.

H_2
H_1

2-91 He adopts this conservative strategy because he wishes to avoid making an ______ claim.

unwarranted

2-92 An unwarranted claim is a type ______ error. It occurs whenever ______ is rejected while ______ is actually true.

I
H_1; H_1

2-93 He plans to determine whether his data could have occurred when H_1 is really true. His statistical analysis will tell him the ______ of getting such data when H_1 is true.

probability

2-94 This probability, to be determined from analysis of the data, is called ______. It will not become known until the data are analyzed.

p_o

2-95 However, he plans in advance to reject H_1 if p_o turns out to be no greater than a certain number. This number, which is called ______, is his ______ probability.

α; *rejection*

2-96 With $\alpha = .05$, H_1 will be rejected only if p_o ______ .05; that is, if the data would occur only ______ times in 100 when H_1 is true.

$\leq$
5

2-97 When α is chosen to be .05, we call .05 the ______ level or the ______ level planned for the experiment.

α
significance

2-98 If $\alpha = .05$ and p_o turns out to be .01, H_1 will be ______, and the results of the experiment will be said to be significant at the .01 ______. They are significant *beyond* the α ______.

rejected
level
level

2-99 The investigator always *plans* to reject H_1 when the probability of a type 1 error is not greater than ______. After he calculates p_o, he knows that the probability of a type I error is ______.

α

p_o

2-100 When an investigator wishes to be very cautious about making a type I error, he will pick a number for α which is relatively (*high; low*).

low

PROBLEMS

1. You are selecting a sample of 10 from a small (finite) population. You do not select the elements sequentially, one at a time, but instead take a handful of 10 elements all at once (a sort of grab-bag procedure). Will the sample meet the conditions for a random sample? Justify your answer.

2. Each of the following sampling procedures is to be classified as producing a random sample or as producing a biased sample. Consider each case, and decide whether the procedure violates the condition of equal chance, the condition of independence, or both, or neither.

a. The population about which inference is to be made is a population of scores in dart throwing, and the question to be decided is whether men achieve higher scores than women. The investigator selects 10 men and 10 women at random; he then obtains five scores from each, making a sample of 100 scores. Is this a random sample?

b. In attacking the same problem, the investigator takes 100 *averages* of five scores, each average coming from a different randomly selected person. Is this sample of average scores a random sample?

c. The population of interest is a population of attitude scores at Alpha College; assume that it is an infinite population. The experimenter wishes to obtain a sample of about 100 scores. He administers the attitude test to four existing groups, in this case four classes selected at random whose total enrollment is 100. Is this a random sample?

d. The same investigator interested in attitude scores sends out the attitude test by mail to a sample of 100 students whose names he has selected by taking every seventy-fifth name in the student directory after randomly choosing a starting point. He receives 71 completed tests; the other 29 students fail to respond. Is the original list a random sample? Is the sample of 71 tests random?

3. An experimenter tested for differences in attitudes toward smoking before and after a film on lung cancer was shown. He found a difference which was significant between the .05 and .02 levels.

a. What is the assumed hypothesis? State the hypothesis in words only.

b. Which level of significance indicates the *greater* degree of significance, .05 or .02?

c. If his α level is .05, will he reject H_1? Will he reject it if he employs the .02 level? In choosing $\alpha = .02$ instead of $\alpha = .05$, he *increases* the risk of making one of the two types of error. Which type?

4. A young scientist is testing an interesting hypothesis, and he wants to draw attention to it. He knows from his training that most scientists will not take seriously any reported difference which is not significant at least at the .05 level. Since he personally would not feel embarrassed to find later that he has made a type I error, should he publish right away his first experiment, which reaches significance at the .10 level, or should he do a second experiment?

LESSON 3 THE CONCEPTS OF PROBABILITY AND CONDITIONAL PROBABILITY

After H_1, H_2, and α have been chosen, the time arrives for collection and statistical analysis of the data. This analysis must produce a statement about the probability p_o that such data would arise when H_1 is really true. The concept of probability is central to data analysis, and you will be helped to understand this concept first in the way it applies to a particular case.

It will serve our purpose best to take a case in which the actual calculations of probability are quite simple. Neither of our previous examples will do because in each of those cases the observations can take on too large a number of values. We need an example in which the observations are as limited as possible, and so we choose a new case whose individual observations can assume only one or the other of two possible values.

Case 3 A woman claims that she can distinguish red paper from blue paper by touch alone, without the use of vision. We are skeptical of this claim, and we arrange a test of her alleged ability. In order to eliminate any possible visual information, we construct a box in which we shall present paper swatches to her hands. The box lies on a table, and she is to reach through a curtain into the box to feel the swatches. We also blindfold her to be doubly sure she can see nothing which might help in the discrimination. We select papers which are matched for texture, size, shape, and any other characteristics except color.

We might proceed in several different ways in the actual test, but let us suppose we choose to present a pair of swatches, one red and one blue, and ask her to choose the red one. This procedure is repeated 10 times, varying the relative positions of the colors in a random way so that position (right or left) does not reveal the color. The number of correct responses K is 9.

Case 3 is typical of a considerable number of research problems. We shall call them COIN-TOSS problems because the special problems of their analysis can be solved by analogy with the mathematical study of coin tossing. Case 3 is like a coin-tossing experiment in which a single coin is tossed 10 times and the face landing uppermost is noted each time. Both experiments have 10 TRIALS. Each trial produces an observation which can have either one or the other of only TWO values (correct or incorrect; heads or tails). What happens on any particular trial cannot affect what happens on any other trial; in other words, the 10 trials are INDEPENDENT of each other. Any experiment which consists of a *series of independent trials with only two outcomes per trial* will be classified here as a coin-toss problem.

All the principles developed in Lesson 2 for the other examples apply also to case 3, the color-touch example. This is also a two-variable problem, but when you start to identify the two variables and state the two hypotheses to be tested, you will encounter some oddities. For one thing, you will find it hard to say just what populations are being compared or what kind of a difference is being evaluated for significance, because in case 3 there is *only one sample.* Section A will help you think through the preliminary analysis of this case.

A. STATISTICAL HYPOTHESES FOR COIN-TOSS PROBLEMS

The color-touch example is like a coin-toss problem in which the question is raised, "Is this coin biased toward heads?" Such a question arises, usually, only if someone thinks the coin may behave "abnormally," giving a disproportionately large number of trials with heads as the outcome. A **FAIR** coin, one which is not biased, is defined as a coin which is equally likely to give heads and tails on any toss. On a large number of tosses, a fair coin is expected to give about the same number of heads as tails. On an infinitely large number of tosses, a fair coin will *by definition* give exactly as many heads as tails.

We can easily apply the terms "population" and "sample" to this coin-toss problem. Our sample is the set of tosses actually observed. If there are N tosses, our sample consists of N observations out of which a certain number K are heads and $N - K$ are tails. The population is the infinitely large number of possible observations that could be made by tossing this coin in exactly the same way each time. Our sample should tell us something about the nature of the population from which it is drawn.

Since there is a question of bias toward heads, we want to know whether the population contains more heads than tails. We can think of two hypotheses: (1) that this coin is a fair one; and (2) that this coin is biased toward heads. By letting P stand for the *proportion* (fraction) of heads in the population, we can write the hypotheses in the form of two equations, $P = 1/2$ and $P > 1/2$. Since only the first of these equations is exact, it becomes our assumed hypothesis. We write

$$H_1: P = 1/2 \qquad \text{and} \qquad H_2: P > 1/2$$

When we have obtained our sample, we shall ask, "What is the probability p_o that we would get a sample of this size containing at least 9 heads when H_1 is true?" If p_o is greater than our α level, we shall decide to accept H_1 and reject the claim that the coin is biased toward heads. But if p_o is not greater than α, we shall reject H_1 and accept H_2. In that case, we shall be saying, "There are so many heads in our sample that we cannot believe it really came from a fair-coin population. It is more reasonable to assume that it came from a population biased toward heads."

In the college-height example, we ask, "Could these *two* samples have come from the same population?" In the coin-toss problem we ask instead, "Could this single sample have come from a population with $P = 1/2$?"

3-1 In the coin-toss problem, each toss is a trial. In the color-touch case, each presentation of a pair of swatches is a ____________.

trial

3-2 The coin-toss trials are independent because the outcome of one trial cannot affect the outcome of another trial. The color-touch trials are also ____________ in this sense.

independent

3-3 But suppose we presented only one swatch on each trial and asked the woman to tell its color. If she knew that we planned to present five blue and five red swatches, would her tenth choice be truly independent of her previous choices? ____________

No

3-4 Such a procedure would *not* produce independent trials, since the knowledge that she had already named red four times and blue five times would make her more likely to name ____________ on the tenth trial.

red

3-5 An experiment is analogous to coin tossing only if it consists of a series of ____________ trials each of which can have only 1 of ________ possible outcomes.

independent
2

3-6 The color-touch example consists of a series of ________ independent trials. Each of these trials can have only 1 of 2 possible outcomes, ____________ and ____________.

10
correct ↔ incorrect

3-7 The experimental question is, "Can she distinguish red from blue by touch alone?" She must either get color information by touch, or she must make her choices by guessing. If she is only guessing, her choices will be like the tosses of a fair coin: there will tend to be equal numbers of ____________ and ____________ choices in a large sample.

correct ↔
incorrect

3-8 The sample in this case is the set of ____________ actually observed. The population is the infinitely large number of possible ____________ that she could make under the same experimental conditions.

choices
choices

3-9 If we let P represent the proportion of correct choices in the population, then the hypothesis that she is only guessing will be written mathematically like the hypothesis that a coin is fair: ________. (Write the hypothesis.)

$P = 1/2$

3-10 The hypothesis that she can get information by touch is the same as a hypothesis that she will make more correct choices than incorrect ones. It can be written mathematically as ________.

$P > 1/2$

In the color-touch case, then, we shall want to determine the probability that a sample of 10 trials would contain at least 9 correct choices when $H_1: P = 1/2$ is true. If p_o is greater than our α level, we shall continue to accept the hypothesis that she is only guessing. If p_o is not greater than α, we shall reject H_1 and accept her claim. The α level we choose should be a low one, such as .01, because her claim is one about which others are likely to be as skeptical as we are. If people can ever distinguish colors by touch alone, this ability is not widely recognized.

We have called this a two-variable problem, and it is time now to identify the two variables. One variable indicates what sort of observations we are studying; it is the variable "choices," which can have only two values (correct and incorrect). We get exactly 10 of these observations, and all of them can be sorted into these two categories. "Choices" is therefore a categorical variable.

Any variable which can take on only two values is called a DICHOTOMOUS variable. "Choices" is a dichotomous variable; its values divide the observations into just two categories—a *dichotomy*—of correct and incorrect choices.

The other variable separates the two populations being compared. These are the "guessing" population, for which information on color is *absent*, and the population with $P > 1/2$, for which some information on color must be *present*. We could name this variable in many ways, no doubt; a satisfactory way is to call it "color information" and name its two values "present" and "absent." Under our conditions, no other value is possible; however, this variable is not inherently dichotomous, since other experimental conditions might allow color information to be conveyed by vision, by use of detecting instruments, by word of mouth, and so on.

If p_o turns out to be no greater than $\alpha = .01$, we shall reject $H_1: P = 1/2$ and accept $H_2: P > 1/2$. In that case, we shall have decided that there is a significant difference between the population to which her actual choices belong and a population which arises through guessing. We shall be saying, "There are so many correct choices in this sample that we do not believe it really came from a population with no color information present. It is more reasonable to assume that some color information was present."

If we are sure that our experimental conditions did indeed rule out all other avenues of information, then our decision that the difference is significant is also a decision that some amount of association exists between red-blue choices and information conveyed by touch. We do not yet know the strength of this association, but it is safe to say that it will command intense interest even if it is weak, since most people will be surprised to hear that it exists at all.

Our particular example of a coin-toss problem is thus also an example of a ONE-SAMPLE problem from which a decision about the association of two variables can be made. Many two-variable problems have at least two samples, but this example shows that one sample can suffice under the proper experimental conditions.

3-11 The hypotheses for the color-touch case are exactly like those for a coin-toss problem in which there is a question of bias toward heads. Instead of testing for a bias toward heads, we are testing for a bias toward ________ choices.

correct

3-12 In each case, the assumed hypothesis states that no bias exists. This hypothesis in mathematical form is ______.

$P = 1/2$

3-13 If we can reject this hypothesis, we plan to accept the hypothesis $P > 1/2$. This hypothesis is our ________ hypothesis.

alternate

3-14 Any coin-toss experiment has a certain number of trials N, of which a certain number K turn up heads. The number of tails is therefore ______.

$N - K$

3-15 Since we observe 9 correct choices in 10 trials, $N =$ ______ and $K =$ ______.

10; 9

3-16 "Choices" is the variable which tells what kind of observations we are sampling. It has only two values; are these values discrete? ________

Yes

3-17 The values of the variable "choices" are category names; it is therefore a ________ variable. Because there are only two values, the observations fall into a dichotomy, and "choices" is called a ________ variable.

categorical

dichotomous

3-18 By statistical analysis we must determine the probability p_o of getting at least nine correct choices on ten trials when ______ is really true.

H_1

3-19 If p_o turns out to be .001, we shall know that at least nine correct responses on ten trials would occur only 1 time in ______ such experiments when H_1 is true.

1,000

3-20 Our statistical decision concerns the difference between two populations of choices: choices made with color information ________ and choices made by guessing when color information is ________.

present

absent

3-21 A significant difference will mean an ________ between two variables, choices and color information.

association

3-22 The college-height and vocabulary examples both involved at least 2 samples. The color-touch case is a ______-sample problem which permits a decision about the association between 2 variables.

1

B. PROBABILITY MEASUREMENT AS ASSIGNMENT OF WEIGHTS

The PROBABILITY of an event is a number which reflects its chance of occurrence. Our statistical analysis must enable us to assign such a number to the experimental outcome "9 correct choices in 10 trials."

Let us begin studying probability measurement by noticing something which you already know about probability numbers. If you are told that the probability of rain is 30 percent (.30), you will be able to tell the probability (according to the same prediction) that it will *not* rain. The very fact that probabilities can be expressed in percent figures implies that the probabilities of *all possible outcomes* of a particular experiment must add up to 1 (or 100 if the numbers are percentages). It will either rain or it will not rain; there are no other possible outcomes. Therefore, if the probability of rain is .30, the probability of no rain is $1 - .30$, or .70.

Thus the number representing the probability of a particular outcome is always one of a set of numbers which add up to 1. There is one number for each possible way the event might turn out, and none of these numbers can be negative (although it is possible for a number to be zero). In dichotomous cases (heads vs. tails, correct vs. incorrect, rain vs. no rain) there are only two members of the set, and it is traditional to call the two probabilities p and q. In other cases the set of probability numbers may have many members; for instance, in dice problems there will be a number for each of the six faces of the die. But whatever the number of elements in this set, the *sum* of the set of numbers must always be 1.

There is a name for the elements of such a set of numbers. Whenever a set of numbers whose sum is 1 is paired, one for one, with a corresponding set of objects, it is possible to call these numbers the WEIGHTS assigned to the objects. The objects in this case are outcomes. Probability measurement is the assignment of weights to the set of all possible outcomes of an experiment.

In a one-trial coin-toss experiment, there are only two possible outcomes. The probability of heads is p, the probability of tails is q, and $p + q = 1$. The value of p may be any number at all from 0 to 1; the value of q will always be $1 - p$. If we assume that the coin is a fair one, then we shall assign *equal* weights to heads and tails, and p will become $1/2$.

If our color-touch experiment had only one trial, we would already have the answer to our question about the value of p_o for the observed outcome. Since H_1 states that correct and incorrect choices are equal in the population, the probability p of a correct choice must equal the probability q of an incorrect choice. Both p and q equal $1/2$. If we observe only one choice and it is correct, we can say that an event has occurred with $p_o = 1/2$ when H_1 is true. With an α level of .01, we could not reject H_1 after only one trial.

3-23 Whenever an experiment has only two possible outcomes, the probability of one outcome is called p and the probability of the other outcome is called ______.

q

3-24 Since the sum $p + q$ must equal ______, q can always be determined when p is known; q always equals ______.

1
$1 - p$

3-25 If the two outcomes are equally probable, then p must equal q, and both must equal ______.

$1/2$

3-26 If the first outcome is *certain* to occur, $p = 1$ and $q =$ ______.

0

3-27 Suppose that a coin is biased so that heads is 3 times as likely to occur as tails. Then $p = 3/4$ and $q =$ ______.

$1/4$

3-28 When a set of numbers is assigned pairwise to a set of objects, and when the sum of the numbers is 1, the numbers are called the weights of those objects. Probabilities are assigned to the set of all possible ______ of an experiment; a set of probabilities must add to 1, and therefore the probabilities can be called ______.

outcomes
weights

3-29 Another kind of experiment requires drawing a bead at random from a bowl containing 10 red beads and 10 white ones. The bead drawn may be either red or white. The probability that it will be red is ______.

$1/2$

3-30 If the bowl contains 10 red, 10 white, and 10 blue beads, the probability that the bead drawn will be red is ______.

$1/3$

3-31 You have answered these two frames by deciding, rightly, that a probability weight may be assigned on the basis of the *relative number* of red beads in the container. If only $1/4$ of the beads are red, then the probability of drawing a red bead is ______.

$1/4$

3-32 We do almost the same thing when we consider the proportion of heads in a population of tosses of a fair coin. Since $P = 1/2$ for such a coin, it is reasonable to assign the probability $p =$ ______ to heads.

$1/2$

3-33 If the population contains 3 times as many heads as tails, we shall write $p =$ ______ and $q =$ ______.

$3/4$; $1/4$

3-34 If $1/4$ of the cards in a deck are spades, the probability that a card drawn at random will be a spade is ______. The probability that it will *not* be a spade is ______.

$1/4$
$3/4$

C. EMPIRICAL PROBABILITIES

In Lesson 4 we shall continue the statistical analysis of the color-touch problem and show how the value of p_0 is determined for the result "9 correct choices in 10 trials." For the moment, as we continue to develop the concept of probability, we shall need another kind of example.

Suppose a box contains 1,000 beads. Some of the beads are yellow, and some are green; some are round, and some are square. We take a random sample of 100 beads, obtaining the results shown in Table 3-1.

Table 3-1 Color and Shape of Beads in a Random Sample of 100 Beads

	Round	Square	Total
Yellow	10	30	40
Green	40	20	60
Total	50	50	100

The numbers in this table describe the frequencies with which yellow or green, round or square beads were found to occur in our sample. If we divide these frequencies by 100, we can describe the RELATIVE frequencies of the four kinds of beads as shown in Table 3-2.

Suppose we now replace all the beads drawn and prepare to draw a single bead at random. Can we state the probability of drawing a yellow bead? Not precisely, of course; we have not observed the entire population of 1,000 beads. However, the relative number of yellow beads in our sample is $40/100 = .4$, and we can *estimate* the number of yellow beads in the population as about 400. If this estimate is correct, the probability of drawing a yellow bead is .4. Because this probability is only an estimate, based on empirical (experimental) observation of a sample, we call it an EMPIRICAL PROBABILITY.

The event "yellow bead" is a SIMPLE event with empirical probability .4; the event "green bead" is a simple event with empirical probability .6. These probabilities appear as the row totals in Table 3-2. The probabilities of the simple events "round bead" and "square bead" appear as the column totals. The event "round yellow bead," on the other hand, is a complex, or JOINT, event. It represents the joint occurrence of "yellow bead" and "round bead," and its empirical probability is .1. If we use Y and G to represent the colors and use R and S to represent the shapes, we can write this joint event as $R \cup Y$ (read "R union Y") where the symbol $\cup$ from mathematical set theory denotes the UNION of two sets of simple events, round beads and yellow beads. The empirical probability of this joint event is then written $p(R \cup Y) = .1$.

Joint events arise when events can be classified in two or more different ways. In Table 3-2, the empirical probabilities of the four possible joint events appear in the four CELLS of the table, while empirical probabilities of the four possible simple events appear as the row and column (or MARGINAL) totals.

Table 3-2 Relative Frequencies of Bead Colors and Shapes in the Random Sample

	Round	Square	Total
Yellow	.1	.3	.4
Green	.4	.2	.6
Total	.5	.5	1.0

3-35 Table 3-1 shows the results of an experiment in which a random sample of ______ beads was drawn from a box containing ______ beads. — *100*; *1,000*

3-36 Out of the 100 beads drawn, ______ were yellow, and ______ were green. These frequencies appear as the (*column; row*) totals in Table 3-1. — *40*; *60*; *row*

3-37 Out of the 100 beads drawn, ______ were round, and ______ were square. These frequencies appear as the ______ totals in Table 3-1. — *50*; *50*; *column*

3-38 The event "round bead" is a (*joint; simple*) event. The event "round yellow bead" is a (*joint; simple*) event. — *simple*; *joint*

3-39 "Green bead" is a ______ event. Its frequency in this experiment was ______. — *simple*; *60*

3-40 "Square green bead" is a ______ event. Its frequency in this experiment was ______. — *joint*; *20*

3-41 When we divide any of the frequencies in Table 3-1 by the number of beads sampled, we obtain a relative ______; $40/100 = .4$ is the ______ of yellow beads. — *frequency*; *relative frequency*

3-42 We do not know how many of the 1,000 beads are yellow, but we estimate that there are about ______ yellow beads on the basis of our empirical observations. — *400*

3-43 We then take the observed relative frequency of yellow beads as the empirical ______ of drawing a yellow bead from this box. — *probability*

3-44 $p(G)$ is the empirical probability of drawing a green bead; $p(G) =$ ______. — *.6*

3-45 $p(S \cup G)$ is the empirical probability of the joint event, "______ bead"; $p(S \cup G) =$ ______. — *square green; .2*

3-46 Because we have information only about a sample, these relative frequencies are called ______ probabilities. — *empirical*

D. THE CONCEPT OF CONDITIONAL PROBABILITY

Suppose a friend has just drawn a bead from our box. He tells us that he has drawn a yellow bead; we must guess whether it is round or square.

The box contains 1,000 beads, and we are estimating that 400 of these are yellow. Of these 400, we think that about 100 are round. On the basis of these estimates, the empirical probability is $100/400 = .25$ that this yellow bead is also round. This probability is dependent upon our knowledge that the bead drawn is yellow; it is not a simple empirical probability, and it cannot be designated as $p(R)$. It is called a CONDITIONAL PROBABILITY because it applies only when a specific condition is met, namely, when the bead drawn is yellow. We say, "Given that a yellow bead is drawn, the probability is .25 that it is also round." We designate this conditional probability as $p(R \mid Y)$, "the probability of R given Y." The condition to be fulfilled *follows* the vertical line which signifies a conditional probability.

Table 3-3 shows the entire set of conditional probabilities which will apply to our example whenever the *color* of the bead is known. If the color is yellow, we consider only the estimated 400 yellow beads. Since we think that about 100 of these are round, $p(R \mid Y) = .25$, and $p(S \mid Y) = .75$. If the color is green, we consider only the estimated 600 green beads. We think about 400 of these are round, so that $p(R \mid G) = .67$, and $p(S \mid G) = .33$.

Table 3-3 Conditional Probabilities of Round and Square Beads Given Knowledge of Color

	Round	Square	Total
Yellow	$p(R \mid Y) = .25$	$p(S \mid Y) = .75$	1.00
Green	$p(R \mid G) = .67$	$p(S \mid G) = .33$	1.00

We could also derive these conditional probabilities directly from the joint- and simple-event probabilities in Table 3-2. For example $p(R \mid Y) = p(R \cup Y)/p(Y) = .1/.4 = .25$. In general, the conditional probability of A given B, $p(A \mid B)$, is the *joint* probability of A and B divided by the probability of the condition B: $p(A \mid B) = p(A \cup B)/p(B)$. This rule is convenient when information is given in the form of probabilities rather than frequencies.

Table 3-4 shows the conditional probabilities which will apply whenever the *shape* of the bead is known. Given that the bead is round, the probability of Y is .2 and the probability of G is .8. Given that the bead is square, the probability of Y is .6 and the probability of G is .4.

The conditional probabilities in Tables 3-3 and 3-4 are all empirical probabilities because they arise from the observation of a random sample from the population of 1,000 beads. If this entire population were examined, a set of *exact* probabilities could be drawn up to replace each of these tables. Because of sampling variability, the empirical probabilities may differ to some extent from the exact probabilities for this population.

Table 3-4 Conditional Probabilities of Yellow and Green Beads Given Knowledge of Shape

	Round	Square
Yellow	$p(Y \mid R) = .2$	$p(Y \mid S) = .6$
Green	$p(G \mid R) = .8$	$p(G \mid S) = .4$
Total	1.0	1.0

3-47 $p(R \mid Y)$ is the probability that the bead drawn will be ________ given that it is ________.

round; yellow

3-48 $p(Y \mid R)$ is the probability that the bead drawn will be ________ given that it is ________.

yellow; round

3-49 $p(R \mid Y)$ and $p(Y \mid R)$ are examples of ________ probabilities. In order for them to apply, the condition which (*follows; precedes*) the vertical line must be fulfilled.

conditional

follows

3-50 When we know that the friend has drawn a round bead, we have only the round beads to consider. We estimate that there are ________ round beads in the box.

500

3-51 Of these 500 round beads, we estimate from our sample that ________ are yellow. We therefore estimate that $p(Y \mid R) =$ ________ and that $p(G \mid R) =$ ________.

100

.2; .8

3-52 When we know that the friend has drawn a square bead, we have only the square beads to consider. We estimate that there are ________ square beads in the box, of which ________ are yellow.

500; 300

3-53 We estimate that $p(Y \mid S) =$ ________ and that $p(G \mid S) =$ ________.

.6

.4

3-54 We could have obtained $p(Y \mid S) = .6$ by the rule $p(Y \mid S) = p(Y \cup S)/p(S)$. Table 3-2 shows that $p(Y \cup S) =$ ________ and $p(S) =$ ________. Then $p(Y \cup S)/p(S) =$ ________.

.3

.5; .6

3-55 We could also use Table 3-2 to obtain $p(Y \mid R)$. To find $p(Y \mid R)$, we divide the joint probability ________ by ________, the probability of the condition to be fulfilled.

$p(Y \cup R)$

$p(R)$

3-56 From Table 3-2, $p(Y \cup R) =$ ________, and $p(R) =$ ________. Therefore, $p(Y \mid R) =$ ________.

.1

.5; .2

3-57 With full knowledge of the population, these probabilities would be exact. Since we are making estimates from a sample, they are ________ probabilities.

empirical

E. STATISTICAL SIGNIFICANCE AS A CONDITIONAL PROBABILITY STATEMENT

Conditional probability statements have many uses. The concept of conditional probability was introduced at this early point so that you might sharpen your understanding of statistical significance by recognizing that *every statement of statistical significance is a conditional probability statement.* Whenever we say that an experimental result is "significant at the .05 level," we are saying, "The conditional probability of this result is not more than .05 given that H_1 is true."

The implications of this fact are very important, well worth dwelling on for a moment. When we choose a rejection probability $\alpha = .01$ before obtaining any data, we make a conditional probability rule for decision making: "If the conditional probability of the result, given H_1, is less than or equal to .01, reject H_1." We then collect the data, examine the result, and calculate its conditional probability given that H_1 is true. When we find that this conditional probability is less than .01, we apply our *decision rule* and reject H_1, saying that this result is "significant *beyond* the .01 level." This phrase is merely a conventional way of stating that the probability of getting this result when H_1 is true is less than .01.

If you like, you can view the phrase "significant beyond the .01 level" as a reminder that a type I error may have been made. The investigator who reports a result "significant beyond the .01 level" is saying, "The probability is less than .01 that I have made a type I error in rejecting H_1." The probability of such an error is greater than 0, but it is small, and he is prepared to live with it. When the significance level is .05 or .10, he may still be prepared to live with it, but increasing numbers of his readers will not agree.

The probability of a type I error is the conditional probability of rejecting H_1 given that H_1 is true. Before data analysis, $p(\text{rejecting } H_1 \mid H_1) = \alpha$. It follows that the probability of accepting H_1 correctly, $p(\text{accepting } H_1 \mid H_1)$, is equal to $1 - \alpha$; the probability that we shall make a *correct* decision when H_1 is true is $1 - \alpha$ as long as we follow our decision rule. We can call the level $1 - \alpha$ our CONFIDENCE LEVEL. When we choose $\alpha = .05$, we arrange a decision rule which will lead to correct decisions in 95 percent of the cases in which H_1 is true.

It is very important to realize that this statement does not tell the absolute probability of a right decision. It merely tells the conditional probability of being right when H_1 is true. An α of .05 will lead to correct decisions in 95 percent of the cases in which H_1 is true and to incorrect decisions in 5 percent of the cases in which H_1 is true. But we have no means of knowing the number of cases we shall actually examine which are "cases in which H_1 is true." This number depends upon our choice of experimental ideas. A brilliant experimenter might have very few such cases to test. Someone else might encounter them quite often. An individual can never know how many times—out of all his statistical decisions—he will be led by his strategy to draw incorrect conclusions.

3-58 The conditional probability of A, given B, is written ______. | $p(A \mid B)$

3-59 In statistical analysis of an experimental result, we calculate the conditional probability of the result, given H_1. If we let A stand for the result, we can write this conditional probability as ______. | $p(A \mid H_1)$

3-60 When we choose a rejection probability α, we plan to reject H_1 if $p(A \mid H_1)$ turns out to be no greater than ______. This plan becomes our ______ rule. | α; *decision*

3-61 If $p(A \mid H_1)$ turns out to be less than α, we shall state that our result A is ______ beyond the level α. | *significant*

3-62 "A is significant at the .05 level" means that ______ is not greater than ______. | $p(A \mid H_1)$; *.05*

3-63 "A is significant beyond the .05 level" means that $p(A \mid H_1)$ is (*less than; equal to; greater than*) .05. | *less than*

3-64 p(rejecting $H_1 \mid H_1$) is the probability of (*a correct; an incorrect*) decision when H_1 is true. | *an incorrect*

3-65 p(rejecting $H_1 \mid H_1$) is the probability of a type ______ error. Before data analysis, the probability of such an error is ______. | *I*; α

3-66 p(accepting $H_1 \mid H_1$) is the probability of a correct decision when ______. Before data analysis, this probability is equal to ______. | *H_1 is true*; $1 - \alpha$

3-67 When $\alpha = .02$, the probability of a type I error is ______. | *.02*

3-68 When $\alpha = .02$, can you state the probability that the decision made will be correct? ______ | *No*

3-69 You can state only the probability of a correct decision given ______. This probability is ______. | *H_1; .98*

3-70 Sometimes .98 is called the "98 percent confidence level." When $\alpha = .02$, one can be 98 percent confident that he is not making (*an incorrect decision; a type I error*). | *a type I error*

3-71 When the probability of a result, given H_1, is *equal* to .05, the result is significant ______ the .05 level. When the same probability is *less than* .05, the result is significant ______ the .05 level. | *at*; *beyond*

REVIEW

3-72 The color-touch example introduced in this lesson is a type of problem which we are calling ________ problems. *coin-toss*

3-73 Coin-toss problems always have a series of ________ trials, and each trial has only ______ possible outcomes. *independent* *2*

3-74 The number of trials is represented by the letter ______ . One of the two outcomes occurs K times; the other outcome occurs ______ times. N $N - K$

3-75 In the color-touch example, $N =$ ______. There are 9 correct choices and ______ choice that is incorrect. *10* *1*

3-76 Since "choices" is a variable with only two values, it is an example of a ________ variable. *dichotomous*

3-77 We may do a coin-toss experiment to find whether a coin is biased toward heads. Since the hypothesis that it *is* biased requires proof, it becomes (H_1; H_2). H_2

3-78 The assumed hypothesis is that the coin is fair. We let P stand for the proportion of heads in an infinite number of tosses of this coin, and we define a fair coin as one for which $P =$ ______. $1/2$

3-79 When we toss the coin one time, we obtain a very small sample from this infinite population. We let p represent the probability of getting a head on this single toss; the probability of getting a tail is called ______. q

3-80 H_1 states that the number of heads in the population is equal to the number of tails. Therefore, if H_1 is true, the probabilities p and q must also be ________. *equal*

3-81 A set of probabilities is a set of weights assigned to the possible outcomes of an experiment. Since the sum of such a set of weights is 1, the sum $p + q$ must be 1 and both p and q must equal ______ for this fair coin. $1/2$

3-82 If a box contains 10 yellow beads and 30 green beads, the probability of drawing a yellow bead at random is ______. This is the probability of a (*joint; simple*) event. $1/3$ *(or .33)* *simple*

3-83 If 6 of the 10 yellow beads are also round, the probability of drawing a round yellow bead at random is ______; this is the probability of a ________ event. $1/5$ *(or .2)* *joint*

3-84 To represent a joint event, we use a symbol called union from mathematical set theory. This symbol is ______. *$\cup$*

3-85 Letting Y represent "yellow" and R represent "round," we write the event "round yellow bead" as follows: ______. *$R \cup Y$*

3-86 $p(R \cup Y) =$ ______, and $p(Y) =$ ______. *$1/5$; $1/3$*

3-87 If we know that a randomly drawn bead is yellow, the probability is ______ that it is also round. This is a conditional probability; it applies only when a condition is fulfilled, namely, that the bead is ______. *.6* *yellow*

3-88 To represent this conditional probability, we write ______ $= .6$. *$p(R \mid Y)$*

3-89 In general, the probability of A given B can be found by the rule, $p(A \mid B) = p(A \cup B)/p(B)$. Applied to this example, $p(R \mid Y) =$ ______ divided by ______; therefore, $p(R \mid Y) =$ ______. *$1/5$; $1/3$* *$3/5$ (or .6)*

3-90 A statement of statistical significance is actually a ______ statement. *conditional probability*

3-91 A result which is "significant at the .01 level" is a result whose probability is not greater than ______ under the condition that ______. *.01* *H_1 is true*

3-92 A result which is "significant beyond the .01 level" is a result whose probability, given ______, is ______ .01. *H_1; less than*

3-93 We let A stand for an observed result. Then $p(A \mid H_1)$ is the ______ probability of getting A, given ______. *conditional; H_1*

3-94 When $\alpha = .05$, one plans to reject H_1 if $p(A \mid H_1)$ is not greater than ______. *.05*

3-95 When $\alpha = .05$, the probability of a type I error is ______. *.05*

3-96 The probability of a type I error is the probability of ______ H_1 when H_1 is true. *rejecting*

3-97 When $\alpha = .05$, the probability of *accepting* H_1 when H_1 is true is equal to ______. *.95*

3-98 If one examines 100 cases in which H_1 is actually true, a decision rule with $\alpha = .05$ will lead to acceptance of H_1 in about ______ of those 100 cases. *95*

PROBLEMS

1. In the color-touch example, H_1 is $P = 1/2$. Is this a null hypothesis?

2. A friend of yours claims that he can tell the difference between instant coffee and fresh-perked coffee. To test his claim, you arrange 10 pairs of samples, each pair containing one sample of each kind of coffee, and ask him to identify the instant coffee sample in each pair.

a. What is your assumed hypothesis?

b. What are the three possible alternate hypotheses?

Do not overlook the possibility that your friend can tell a difference between the two kinds of coffee but thinks that the instant is the fresh-perked coffee and therefore calls the samples by the wrong names.

3. A college has 500 women students and 1,000 men students. The introductory zoology course has 90 students, of whom 50 are women. Suspecting that women tend to elect zoology more frequently than men do, you decide to test your hypothesis statistically with the data from this class.

a. What is the probability that any student drawn at random from this college population will be a woman? What is the probability of drawing a man at random?

b. State your assumed hypothesis verbally. What is your alternate hypothesis?

c. Is this a one-sample or a two-sample case?

d. What are the two variables whose association is being tested?

4. In the following problems assume a standard deck of playing cards, with 13 cards from each of the four suits.

a. A single card is drawn at random. What is the probability that the card drawn is a spade?

b. A second card is drawn from the same deck after the first card drawn was replaced in the deck and the deck was reshuffled. What is the probability that the new card drawn is a spade?

c. A spade has been drawn from the deck and *laid aside*. Now still another card is drawn from those cards remaining in the deck. What is the probability that the new card drawn is a spade? What is the probability that it is a club?

d. What is the probability of drawing a card from a black suit in a random draw from the complete deck? What is the probability of getting either a heart or a diamond?

e. What is the probability of getting the queen of hearts on a single random draw? What is the probability of getting a queen, regardless of suit? What is the probability of getting either a queen or a jack?

f. What is the probability of getting a face card (king, queen, or jack) on a single random draw?

5. A die has six faces. When it is unbiased, what is the probability of getting a ⚀ on a single toss of the die?

6. Suppose a die has been loaded, so that the ⚀ face lands uppermost 3 times as often as any other face, while all the other faces occur equally often. What is the probability of a ⚁ on a single toss? What is the probability of a ⚀ ?

LESSON 4
CONDITIONAL PROBABILITY DISTRIBUTIONS FOR COIN-TOSS PROBLEMS

A probability is a number assigned to a particular outcome, reflecting its chances of occurrence. A PROBABILITY DISTRIBUTION is a set of such numbers, one for each possible outcome that might occur.

In Lesson 3B we have already studied a very simple probability distribution, the distribution for one-trial coin-toss problems. Such an experiment has only two outcomes. If we assume that $P = 1/2$, then each outcome is assigned the probability $1/2$. If we assume that $P = 3/4$, then one outcome has probability $3/4$ and the other has probability $1/4$. Each of these assumptions leads to a probability distribution which is conditional upon that assumption. Each distribution is therefore a *conditional* probability distribution.

In Lessons 4 and 5 we are going to study the conditional probability distributions for coin-toss problems in detail. We shall proceed as if our only object were to make a statistical decision about the color-touch case. However, the study of coin-toss problems will introduce the principles of probability and several other concepts which are basic to statistical analysis.

As soon as we increase the number of trials from one to two, we increase the number of ways our experiment can turn out. The total number of different outcomes will be four, and a probability distribution must have a number for each outcome. A TREE DIAGRAM will show how the four outcomes of a two-trial coin-toss experiment can be visualized. The choice on trial 1 may be

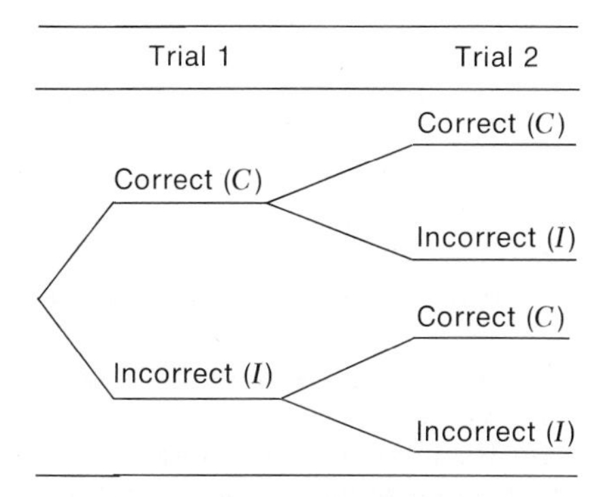

correct (C), and this choice may be followed by a correct or by an incorrect (I) choice on trial 2. These two outcomes can be written as CC and CI. The two outcomes in which the trial 1 choice is incorrect will be written IC and II.

A. FINDING THE TOTAL NUMBER OF OUTCOMES

Imagine, or draw for yourself, a tree diagram for a three-trial experiment. Each of the four branches on trial 2 will now have two new branches under trial 3. There will be eight branches for trial 3.

Every time you add an extra trial, you double the previous number of possible outcomes. For one trial there are two outcomes; for two trials, four; for three trials, eight outcomes. The rule can be put another way: Multiply together as many 2s as there are trials in the experiment. For one trial, take 2 alone; for two trials, multiply 2 times 2; for three trials, multiply 2 times 2 times 2; and so on. If there are N trials, you will need to multiply N 2s together.

Of course, the simplest statement of this rule is in terms of exponents of the number 2: the number of outcomes equals 2 to the Nth power, or 2^N.

When we want to assign a probability to each of the possible outcomes, we must examine our assumed hypothesis. When H_1 is $P = 1/2$, the probability of a correct choice on any single trial is $1/2$; for H_1: $P = 3/4$, the probability of C on a single trial is $3/4$. Since the value of p is the same as the population proportion P, we shall refer to these hypotheses hereafter as $p = 1/2$ and $p = 3/4$.

Suppose we do a large number of two-trial experiments. If $p = 1/2$, half of them should have C on trial 1. Half of these experiments with C on trial 1 should have C also on trial 2. Therefore, the proportion of experiments with CC as their outcome should be half of $1/2$, or $1/4$. The probability of the joint event CC is $1/2$ times $1/2$, or $1/4$. This is the product of the probabilities found on the pathway leading to outcome CC in the diagram below.

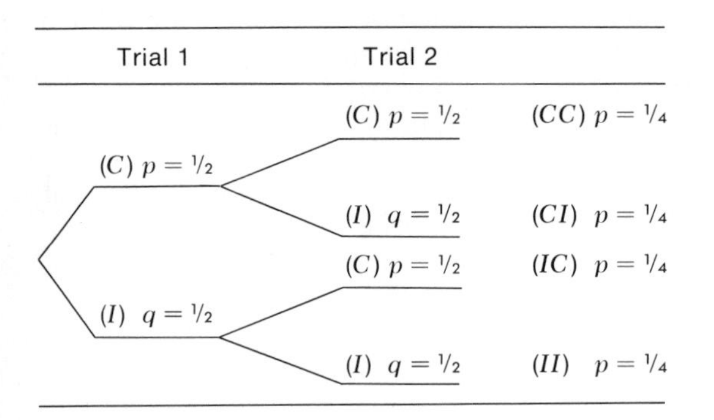

If we assume instead that $p = 3/4$, the product of the probabilities on pathway CC will be $3/4$ times $3/4$, or $9/16$. The product of the probabilities on pathway CI will be $3/4$ times $1/4$, or $3/16$.

We shall call this the PRODUCT RULE for joint probabilities of independent events. Whenever the events A and B are independent, the probability of the joint event AB is equal to the product of the probabilities of the simple events. A coin-toss experiment has independent trials; therefore, the events "C on trial 1" and "C on trial 2" are independent, and we can obtain the probability of CC by multiplying p times p.

4-1 In a 1-trial experiment, $N =$ ______, and the total number of different possible outcomes is ______. In this case $2^N =$ ______. *1* *2* *2*

4-2 In a 2-trial experiment, $N =$ ______, and the total number of different possible outcomes is $2^N =$ ______. *2* *4*

4-3 Fill in the remaining values:

N	Number of Different Outcomes	
1	2	
2	4	
3	______	*8*
4	______	*16*
5	______	*32*
6	______	*64*
7	______	*128*
8	______	*256*
9	______	*512*
10	______	*1,024*

4-4 In the color-touch example, $N =$ ______, and the total number of different possible outcomes is ______. This number is 2 to the ______ power. *10* *1,024* *10th*

4-5 If we can assign to each of these 1,024 outcomes a weight reflecting its chances of occurrence, we shall have a ______ distribution for the color-touch example. *probability*

4-6 We can already draw up a probability distribution for any experiment with $N = 2$. It must be a conditional distribution, and we must first decide what value of p to ______. *assume*

4-7 If we assume $p = 3/4$, the outcome CI has an event on the first trial whose probability is ______ and an event on the second trial whose probability is ______. $3/4$ $1/4$

4-8 The probability of CI is the ______ of the probabilities of its component events C and I, because these events are ______ of each other. *product* *independent*

4-9 The probability of CI is ______. The probability of IC is ______. The probability of II is ______. The probability of CC is ______. $3/16$ $3/16$; $1/16$ $9/16$

4-10 The sum of the probabilities of the four outcomes is ______. *1*

B. PARTITIONING THE OUTCOMES INTO SUBSETS

It is important to know that there are 2^N different outcomes of an N-trial coin-tossing experiment, but we may not actually need to know the probability for every single one. In the color-touch case a record is kept of the number K of correct choices, but no particular attention is paid to the *order* in which the correct and incorrect choices occur.

The outcomes CI and IC are actually different outcomes in a two-trial experiment, yet both have $K = 1$. If we are counting the number of correct choices without regard to order, there are only three values of K that can occur in a two-trial experiment: $K = 0$, for which there is just one outcome II; $K = 1$, for which there are the two outcomes CI and IC; and $K = 2$, for which there is one outcome CC. Since K is the statistic we are interested in studying, we can partition the total set of outcomes into a series of SUBSETS, each with a different value of K, and then count up the number of outcomes in each subset. A little reflection will show that for N trials there must always be $N + 1$ subsets because K may take any whole-number value from 0 to N.

Once we know the assumed values of p and q, we can find the probability of any outcome with a given value of K. In a two-trial experiment the probability of any outcome with $K = 2$ is $p \cdot p$; the probability of any outcome with $K = 1$ is $p \cdot q$; and the probability of any outcome with $K = 0$ is $q \cdot q$.

Let us write these facts in a different way, using the exponents 0, 1, and 2. Remember that $p^1 = p$ and that $p^0 = 1$.

Outcome	K	$N-K$	Probability
CC	2	0	p^2q^0
CI	1	1	p^1q^1
IC	1	1	p^1q^1
II	0	2	p^0q^2

From this analysis you can derive a rule: the probability of a particular outcome with K correct is p^Kq^{N-K}. This rule will hold for all values of N and K.

However, we do not yet know the probability of $K = 1$, which has two orders in which it can occur. Either CI or IC will count as $K = 1$, and we have a SUMMATION RULE for such either-or cases: when events A and B are mutually exclusive, the probability of getting *either* A *or* B is equal to the sum of the probabilities of A and B. Since the probability of CI is p^1q^1 and the probability of IC is also p^1q^1, the probability of $K = 1$ is $2p^1q^1$.

We can now draw a picture of the probability distributions for $N = 2$ when $p = 1/2$ and when $p = 3/4$ (Fig. 4-1). We put the discrete values of K on the horizontal axis and draw a vertical bar for each value, letting the height of the bar represent the probability of getting that value of K. Such a bar graph is called a HISTOGRAM. We think of the width of each bar as 1 unit; since the bar height is the probability of K, the area of the bar equals 1 times $p(K)$. Both the bar's height and its area thus represent its probability $p(K)$. The total area in *all* the histogram bars will always be 1.

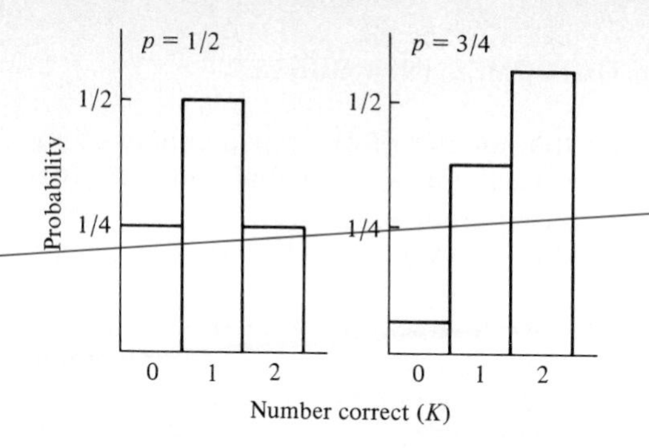

Figure 4-1: Histograms of the probability distributions for $N = 2$, $p = 1/2$ (*left*) and $p = 3/4$ (*right*).

4-11 When we partition a 2-trial experiment into subsets according to K, we obtain 3 such subsets. The number of subsets is always equal to N plus ______.

1

4-12 The probability of any particular outcome can be found by taking $p^K q^{N-K}$. When $N = 2$, the probability of an outcome in subset $K = 1$ is ______.

$p^1 q^1$ (*or* pq)

4-13 According to the summation rule, the probability of getting either one or the other of two mutually exclusive events is equal to the ______ of the probabilities of the separate events.

sum

4-14 When $N = 2$, there are two outcomes with $K = 1$. The probability of getting a *particular* one of these outcomes is ______. The probability of getting either one or the other of these outcomes is ______.

pq
$2pq$

4-15 Write the probabilities for the 8 outcomes of a 3-trial experiment, assuming $p = 3/4$.

Outcome	K	Probability	
CCC	3	$p^3q^0 = 27/64$	
CCI	2	______	$9/64$
CIC	2	______	$9/64$
ICC	2	______	$9/64$
IIC	1	______	$3/64$
ICI	1	______	$3/64$
CII	1	______	$3/64$
III	0	______	$1/64$

4-16 With $N = 3$, the number of subsets according to K is ______. The number of outcomes in subset $K = 2$ is ______. The probability of getting $K = 2$ is ______.

4
3; $27/64$

C. COUNTING THE NUMBER OF OUTCOMES IN A SUBSET

Our next task is to learn how to count the number of outcomes contained in each of the $N+1$ subsets with different values of K. We shall illustrate the method first by a triangle which gives these numbers for experiments of up to five trials. Figure 4-2 has five rows, one for each value of N from 1 to 5. There are six diagonals, one for each value of K from 0 to 5.

Each cell in the triangle is located at the intersection of a row and a diagonal. These cells have been left empty so that you can write a number in each one; the number you write should represent the number of outcomes in the subset K for the row N. Start with the top row, which has only two cells. There are only two possible outcomes when $N = 1$; for one of these $K = 0$, and for the other $K = 1$. Write the number 1 in each of these two cells.

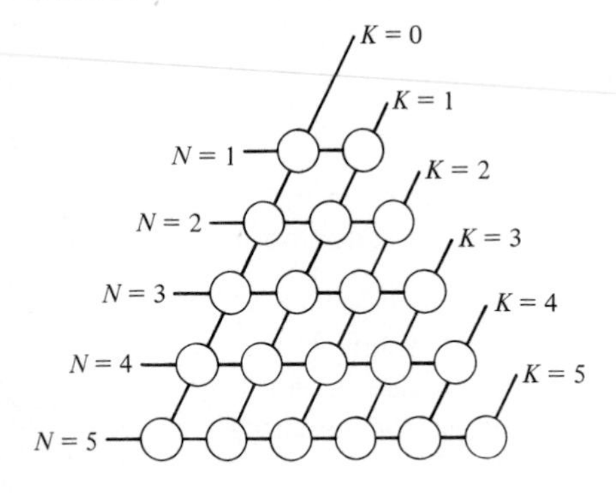

Figure 4-2: How to count the outcomes in each subset for experiments of 1, 2, 3, 4, and 5 trials. Number of trials is represented by N; number of correct responses is represented by K.

Now fill in the remaining cells by following these steps:

1 The diagonal for $K = 0$ contains cells which can only have the number 1, since there is always only 1 outcome with 0 correct no matter how many trials the experiment includes.

2 The second cell in each row, which is always on the diagonal $K = 1$, will always contain a number equal to N for that row. The result "1 correct" can always occur in just N outcomes, since there will always be exactly N positions (trials) in which that single correct choice can appear.

3 The last cell in each row must always contain the number 1. It is always the cell for $K = N$, and the result "all correct," like the result "0 correct," can occur in only one way.

4 The next to last cell in each row is the cell for $K = N - 1$, the subset "1 incorrect." Like the result "1 correct," this result can occur in just N outcomes; there will always be exactly N trials on which that single incorrect choice can appear.

5 The number in any cell is exactly equal to the sum of the two numbers directly above it in the preceding row.

4-17 Take the row for $N=2$. The middle cell is for $K=$ _______, and the number in the cell is _______. *1; 2*

4-18 Whatever the value of N, the cells in the diagonal $K=1$ must always have numbers equal to _______. *N*

4-19 The second cell in row 5 shows the number of outcomes giving $K=$ _______ when $N=5$. The number is _______. *1; 5*

4-20 The last cell in each row is for values of K which are equal to N. The next to last cell is for values of K equal to _______. *N − 1*

4-21 When $K=N-1$, there are $N-1$ correct choices and _______ incorrect. The number in the next to last cell of each row must equal the number _______ for that row. *1* *N*

4-22 In the third row, $N=$ _______. The number in the second cell is for $K=$ _______, and the number is _______. *3* *1; 3*

4-23 The number in the next to last cell in the third row is for $K=N-1$, and it is also equal to _______. *3*

4-24 The sum of all four numbers in the third row is _______. The total number of possible outcomes when $N=3$ is $2^3=$ _______. *8* *8*

4-25 The numbers in row 4 are, in order from left to right, _______, _______, _______, _______, and _______. *1; 4; 6; 4; 1*

4-26 The number 6 in the third cell is the sum of the two numbers nearest to it in the preceding row; these numbers are _______ and _______. *3; 3*

4-27 The total number of outcomes when $N=4$ is _______. The sum of the five numbers in row 4 is _______. *16* *16*

4-28 The numbers in row 5 are, in order from left to right, _______, _______, _______, _______, _______, and _______. *1; 5; 10; 10; 5; 1*

4-29 The number of outcomes in the subset for $K=2$ and $N=5$ is _______. *10*

4-30 When $N=5$, the probability of any *particular* outcome with two correct is _______. The probability of getting *some* outcome in the subset $K=2$, no matter which one, is _______. p^2q^3 $10p^2q^3$

4-31 When $N=5$, the probability of getting some outcome in the subset $K=1$ is _______. The probability of getting $K=5$ is _______. The probability of $K=0$ is _______. $5pq^4$ p^5; q^5

As we extend this table, called the PASCAL TRIANGLE, to experiments of up to 10 trials, it will be convenient to have a shorter way of referring to the cells. Instead of saying, "The cell for $N = 5$ and $K = 2$," we shall give that cell the label $\binom{5}{2}$, in which the upper number is understood as N and the lower number is understood as K. Expressions of the form $\binom{N}{K}$ will mean "the number of outcomes of N-trial experiments in which there are exactly K correct choices." The expression $\binom{N}{K}$ always stands for a single number, and we can speak of "the number $\binom{N}{K}$."

Any number $\binom{N}{K}$ can be found by means of a simple rule making use of FACTORIAL NUMBERS. A factorial number is written with the symbol ! immediately following it; factorial 2 is written 2!, factorial 4 is written 4!, and so on. Factorial 2 means the product of 2 and all the positive whole numbers less than 2; in other words, $2! = 2 \cdot 1 = 2$. Similarly, $3! = 3 \cdot 2 \cdot 1 = 6$, and $4! = 4 \cdot 3 \cdot 2 \cdot 1 = 24$. The expression 2! is usually read "2 factorial."

The rule † for finding any number $\binom{N}{K}$ is stated in terms of factorial numbers.

$$\binom{N}{K} = \frac{N!}{K!(N-K)!}$$

To find the number $\binom{5}{2}$, substitute in this expression, writing $5!/[2!(5-2)!]$. When you write out the factors in these factorial numbers, you find that the task of multiplication is not arduous:

$$\frac{5 \cdot 4 \cdot \not{3} \cdot \not{2} \cdot \not{1}}{(2 \cdot 1)\,(\not{3} \cdot \not{2} \cdot \not{1})} = \frac{5 \cdot 4}{2} = 10$$

If you look back now at your Pascal triangle on page 56, you will observe that 10 is indeed the number you have entered for $N = 5$ and $K = 2$.

By means of this rule you will be able to supply the number for any cell in a Pascal triangle without knowing the numbers in the surrounding cells. You will meet only one kind of puzzle: in evaluating $\binom{N}{K}$ where $K = 0$ or $K = N$, you may not know what to do about factorial zero, 0!. Factorial 0 is simply defined as 1. Therefore, $\binom{5}{0}$ will be found by $5!/0!5!$ and equals 1, since 5! appears in both numerator and denominator.

---

† The rule for evaluating $\binom{N}{K}$ is often called the COMBINATIONS rule. In the most general sense, it is the number of different combinations of N things taken k at a time.

4-32 In the context of the color-touch case, the N in $\binom{N}{K}$ is the number of ________ in an experiment and K is the number of ________ choices.

trials
correct

4-33 When we want to find the number of outcomes in a 10-trial experiment which have exactly 8 correct choices, we need to evaluate the number ________.

$\binom{10}{8}$

4-34 $\binom{N}{K} = \frac{N!}{K!(N-K)!}$. Therefore, $\binom{10}{8}$ = (*write the factorial fraction*) ________.

$\frac{10!}{8!2!}$

4-35 The value of the number $\binom{10}{8}$ is ________.

45

4-36 The probability of getting a *particular* one of these 45 outcomes is ________.

p^8q^2

4-37 The probability of getting some member of the subset $K = 8$, no matter which one, is ________.

$45p^8q^2$

4-38 If $p = 1/2$, $p^8q^2 = (1/2)^8(1/2)^2 = (1/2)^{10}$. This quantity is the same as $1^{10}/2^{10}$, or 1 divided by 2^{10}. You found the value of 2^{10} on page 53; it is ________. The probability p^8q^2 equals ________ when $p = 1/2$, and the probability of getting $K = 8$ is $45p^8q^2$ = ________.

1,024; 1/1,024

45/1,024

4-39 Now consider the subset $K = 5$. The probability of getting a *particular* outcome in this subset is p^5q^5, or $1/2$ to the ________ power. The probability p^5q^5 equals ________ when $p = 1/2$.

10th; 1/1,024

4-40 The number of outcomes with $K = 5$ is found by the number $\binom{N}{K}$ = (________). The value of this number is ________.

$\binom{10}{5}$; *252*

4-41 Complete the following table of probabilities in the upper half of the distribution for $N = 10$ and $p = 1/2$.

K	$\binom{N}{K}$	Probability of Getting an Outcome in Subset K
5	252	252/1,024
6	________	________
7	________	________
8	45	45/1,024
9	________	________
10	________	________

210; 210/1,024
120; 120/1,024

10; 10/1,024
1; 1/1,024

D. STATISTICAL TESTS

You have now seen how a conditional probability distribution can be derived for coin-toss problems, and you have calculated part of the distribution needed to make a decision about the color-touch case. This is the distribution for experiments of $N = 10$, given that H_1: $p = 1/2$ is true. Figure 4-3 is its graph.

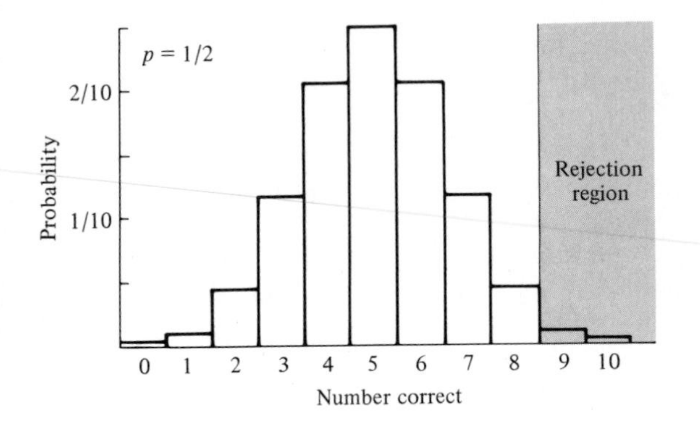

Figure 4-3: Probability distribution for $N = 10$, $p = 1/2$. Shaded area represents rejection region for H_1: $p = 1/2$ when H_2 is $p > 1/2$ and $\alpha = .01$.

At the right end of this distribution a shaded region has been marked REJECTION REGION. This is the region of rejection of H_1 and acceptance of H_2. If the result of the color-touch experiment falls into this region, H_1 may be rejected at the .01 level of significance.

To understand how this region is located, remember that an α level of .01 was chosen because of the unexpectedness of the woman's claim. We do not want the probability of a type I error in this case to be greater than .01. We therefore made a decision rule: if p_o, given H_1, is not greater than .01, reject H_1. Now suppose we get 9 correct. The probability of *exactly* 9 correct is .010, given H_1. But there is also a chance of getting an even more extreme result, $K = 10$; its probability is .001. Such a result would be even more damaging to H_1 and even more in line with H_2, and so we must consider its probability also. The probability of getting *either 9 or 10* correct is the sum of the two separate probabilities, .011. If we get $K = 9$, we can therefore say that the probability of getting a result *at least this extreme* under H_1 is not greater than .011.

Our α level is described to only two decimal places, and there is no difference between .010 and .011 when we round off to two places. Therefore, we mark a boundary for our rejection region at $K = 9$; the probability of getting a result lying anywhere in this region is .01.

The value $K = 9$, which marks the edge of this rejection region, is the CRITICAL VALUE of K for our example. If the experiment yields $K < 9$, H_1 will be accepted. If it yields $K \geqslant 9$, the probability p_o of the observed result will not be greater than α and H_1 will be rejected.

4-42 Someone planning a color-touch experiment will choose N, H_1, H_2, and α before he begins to collect his data. He can calculate the conditional probability distribution from a knowledge of just two of these, ______ and ______.

N; H_1

4-43 He can then designate a rejection region whose probability is not greater than ______. If he gets a value of K lying in this rejection region, he will reject ______.

α
H_1

4-44 The critical value of K is the value at the boundary of the ______ region.

rejection

4-45 In our example the value of α is .01, and 9 is the ______ value of K.

critical

4-46 Any value of K which is equal to or greater than 9 will lie in the ______.

rejection region

4-47 The probability of $K = 9$ is .010; the probability of $K = 10$ is .001. The probability of getting a result which lies somewhere in the rejection region, given H_1, is ______.

.011

4-48 We reject H_1 when we get $K = 9$ because the probability of (*at least; exactly*) 9 correct is not greater than α.

at least

4-49 If we had chosen $\alpha = .001$, the critical value of K would be ______.

10

4-50 The probability of $K = 8$ is 45/1,024, or .044. The probability of at least 8 correct is ______.

.055

4-51 If we had chosen $\alpha = .05$, the probability of at least 8 correct would be almost exactly equal to α. We might choose to regard it as close enough; our critical value would then be ______, and the rejection region would then include the bars for $K =$ ______, ______, ______. (Give the values of K.)

8
8; 9; 10

4-52 With $\alpha = .01$, we know that we shall reject H_1 if we get a value of K which is at least ______. We also know that the probability of getting such a large value of K is ______ when H_1 is actually true.

9
.011

4-53 Thus we are deliberately planning to reject H_1 on the grounds of a result which *could* happen when H_1 is true. We know our plan might lead to a type I error, rejecting ______ falsely.

H_1

4-54 But we know that the probability of such an error is not great; it is exactly ______.

.01 (or α)

The probability distribution you have used in drawing conclusions about the color-touch example has served as a STATISTICAL TEST of H_1: $p = 1/2$. You will learn to use many other kinds of statistical tests as you study this book. All of them operate in the same manner as this one, and most of them are named for the probability distribution they employ. Lesson 5 will explain that the one you have just used is a BINOMIAL test.

A statistical test always starts with some statistic which can be calculated from the sample data. Our statistic was K. The statistical test then produces, by logical deduction from H_1, a set of probabilities for the values which the sample statistic might assume. The test then identifies a rejection region in the probability distribution. The boundary of the rejection region constitutes a critical value for the statistic; when the statistic surpasses that value, H_1 can be rejected at the α level of significance.

The location of this rejection region depends on α, as you already know. We shall now show that it also depends on H_2.

In our example we chose H_2: $p > 1/2$ to represent the claim that a discrimination between red and blue can be made by touch alone. This alternate hypothesis is directional. Since we felt justified in looking for a difference in a specific direction, we did not choose H_2: $p \neq 1/2$. We also did not consider $p < 1/2$, since a difference in this direction would not coincide with the woman's claim.

The hypothesis $p > 1/2$ would also be chosen in a coin-toss experiment testing whether a coin is biased toward heads, and the binomial test would be used in exactly the way we have used it for our example. If we had to test for a bias toward *tails*, H_2 would be different, and the rejection region would lie at the opposite end of the distribution. This case requires H_2: $p < 1/2$, and the rejection region for $\alpha = .01$ would include $K = 0$ and $K = 1$. The critical value would be 1, and we would say that the probability is not greater than .01 of getting a value of K which is at least as *low* as 1.

Finally, we might have to test simply whether the coin is biased. We would then use the nondirectional hypothesis $p \neq 1/2$. In this case a value of K might be either too large or too small, and the rejection region would be divided between the two extreme ends of the distribution. Since the distribution is symmetrical, each part of the rejection region would contain values of K with probability not greater than $\alpha/2$. For $\alpha = .01$, each part of the rejection region could contain only $p \leqslant .005$; these parts would be $K = 10$ and $K = 0$. These two values would both be critical values of K.

When H_2 is directional, the rejection region has only one part. It lies at one extreme, or TAIL, of the probability distribution, and the statistical test is said to be a ONE-TAILED TEST. When H_2 is nondirectional, the rejection region has two parts, one in each tail, and the test is said to be a TWO-TAILED TEST. Figure 4-4 shows the rejection region for a two-tailed test in our example, but with α taken as .05.

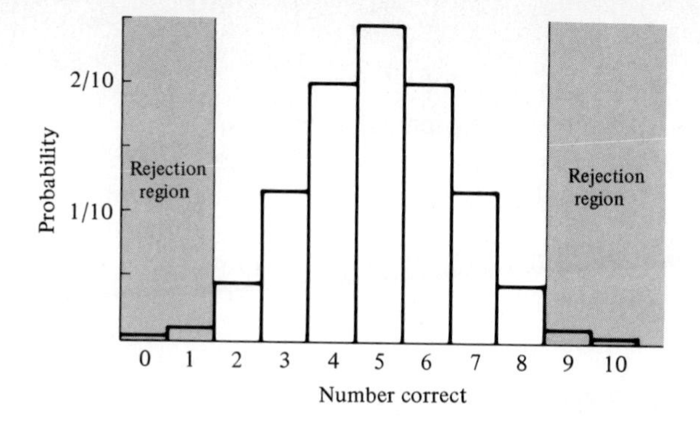

Figure 4-4: Rejection region for H_1: $p = 1/2$ when $N = 10$, $\alpha = .05$, and H_2 is $p \neq 1/2$.

4-55 If H_2 is $p < 1/2$, it would be reasonable to accept H_2 when K is (*larger; smaller*) than a critical value. This H_2 is a (*directional; nondirectional*) hypothesis.

smaller
directional

4-56 When we think K might be quite extreme in either direction, we shall choose a ________ H_2, such as $p \neq 1/2$.

nondirectional

4-57 When H_2 is nondirectional, the rejection region will have one part in each tail of the distribution. A nondirectional H_2 requires a ________-tailed statistical test.

two

4-58 The probability of getting a result in either one part or the other of this rejection region must not be greater than ______.

α

4-59 The two parts are identical in size and shape. If the sum of their two probabilities is α, the probability of either part by itself must be ______.

$\frac{\alpha}{2}$

4-60 Figure 4-4 shows the rejection region for $\alpha = .05$ and H_2: $p \neq 1/2$. The rejection region has two parts because of the nature of (α; H_2).

H_2

4-61 The boundary of the upper part is drawn at $K =$ ______. The boundary of the lower part is drawn at $K =$ ______. The probability of a result which falls in either one part or the other is ______, since the probability of $K \geqslant 9$ is .011.

9
1
.022

4-62 The probability .022 is less than α. If the boundaries had been drawn at 2 and 8, the probability of getting a result inside the rejection region would have been (*equal to; greater than*) α.

greater than

4-63 A two-tailed test is required when H_2 is ________; a one-tailed test is required when H_2 is ________.

nondirectional
directional

REVIEW

4-64 This lesson teaches a particular statistical test which is appropriate for experiments which have ________ trials with only two possible ________ per trial.

independent
outcomes

4-65 A statistical test always starts with a statistic which will be obtained from the data. The statistic used in our test has been ______.

K

4-66 A statistical test is always a test of some hypothesis. This hypothesis is designated (H_1; H_2), and it becomes the basis for a conditional probability distribution.

H_1

4-67 Given any particular H_1, there is a different distribution for every value of N in a coin-toss problem. N determines the total number of outcomes which the experiment can have; the number is always ______.

2^N

4-68 N also determines the number of values which the statistic K can have. This number is always ______.

$N + 1$

4-69 We learned how to find the probability of any particular one of the 2^N outcomes. First we find values for p and q from our knowledge of ______, letting p represent the probability of a ________ choice on any *single* trial.

H_1
correct

4-70 Then we apply the product rule for joint probabilities of independent events. This rule states that the joint probability is equal to the ________ of the probabilities of the component events.

product

4-71 Any particular outcome consists of K correct choices, each with probability equal to ______, and ______ incorrect choices, each with probability equal to ______.

p; $N - K$
q

4-72 Thus the probability of any particular outcome, by the product rule, is ______.

$p^K q^{N-K}$

4-73 When $N = 3$ and H_1 is $p = 1/2$, the probability of a particular outcome in which $K = 1$ is ______.

$1/8$

4-74 When $N = 3$, the probability of a particular outcome with $K = 1$ (*is; is not*) the same as the probability of getting $K = 1$.

is not

4-75 These probabilities differ because $K = 1$ can occur in more than one way. The one correct (C) and two incorrect (I) choices may occur in any one of three different orders, namely, ______, ______, and ______.

CII; ICI; IIC

4-76 To determine the probability of $K = 1$, we apply the summation rule for either-or cases. This rule can be applied only when the events are mutually ________.

exclusive

4-77 The summation rule states that the probability of getting either one or the other of two (or more) mutually exclusive events is equal to the ________ of the probabilities of the separate events.

sum

4-78 Thus when $p = 1/2$ and $N = 3$, the probability of getting $K = 1$ is ________.

$3/8$

4-79 The number $\binom{6}{4}$ is the number of outcomes of experiments having ________ trials in which K equals ________.

6; 4

4-80 The number $\binom{6}{4}$ is evaluated by means of the factorial fraction, ________. Its value is ________.

$\frac{6!}{4!2!}$; *15*

4-81 Once we have the proper conditional probability distribution, we can locate a rejection region. If we obtain a statistic inside this region, we intend to ________ H_1 and ________ H_2.

reject; accept

4-82 The rejection region will have two parts if H_2 is ________. The statistical test is then said to be ________.

nondirectional
two-tailed

4-83 The rejection region will have one part if H_2 is ________, and the statistical test will be ________.

directional; one-tailed

4-84 The boundary, or boundaries, of the rejection region are the ________ values of the statistic K.

critical

4-85 We choose the critical values so that the probability of getting a result inside the rejection region is not greater than ________.

α

4-86 When the test is two-tailed, the probability of getting a result inside a *particular part* of the rejection region must not be greater than ________.

$\frac{\alpha}{2}$

4-87 For $H_2: p > 1/2$ the rejection region will include only values of K which are extremely (*high; low*). For $H_2: p < 1/2$ it will include only extremely (*high, low*) values of K.

high
low

4-88 The α level determines how extreme the critical value must be. In general, a lower number for α means that the critical value must be (*less; more*) extreme.

more

PROBLEMS

1. Consider an experiment with N independent trials in which each trial can have any one of *three* outcomes, A, B, or C. Draw a tree diagram for the first two trials, and answer the following questions. How many different outcomes are there when $N = 2$? When $N = 3$? What rule, analogous to 2^N, will give the total number of outcomes for any value of N?

2. A coin is so biased toward tails that the probability of heads is $1/10$. What is the probability on two tosses of this coin of obtaining 2 heads? Of obtaining 2 tails? Of obtaining 1 head and 1 tail?

You are testing this coin's bias. Is there *any* value of K which would enable you to reject H_1: $p = 1/2$ in favor of H_2: $p < 1/2$ at the .01 level? Look again at the probability distribution on page 55, and consider whether there is any place for the necessary critical value of K. Now consider the probability distribution for H_1: $p = 1/10$, which you have just calculated. Is there any value of K which would enable you to reject *this* H_1 in favor of H_2: $p > 1/10$ at the .01 level? What value?

3. A coin is tossed nine times. What is the total number of possible outcomes of the nine-toss experiment? How many subsets of outcomes are there, according to "number of heads"? Assuming that the coin is unbiased, what is the probability of getting any one of the possible outcomes? How many elements are in the subset "6 heads and 3 tails"? What is the probability of getting exactly 6 heads and 3 tails in nine tosses of this unbiased coin?

4. A die is tossed one time. If it is unbiased, any one of its six faces is equally likely to fall uppermost. What is the probability that the face will be ⚀? What is the probability that it will be either ⚀ or ⚁?

5. The same die is tossed twice. What is the probability of getting ⚀ on the first toss and ⚁ on the second toss? What is the probability of getting ⚀ on both tosses? What is the probability of getting ⚀ on one toss (either one) and ⚁ on the other? What is the probability of getting the same face (any face) on both tosses?

6. Suppose you are trying to predict the guesses which will be made by a newcomer on two tosses of a coin. From your viewpoint, the probability of his calling heads on the first toss can be considered to be $1/2$; you have no reason to assign a different probability. But it would be incorrect to say, "The probability of his calling heads on *both* tosses is $1/4$." Why?

7. A deck of cards can be dichotomized into black cards and red cards. If p is the probability of a black card on a single draw and q the probability of a red card, $p = 1/2$ and $q = 1/2$. Six cards are sampled with replacement. What is the probability, on six draws, of getting 4 black and 2 red cards? Of getting all black cards?

8. If the deck is dichotomized into hearts and all other cards, what is the probability p of getting a heart on a single draw? What is the probability q of getting a spade, club, or diamond? When 7 cards are sampled with replacement, what is the probability of getting no hearts at all? What is the probability of getting 4 hearts? What is the probability of getting 2 hearts out of the first 4 draws and then 2 hearts out of the next 3? Is this result more or less probable than "4 hearts out of 7"? Why?

LESSON 5 THE POWER OF A STATISTICAL TEST

In setting up a statistical test, several choices have to be made. H_1 and H_2 must be formulated, and the α level must be chosen. Usually an investigator is also able to choose the size of his sample. All these choices affect the POWER of the test. In Lesson 5 we shall study the concept of power as it applies to statistical tests in coin-toss problems.

Two tests differ in power when one of them is more likely than the other to lead to a correct decision between H_1 and H_2, even though both are applied with the same decision rule. The value α is part of the decision rule.

Table 5-1 describes the possible outcomes of a decision based on such a rule. There are two kinds of error we can make, and we know that the probability of a type I error is α. We shall call the probability of a type II error (whose value we do not yet know how to calculate) simply β (beta). We are interested in the probability of a correct decision between H_1 and H_2 using a fixed decision rule in which α plays an important part. The diagram shows two ways in which we can be correct: by accepting H_1 when H_1 is true, and by accepting H_2 when H_2 is true. Just as $p(\text{accepting } H_1 \mid H_1) = 1 - \alpha$, the probability of the other correct decision $p(\text{accepting } H_2 \mid H_2) = 1 - \beta$.

Table 5-1. Possible Outcomes of a Statistical Decision

Decision	H_1 Is True	H_2 Is True
Accept H_1 and reject H_2	*Correct* $p(\text{accepting } H_1 \mid H_1) = 1 - \alpha$	*Type II error* $p(\text{rejecting } H_2 \mid H_2) = \beta$
Accept H_2 and reject H_1	*Type I error* $p(\text{rejecting } H_1 \mid H_1) = \alpha$	*Correct* $p(\text{accepting } H_2 \mid H_2) = 1 - \beta$

The probability $1 - \alpha$ is fixed by the decision rule. For this reason we take the value of $1 - \beta$ as indicating the power of a statistical test. As $1 - \beta$ becomes larger, the test is said to become more powerful.

To find $1 - \beta$, we must find $p(\text{accepting } H_2 \mid H_2)$, and for this we need a probability distribution conditional upon H_2. Here we run onto what at first seems an insurmountable difficulty. H_2 is not usually an exact hypothesis; in our color-touch example, H_2 is $p > 1/2$. To calculate a conditional probability distribution, we must have an *exact* H_2. We can resolve this difficulty by assuming, in turn, several exact hypotheses in which $p > 1/2$ and calculating $p(\text{accepting } H_2 \mid H_2) = 1 - \beta$ for each of them.

A. FINDING THE PROBABILITY OF A TYPE II ERROR

Figure 5-1 shows a graph of our decision-making distribution. It is conditional upon H_1: $p = 1/2$; in it we locate α and $1 - \alpha$. With a decision rule which requires us to reject H_1 when $\alpha = .01$, we have found the critical value of K to be 9. In Fig. 5-1 a line is drawn at this cutoff point, and the region to its right has been labeled α. We know that the region labeled α has an area equal to .01 and that the remaining area, $1 - \alpha$, must equal .99.

To find the probability β, we must have an exact H_2. Let us first assume H_2: $p = 3/4$ and then assume H_2: $p = 9/10$, comparing the values of β which arise from these assumptions. Figure 5-2 shows the distributions which are conditional upon these two assumptions; the calculations for these distributions are shown in Table 5-5 (page 82). If $p = 3/4$ is true, then our cutoff at $K = 9$ will have the consequences shown in Fig. 5-2*a*. When $K < 9$, H_1 will be accepted and H_2 will be rejected; if $p = 3/4$ is true, this decision will be a type II error. The probability of a type II error is therefore represented by the region to the left of the cutoff point; it is labeled β. When $K \geqslant 9$, H_2 will be accepted; if $p = 3/4$ is true, this will be a correct decision. The probability $1 - \beta$ is therefore represented by the region to the right of the cutoff point. For H_2: $p = 3/4$, $\beta = .756$, and $1 - \beta$, the measure of power, equals .244.

The same reasoning applies to Fig. 5-2*b*, where H_2: $p = 9/10$ is assumed. Again using the cutoff point at $K = 9$, we find that the region β to the left of the cutoff point is now considerably smaller. The probability that we shall mistakenly reject this more extreme hypothesis is only .264. Here $1 - \beta = .736$.

Of course, we have no way of knowing what value of p is the actually true value. But we can often state the lowest value of *p which we must be sure to detect.* If we want to avoid a missed discovery when p actually equals $3/4$, our test will have very limited power. When we shall be satisfied to miss the discovery unless p is at least $9/10$, our test will have greater power. We can best put this point in terms of the difference between H_1 and the value of p to be detected: The smaller this difference, the lower the power of any test.

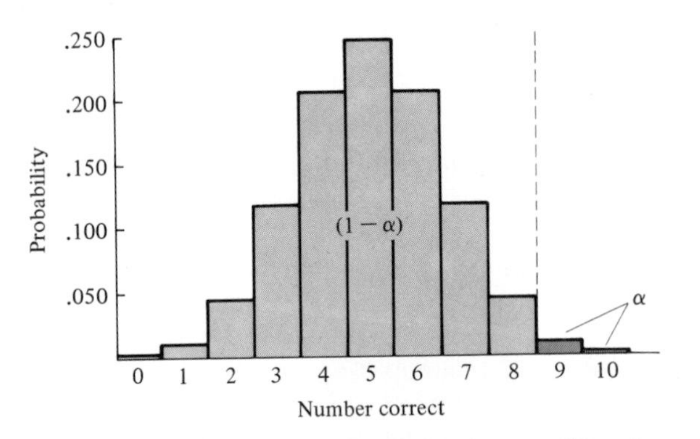

Figure 5-1: Decision-making distribution conditional upon H_1: $p = 1/2$, with $N = 10$, $\alpha = .01$, and H_2: $p > 1/2$.

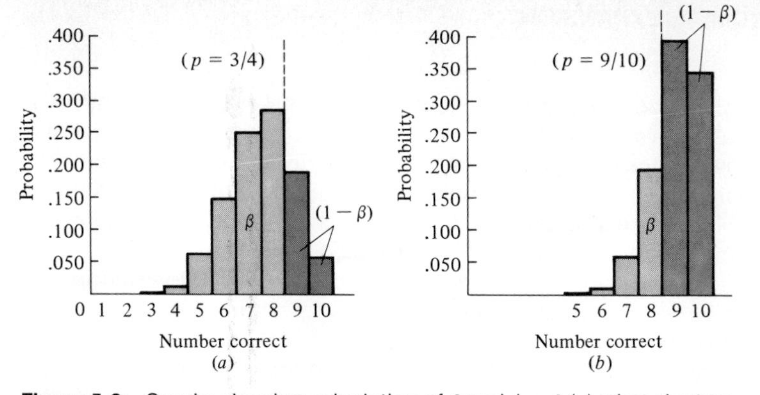

Figure 5-2: Graphs showing calculation of β and $1-\beta$ (*a*) when the true value of p is $3/4$ and (*b*) when the true value of p is $9/10$.

5-1 We use a distribution conditional upon (H_1; H_2) to find a rejection region with area approximately equal to α.

H_1

5-2 Since we are assuming H_1: $p = 1/2$, we use the distribution for $N = 10$ and $p = 1/2$ (Fig. 5-1). With $\alpha = .01$, the critical value of K is ______.

9

5-3 Whatever the true value of p may be, we have now decided to reject H_1 and accept H_2: $p > 1/2$ if the observed value of K is ______ 9.

at least

5-4 We do not know the true value of p, but there is a possibility that it may be $3/4$. If it is, the probability that $K < 9$ is .756 and the probability that $K \geqslant 9$ is ______.

.244

5-5 But if $K \geqslant 9$, we shall reject ______ and accept ______ under our decision rule. When $p = 3/4$ is true, this will be (*a correct; an incorrect*) decision.

H_1; H_2

a correct

5-6 Therefore, if $p = 3/4$ is true, the probability of a correct decision using our decision rule is ______. We designated this probability in Table 5-1 as ______.

.244
$1-\beta$

5-7 If $p = 9/10$ is true, the probability of $K < 9$ is .264 and the probability of $K \geqslant 9$ is ______. In this case, $1 - \beta =$ ______.

.736; .736

5-8 When we compare these two alternate possibilities ($p = 3/4$ and $p = 9/10$), we find that the power of our test is ______ when $p = 3/4$ is true and its power is ______ when $p = 9/10$ is true.

.244
.736

5-9 Our test has *greater* power when the hypothesis to be detected is (*less; more*) similar to H_1.

less

B. THE BINOMIAL EXPANSION

The values of β for our example turn out to be surprisingly high, and in a moment we shall return to give them careful consideration. In doing so, however, we shall need to calculate more probability distributions, and we can help ourselves by simplifying the way we think of probability in coin-toss problems.

The probabilities in distributions like Figs. 5-1 and 5-2 are actually terms in the expansion of $(p+q)^N$. In algebra $p+q$ is called a BINOMIAL because it has two variables; the expansion of $(p+q)^N$ is called a BINOMIAL EXPANSION. A probability distribution whose elements are terms in a binomial expansion is a BINOMIAL PROBABILITY DISTRIBUTION.

In order to see that we have really been expanding the binomial $(p+q)^N$, let us review a little of the algebra involved, using small values of N. When $N=2$, $(p+q)^N=(p+q)^2=p^2+2pq+q^2$. There are $N+1$ terms, one for each of the $N+1$ subsets into which we partition the 2^N possible outcomes of an N-trial experiment. The coefficients of these terms are 1, 2, and 1, the numbers appearing in the row for $N=2$ of our Pascal triangle. These numbers are obtainable by the factorial rule $\binom{N}{K}=\frac{N!}{K!(N-K)!}$. The factors p^2, pq, and q^2 are the probability weights for *individual* outcomes in each of the subsets. Our rule for finding these weights is $p^K q^{N-K}$; the rule clearly fits the successive terms of this binomial expansion, with K going from 2 in the first term to 0 in the last term.

In practice, the two rules we have been using (one for finding the probability of an individual outcome in a subset and one for finding the number of outcomes in a subset) are the standard rules for determining the series of terms in expanding $(p+q)^N$. If you have to expand $(p+q)^3$, the easiest way to find the terms required is by means of these rules.

$$\begin{aligned}(p+q)^3 &= \frac{3!}{3!0!}\,p^3q^0+\frac{3!}{2!1!}\,p^2q^1+\frac{3!}{1!2!}\,p^1q^2+\frac{3!}{0!3!}\,p^0q^3\\ &= p^3+3p^2q+3pq^2+q^3\end{aligned}$$

The coefficients (1, 3, 3, 1) are indeed the numbers in row $N=3$ of the Pascal triangle on page 56. When we use that triangle, it is very easy to write the expansion of $(p+q)^5$:

$$(p+q)^5=p^5+5p^4q+10p^3q^2+10p^2q^3+5pq^4+q^5$$

The three probability distributions shown on pages 68 and 69 are all expansions of the binomial $(p+q)^{10}$. For Fig. 5-1, $p=q=1/2$; for Fig. 5-2a, $p=3/4$ and $q=1/4$; and for Fig. 5-2b, $p=9/10$ and $q=1/10$. Each vertical bar in Figs. 5-1 and 5-2 stands for a single term in one of these binomial expansions. In Table 5-5 (page 82) the coefficients and the probability weights are written out separately, term by term.

A statistical test based on a binomial distribution is called a BINOMIAL TEST. The tests discussed in Lesson 4E were therefore binomial tests. Presently we shall use the binomial expansion to make a binomial test when $N=20$.

5-10 If $p = q = 1/2$, then the four probabilities in a distribution for $N = 3$ are the four ________ of the binomial $(1/2 + 1/2)^3$ when it is expanded.

terms

5-11 If $p = 3/4$ and $q = 1/4$, the probabilities for $N = 3$ are terms of the binomial ________.

$(3/4 + 1/4)^3$

5-12 The first term of the binomial $(p + q)^3$ is $\binom{N}{K} p^K q^{N-K}$ for $K =$ ________. The coefficient of this term is 1, because $\binom{N}{K} = 1$ when $N = 3$ and $K =$ ________.

3
3

5-13 The second term of this binomial is $\binom{N}{K} p^K q^{N-K}$ for $K =$ ________. The value of $\binom{N}{K}$ is ________; the term is therefore ________.

2; 3
$3p^2q$

5-14 If $p = 3/4$, the second term $3p^2q$ has the value ________.

$27/64$

5-15 When we know that $3p^2q = 27/64$, we know the probability that K will be exactly ________ in an experiment with $N = 3$ and $p = 3/4$.

2

5-16 What is the probability that K will be exactly 1 in such an experiment? This probability is given by the term ________, and its value is ________.

$3pq^2$
$9/64$

5-17 Summarize the probability distribution for $N = 3$ and $p = 1/2$:

K	$\binom{N}{K}$	$p^K q^{N-K}$	$p(K)$
3	________	________	________
2	________	________	________
1	________	________	________
0	________	________	________

1; 27/64; 27/64
3; 9/64; 27/64
3; 3/64; 9/64
1; 1/64; 1/64

5-18 What is the sum of the four probabilities in this distribution? ________

1

5-19 When $N = 20$, the probability distribution is the set of terms in the binomial, ________. The set has ________ terms.

$(p + q)^{20}$; *21*

5-20 In order to find $p(K \geq 16)$, the terms needed are those with $K =$ ________, ________, ________, ________, and ________.

16; 17; 18; 19; 20

C. POWER AND SAMPLE SIZE

In experiments with $N = 10$ we have found $\beta = .756$ when the true hypothesis is $p = 3/4$ and $\beta = .264$ when the true hypothesis is $p = 9/10$. The chance that we shall make a type II error is much greater if the true value of p is only $3/4$ than if the true value is as high as $9/10$. In order to grasp the meaning of this power difference, let us spell it out carefully as it applies to the color-touch example, a case in which we certainly want to avoid a type I error, yet at the same time the sort of case in which it would be very gratifying to be able to accept H_2 and report an important discovery.

The woman claims that she can discriminate colors by touch alone, but she may not expect to be correct *all* the time. Many difficult discriminations have a less than 100 percent accuracy. By proposing a 10-trial experiment to test her claim, we make it necessary for her to get at least 9 correct in order to meet our α level of .01. Will she be happy with the proposed experiment? If she thinks that her ability will permit her to be right at least 90 percent of the time, she may regard the proposal as fair. On the other hand, she will hesitate if she thinks that her ability, while genuine, is effective only about 75 percent of the time.

She can calculate her chances of proving her claim on our terms. If she believes she has 90 percent accuracy, she thinks $p = 9/10$ is true, and she calculates the probability distribution for $N = 10$ based on $p = 9/10$. Since this calculation leads to a probability $1 - \beta = .736$ that we shall accept her claim, she may be content. If she believes she has only 75 percent accuracy, she will calculate the probability for $N = 10$ based on $p = 3/4$, and she will find the probability that we will accept her claim to be only .244. Probably she will not want to agree to these odds.

From our own point of view, these calculations are also not comforting. We picked $\alpha = .01$ in order to avoid a type I error in testing this unusual claim. Yet the claim is an interesting one, and we find it somewhat disquieting that the probability of a type II error is quite large in both the distributions we have calculated. The value of β is .756 when $p = 3/4$; thus, if the true value of p is $3/4$, we stand a good chance of missing an interesting discovery. Even if her ability is marked enough for the true value of p to be $9/10$, β is still .264 and there is better than 1 chance in 4 that we shall miss the discovery.

Thus, from anyone's point of view, this is a bad situation. We have an interesting claim to investigate, one which is so unexpected that it is safe to say the world would be astonished to find that the true value of p is anything in excess of $1/2$. Yet even if the true value is as high as $9/10$, we stand a good chance of missing its discovery. What can we do about this predicament? Obviously, we need a *more powerful test,* one for which the value of $1 - \beta$ is higher even though we continue to use the same decision rule.

Before going on to such a test, we shall pause to consider how the choice of α affects a test's power. The probability distributions you will need in the following frames are found in Table 5-5 on page 82.

5-21 The probability, on H_1, of $K \geqslant 8$ is .055. If α is .055, the critical value of K would be ______. | *8*

5-22 Suppose that the true value of p is 3/4. When 8 is the critical value of K, a type II error will be made whenever K_o turns out to be ______. | *<8*

5-23 If $p = 3/4$, a correct decision will be made whenever K is ______. The probability of such a correct decision is ______. Use Table 5-5, and round off to three decimals. | *≥8; .526*

5-24 The probability of a correct decision is $1-\beta$. The probability β will be ______ in this case. | *.474*

5-25 Determine $1-\beta$ for $p = 3/4$ and $\alpha = .001$. The critical value of K is 10; $1 - \beta =$ ______. | *.056*

5-26 Fill in the values of $1-\beta$:

	$\alpha = .055$	$\alpha = .01$	$\alpha = .001$
$p = 3/4$	______	.244	______
$p = 9/10$	______	.736	______

.526; .056
.930; .349

5-27 The value of $1-\beta$ measures the ______ of a statistical test. As $1-\beta$ increases, β ______. | *power* / *decreases*

5-28 Write β for each H_2 and α:

	$\alpha = .055$	$\alpha = .01$	$\alpha = .001$
$p = 3/4$	______	.756	______
$p = 9/10$	______	.264	______

.474; .944
.070; .651

5-29 The value of α is the probability of a type ______ error; the value of β is the probability of a type ______ error. | *I* / *II*

5-30 The value of β cannot be calculated from a knowledge of α alone; we also have to have the probability distribution conditional upon (H_1; H_2). But we can say that an increase in α means (*a decrease; an increase*) in β. | H_2 / *a decrease*

5-31 As we *reduce* the probability of a type I error, we automatically (*decrease; increase*) the probability of a type II error. | *increase*

5-32 As α decreases, $1 - \beta$ (*decreases; increases*). Therefore, as we reduce the probability of a type I error, we (*decrease; increase*) the power of our test. | *decreases* / *decrease*

We must find a way to get a more powerful test. For a clue, recall the probability distributions calculated for $N = 2$ (page 55). The smallest probability in the distribution was $1/4$, given $p = 1/2$, and therefore we could not possibly have obtained any outcome reaching the .01 level with only 2 trials. Perhaps, then, 10 trials are still too few for a satisfactory experiment. We can consider a larger experiment of $N = 20$ as a tentative answer to the difficulty.

For a binomial test with $N = 20$, we must begin by finding the critical value of K for $\alpha = .01$ when we assume H_1: $p = 1/2$. Since we want to stop calculating as soon as the sum of the probabilities reaches .01, we write the distribution starting with $K = 20$ and include a column for the *cumulative* probability. This column is the probability of at least K correct, for each value of K; it is obtained by adding the probability of *exactly* K correct to the probabilities already calculated for higher values of K.

Table 5-2 Upper End of Binomial Expansion for $N = 20$, $p = 1/2$

K	$\binom{N}{K}$	$p(K)$	Cumulative Probability
20	1	$\frac{1}{2^{20}}$	.0000
19	20	$20\frac{1}{2^{20}}$	.0000
18	190	$190\frac{1}{2^{20}}$	.0002
17	1,140	$1{,}140\frac{1}{2^{20}}$	.0013
16	4,845	$4{,}845\frac{1}{2^{20}}$	.0059
15	15,501	$15{,}501\frac{1}{2^{20}}$	.0207

No further terms are needed, for we can now state that the critical value of K is 16. The probability of at least 16 correct in 20 trials is .0059; the probability of at least 15 correct is .0207, too high to meet $\alpha = .01$. We now make a new decision rule: reject H_1 if the observed value of K is at least 16.

It already seems that the addition of 10 trials may have improved our experiment. Instead of having to be correct on at least 9 out of 10 trials (90 percent), the woman will now have to be correct on only 16 out of 20 trials (80 percent) in order to meet our α level.

To find the value of β, we must now calculate the first five terms for each of the conditional probability distributions we are studying, those with $p = 3/4$ and with $p = 9/10$. Table 5-3 shows the terms and cumulative probabilities.

Following our decision rule, we must reject H_1 and accept H_2 if K is at least 16. If $p = 3/4$ is true, this decision will be correct. Given $p = 3/4$, the probability of getting $K \geq 16$ is $.4148 = 1 - \beta$. With $N = 10$, $1 - \beta$ is only .244 for this same H_2. *Increasing the sample size always increases the power of a test.*

Table 5-3 Upper End of Binomial Expansions for $N = 20$, $p = 3/4$ and $p = 9/10$

K	$p(K)$ for $p = 3/4$	Cumulative Probability	$p(K)$ for $p = 9/10$	Cumulative Probability
20	$\frac{3^{20}}{4^{20}}$	.0032	$\frac{9^{20}}{10^{20}}$	.1216
19	$20\frac{3^{19}}{4^{20}}$	.0243	$20\frac{9^{19}}{10^{20}}$	.3917
18	$190\frac{3^{18}}{4^{20}}$	.0912	$190\frac{9^{18}}{10^{20}}$	.6768
17	$1{,}140\frac{3^{17}}{4^{20}}$	.2251	$1{,}140\frac{9^{17}}{10^{20}}$	.8669
16	$4{,}845\frac{3^{16}}{4^{20}}$	.4148	$4{,}845\frac{9^{16}}{10^{20}}$	.9567

5-33 To determine the cutoff point for our decision rule, we calculate the probability distribution conditional upon (H_1; H_2).

H_1

5-34 Table 5-2 shows the upper end of this probability distribution, calculated as the expansion of the binomial $(p + q)^N$ with $N =$ ______ and $p =$ ______.

20; 1/2

5-35 The probability of $K = 16$ is $\binom{N}{K} p^K q^{N-K}$. For this term $\binom{N}{K} =$ ______, and $p^K q^{N-K} =$ ______.

4,845; $\frac{1}{2^{20}}$

5-36 Table 5-2 shows that the probability is .0059 of getting ($K = 16$; $K \geqslant 16$) when $p = 1/2$.

$K \geqslant 16$

5-37 With $\alpha = .01$, our cutoff point is $K = 16$. If α were .001, our cutoff point would be $K =$ ______; if α were .02, our cutoff point would be $K =$ ______.

17
15

5-38 To find $1 - \beta$ for H_2: $p = 9/10$, we expand the binomial $(p + q)^{20}$ with $p =$ ______ and $q =$ ______. We start with the term for $K =$ ______, and we stop with $K =$ ______.

9/10; 1/10
20; 16

5-39 Suppose that $p = 9/10$ is actually true. By our decision rule we shall reject H_2: $p > 1/2$ if K is less than 16. This decision will be a type ______ error, and its probability is called ______.

II; β

5-40 The probability of $K \geqslant 16$ is ______ when $p = 9/10$ is true; the probability of $K < 16$ is ______ when $p = 9/10$ is true.

.9567
.0433

5-41 When $p = 9/10$ is true, $\beta =$ ______, and $1 - \beta =$ ______. With $N = 20$, the power is ______ when $p = 9/10$ is true.

.0433; .9567
.9567

5-42 With $N = 10$, the power is .736 when $p = 9/10$ is true. Increasing sample size always (*decreases; increases*) power.

increases

D. DRAWING A POWER FUNCTION

We have been calculating the power of our binomial test by assuming, in turn, two different hypotheses as H_2. These calculations do not of course give a complete answer to questions about power in the color-touch example. The most complete way to answer such questions in any particular case is by drawing a POWER FUNCTION.

A power function is a graph which shows how the value of $1-\beta$ varies with different assumptions concerning the true hypothesis, all other things remaining the same. We have actually begun to draw two different power functions for our case, one in which $N=10$ and another in which $N=20$. Table 5-4 gives the values for these power functions. Two new alternate hypotheses have been added in this table so that each of the functions now has four points, one for each H_2 assumed.

Figure 5-3 shows these four points in the power function for $N=10$ as a graph, with the assumed true value of p on the horizontal axis and the value of $1-\beta$ on the vertical axis. When these four points are connected, they suggest how the power of this test varies when the true value of p changes from $2/3$ to $99/100$. If we calculated $1-\beta$ for many more values of p, we would find that the curve becomes smoother and more continuous as we add these points.

Before beginning the frames on page 77, plot the power function for $N=20$ in Fig. 5-3. Take the values of $1-\beta$ from Table 5-4.

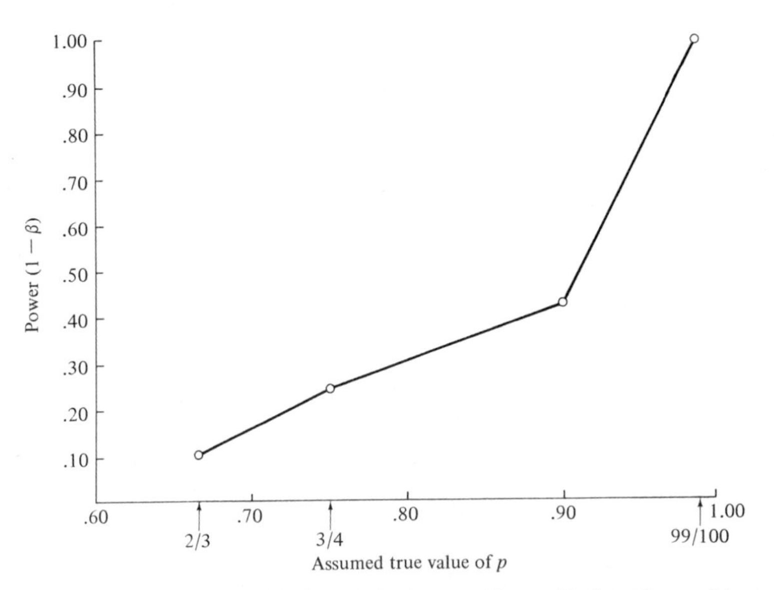

Figure 5-3: Power function for a coin-toss problem with $N=10$, $\alpha=.01$, and H_1: $p=1/2$.

Table 5-4 Power Functions for Coin-Toss Problems in Which $\alpha = .01$ and H_1 Is $p = 1/2$

Assumed True Value of p	$1 - \beta$	
	$N = 10$	$N = 20$
2/3	.1040	.1523
3/4	.2440	.7361
9/10	.4148	.9567
99/100	.9959	1.0000

5-43 At every one of the four points, the power function for $N = 20$ is (*above; below*) the power function for $N = 10$. This difference is *least* when the true value of p is ________.

above
99/100

5-44 The value of p in the assumed hypothesis (H_1) is ________. The power of the test is *least* when the true value of p is (*very close to; very different from*) this assumed value.

1/2
very close to

5-45 What will happen to these power functions if you relax your α level from .01 to .055? The values of $1 - \beta$ will tend to (*decrease; increase*).

increase

5-46 Some values of $1 - \beta$ in Fig. 5-3 cannot increase very much; $1 - \beta$ can never be greater than ________.

1

5-47 If you keep $\alpha = .01$ but change H_1 so that H_1 asserts $p = 3/5$, all the values of $1 - \beta$ will (*decrease; increase*) because the test must now discriminate between hypotheses which are (*less; more*) different than before.

decrease
less

5-48 Suppose that you want to be sure, *at the .05 level*, that you will detect a true p of 9/10, but you are not interested in any color-perceiving ability that works less than 90 percent of the time. You will need a test for which β is not greater than ________ when the true value of p is 9/10.

.05

5-49 If β must not be greater than .05, then $1 - \beta$ must be at least ________. Will an experiment of 10 trials have sufficient power? ________

.95
No

5-50 To detect a true p of 9/10 at the .95 *confidence* level, you will need an experiment with $N =$ ________.

20

5-51 To detect a true p of 99/100 at the .95 confidence level, you do not need more than ________ trials in the experiment.

10

5-52 To detect a true p of 3/4 at the .95 confidence level, you will need more than ________ trials in the experiment.

20

E. THE FAMILY OF BINOMIAL DISTRIBUTIONS

We have now examined quite a number of different binomial probability distributions: we have considered $N = 1$, 2, 10, and 20 when H_1: $p = 1/2$ is assumed to be true, and we have looked at parts of the distributions for $p = 3/4$ and $p = 9/10$ for experiments with $N = 10$ and 20. In Sec. A we examined graphs of three of these distributions, and in Sec. B we learned to think of their probability elements in terms of the binomial $(p + q)^N$.

All probability distributions which are obtained from the terms of $(p + q)^N$ are members of the family of binomial probability distributions. There is a different distribution for every possible combination of values of N and p; these values are called the PARAMETERS of the distribution. There is no limit to the number of members of this family; it is infinitely large.

Before concluding this lesson, we want to observe how the members of this family change their appearance as p remains the same and N increases. Figure 5-4 shows four binomial probability distributions in which $p = 3/4$; these distributions differ only in the value of N. Notice the following differences; we shall be discussing each of them in Lessons 6 to 8.

1 The LOCATION OF THE PEAK. When $N = 10$, the peak is at $K = 8$; when $N = 200$, it is at 150.

2 The DISPERSION (spread) of the distribution. When $N = 10$, the dispersion is clearly less than for any higher value of N. As N increases, the distribution becomes more spread out along the K axis.

3 The HEIGHT of the peak, which becomes less as N increases.

4 The SYMMETRY of the distribution. As N increases, the distribution becomes more nearly symmetrical.

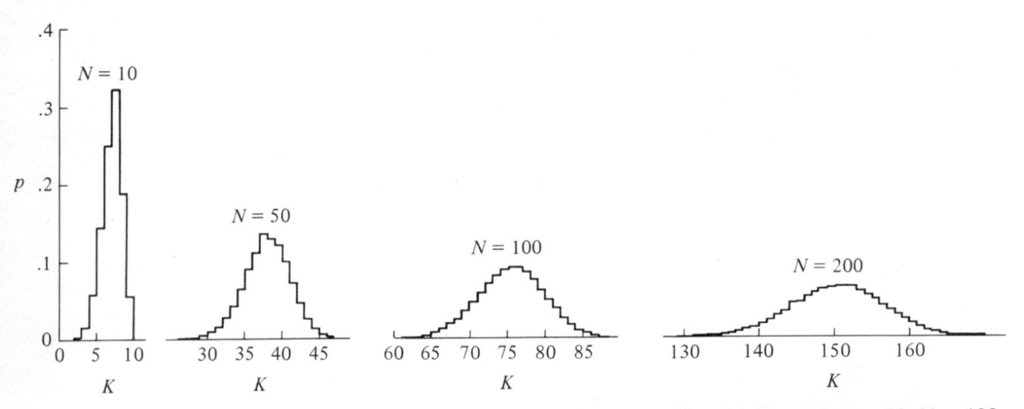

Figure 5-4: Binomial probability distributions for $p = 3/4$, with $N = 10$, $N = 50$, $N = 100$, and $N = 200$.

5-53 Every one of the four distributions shown in Fig. 5-4 is a ________ probability distribution. *binomial*

5-54 Each combination of the parameters N and p gives a different distribution. For all the distributions in Fig. 5-4, $p =$ ________. *3/4*

5-55 As N increases, the number of terms in the expansion of $(p + q)^N$ increases; there are always ________ terms. *$N + 1$*

5-56 For each term whose value is greater than 0, there is a vertical bar in the graph of the distribution. Figure 5-4 shows that ________ of the 11 terms have nonzero values when $N = 10$. *8 (including $K = 3$)*

5-57 The height of each bar in the graph for $N = 10$ is equal to $p(K)$; the width of each bar is equal to 1. The *area* inside any particular bar is therefore equal to ________. *$1 \cdot p(K) = p(K)$*

5-58 If we add together all the values of $p(K)$ in this distribution, the sum will be ________. Therefore, the total *area* inside this distribution is ________. *1* *1*

5-59 The area inside the distribution for $N = 50$ is also equal to ________. *1*

5-60 In which distribution, $N = 10$ or $N = 50$, is the maximum value of $p(K)$ highest? It is highest for $N =$ ________. *10*

5-61 Which of these two distributions is most spread out? $N =$ ________. The degree of spread is referred to as the ________ of the distribution. *50* *dispersion*

5-62 The area in $N = 50$ has to be distributed among a (*larger; smaller*) number of nonzero terms. This difference makes the distribution both wider and flatter. *larger*

5-63 The area in the distribution for $N = 100$ is also equal to ________. Since it has still more nonzero terms than $N = 50$, its dispersion is still ________ and its maximum $p(K)$ is still ________. *1* *greater* *smaller*

5-64 The most symmetrical of the four distributions is the distribution for $N =$ ________. *200*

5-65 This distribution also has the greatest degree of ________ and the lowest maximum ________. *dispersion; $p(K)$*

REVIEW

5-66 The probability of a type I error is α; it is the probability of rejecting H_1 when ______ is true. *H_1*

5-67 The probability of a type II error is ______; it is the probability of rejecting H_2 when ______ is true. *β* *H_2*

5-68 $1 - \beta$ is the probability of a correct decision when ______ is true. This probability indicates the ______ of the statistical test. *H_2; power*

5-69 To calculate β or $1 - \beta$, the probability distribution conditional upon (H_1; H_2) must be calculated. This hypothesis is ordinarily (*exact; inexact*). *H_2* *inexact*

5-70 We can assume an exact H_2 for the purpose of calculating this conditional probability distribution. The power of the test (*is; is not*) the same for each H_2 we might assume. *is not*

5-71 Consider H_1; $p = 1/2$, H_2: $p = 1/4$, and H_2': $p = 1/10$. $1 - \beta$ will be greater for alternate hypothesis (H_2; H_2'). *H_2'*

5-72 A series of exact H_2's can be considered and $1 - \beta$ calculated for each. The set of values of $1 - \beta$ can be drawn as a graph called a ______ function. *power*

5-73 When $N = 20$ in a binomial test, the power function lies (*above; below*) the function for a binomial test with $N = 10$. *above*

5-74 If everything else remains the same, increasing the sample size always ______ the power of a test. *increases*

An experimenter is considering a coin-toss experiment. When $\alpha = .01$ and H_1 is $p = 1/2$, the power $1 - \beta$ for detecting H_2: $p = 3/4$ is .9567 for a 20-trial experiment. He thinks, however, that he might like to do the following things:

5-75 Change α to .05. If he does this, he will need (*fewer; more*) than 20 trials to achieve the same power. *fewer*

5-76 Detect H_2: $p = 2/3$ at the same power, that is, .95. If he wishes to do this, he will need (*fewer; more*) than 20 trials. *more*

5-77 Decrease the probability of a type II error. In order to do this, he must (*decrease; increase*) the power of his test, and this will require (*fewer; more*) than 20 trials in the experiment. *increase* *more*

5-78 Each value of $p(K)$ in a binomial probability distribution is a term in the expansion of the binomial ______. $(p+q)^N$

5-79 Since $q = 1 - p$, there are only two parameters needed in order to determine a complete binomial probability distribution. These parameters are ______ and ______. $N; p$

5-80 The expansion of $(p+q)^N$ has ______ terms. $N+1$

5-81 If $K = 5$, the propability $p(K)$ is calculated from the rule $p(5) =$ ______. $\binom{N}{5}p^5q^{N-5}$

5-82 If $K = 5$ and $N = 7$, $p(K) =$ ______. $\binom{7}{5}p^5q^2$

5-83 Determine the coefficient $\binom{7}{5}$ for this term. It is ______. 21

5-84 When $p = 1/2$, and $N = 7$, the probability of $K = 5$ is (*fraction*) ______. $21/128$

5-85 In a histogram of the probability distribution for $N = 7$ and $p = 1/2$, the height of the bar for $p(5)$ would be ______. $21/128$

5-86 The width of the same bar is equal to ______. Therefore, the area inside the bar is ______. 1; $21/128$

5-87 In the binomial probability histogram, $p(K)$ is represented by both the ______ and the ______ of the bar at K. *height; area*

5-88 The sum of all $K+1$ values of $p(K)$ is equal to ______. The total area inside a probability histogram is ______. 1; 1

The following questions refer to a 4-trial coin-toss experiment with H_1: $p = 1/2$ and $\alpha = 1/16$. H_2 is $p > 1/2$.

5-89 State the decision rule in terms of α: reject ______ if p_o is ______. H_1; $\leq \alpha$

5-90 State the decision rule in terms of a critical value of K. To do this, you need to calculate part of a binomial probability distribution; it is the distribution for $N =$ ______ and $p =$ ______. 4; $1/2$

5-91 The critical value of K is ______. The decision rule is: reject H_1 if ______. 4; $K = 4$

5-92 Calculate the power of this test for detecting H_2: $p = 3/4$. It is the probability of $K =$ ______ when $N =$ ______ and $p =$ ______. The power is ______. 4; 4; $3/4$; $81/256$

PROBLEMS

1. Consider a binomial probability distribution with $N = 12$ and $p = 1/4$. What is the probability of *exactly* $K = 11$?

2. On a multiple-choice test with five alternatives given for each question, the probability of a correct answer by guessing is $1/5$ on each question. How many questions must a student get correct in order for you to reject the hypothesis that $p = 1/5$ (that he is merely guessing) at the .001 level? Answer this question for $N = 5$, $N = 10$, and $N = 20$, that is, for a 5-, a 10-, and a 20-question test.

3. In a coin-toss type of experiment, H_1 is $p = 1/3$. Calculate the power of a binomial test with $\alpha = .05$ and $N = 10$ for detecting H_2: $p = 1/2$ and H_2: $p = 2/3$. Calculate the power of the test for detecting these same H_2's when $N = 20$.

4. In designing an experiment of the coin-toss type, an investigator must make a decision about the number of trials to include in the experiment. What *three* decisions must he make first, before he can determine the minimum number of trials he must include in order to realize his aims?

5. Calculate the binomial probability distributions for $N = 10$ and $p = 1/8$, $1/4$, $3/8$, $5/8$, $3/4$, and $7/8$. Compare these distributions with the distribution for $N = 10$ and $p = 1/2$. Do they differ in degree of dispersion? Position of peak? Maximum height of peak? Degree of symmetry? Are any of these distributions mirror images of each other?

6. You are considering a color-touch experiment with $N = 14$. Determine the critical value of K for $\alpha = .01$. Find $1 - \beta$ for H_2: $p = 3/4$. How much increase of power can be obtained by increasing sample size from 10 to 14? From 14 to 20?

Table 5-5 Binomial Expansions for $N = 10$, $p = 3/4$, and $p = 1/10$

K	$\frac{N!}{K!(N-K)!}$	Probability of K: Given $p = 3/4$	Probability of K: Given $p = 9/10$
3	120	$120\frac{3^3}{4^{10}} = .0034$	$120\frac{9^3}{10^{10}} = .0000$
4	210	$210\frac{3^4}{4^{10}} = .0162$	$210\frac{9^4}{10^{10}} = .0001$
5	252	$252\frac{3^5}{4^{10}} = .0584$	$252\frac{9^5}{10^{10}} = .0015$
6	210	$210\frac{3^6}{4^{10}} = .1460$	$210\frac{9^6}{10^{10}} = .0112$
7	120	$120\frac{3^7}{4^{10}} = .2502$	$120\frac{9^7}{10^{10}} = .0574$
8	45	$45\frac{3^8}{4^{10}} = .2815$	$45\frac{9^8}{10^{10}} = .1937$
9	10	$10\frac{3^9}{4^{10}} = .1877$	$10\frac{9^9}{10^{10}} = .3874$
10	1	$\frac{3^{10}}{4^{10}} = .0563$	$\frac{9^{10}}{10^{10}} = .3487$

LESSON 6
THE MEAN AND OTHER INDICATORS OF CENTRAL TENDENCY

In Lesson 5E we compared four binominal probability distributions for $p = 3/4$ and observed that the peak shifts to higher values of K as N increases. This peak is a striking feature of any binomial distribution; when we describe its location, we make a statement about the central point of the distribution. However, the peak is not the only kind of central point; Lesson 6 explains three different kinds, and all of them can be called INDICATORS OF CENTRAL TENDENCY.

Not all distributions are probability distributions. When a test is given to a group of N persons, the N scores obtained also form a distribution; it is called an empirical (rather than theoretical) distribution. Figure 6-1 shows an empirical distribution of 50 scores in the familiar form of a histogram. There is no score below 1 or above 11, and the number of persons obtaining each score is shown by the height of the vertical bar for that score. The height of the bar thus represents the FREQUENCY with which its score occurs, and the distribution is called a FREQUENCY DISTRIBUTION.

In a frequency distribution, the sum of the heights of all the bars equals N. This sum is an important difference between frequency histograms and probability histograms, in which the sum of the bar heights must always equal 1.

The bars in Fig. 6-1 are considered to be 1 score unit wide. The first bar occupies the space between .5 and 1.5, centered at the score value 1 as its midpoint. Even though the scores in this case are discrete, it is convenient to think of them as being superimposed on a horizontal axis which is continuous. This way of thinking will simplify our graphical and arithmetical analysis.

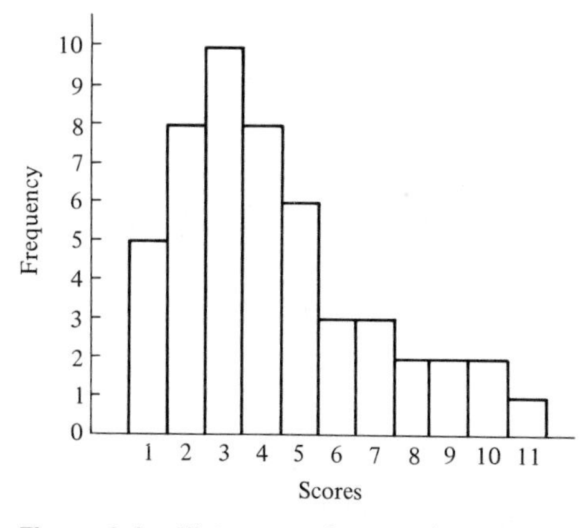

Figure 6-1: Histogram of scores ($N = 50$).

A. THE MEAN AS THE ARITHMETIC AVERAGE

Figure 6-2 shows two frequency distributions of 50 test scores each; B is the same distribution shown in Fig. 6-1, but it appears here as a FREQUENCY POLYGON rather than a histogram. In Fig. 6-2, A is *symmetrical,* and B is *skewed.* Since its longer tail is toward the right, B is said to be skewed to the right.

A frequency polygon is a many-sided figure drawn by connecting the midpoints of the tops of the histogram bars. Table 6-1 shows how to find the average score for B; you will later find the average score for A. The "average" (4.34 for B) is technically called the ARITHMETIC MEAN. We shall usually call it the mean, since it is the only mean in common use for statistical purposes. It is obtained by taking the sum of the 50 scores and dividing this sum by 50. In order to define the mean precisely, we adopt certain conventional symbols.

1 Instead of using many plus signs, we use the Greek capital letter S, Σ (sigma), to indicate the *operation* of adding (summation). We call Σ the SUMMATION SIGN. Whenever you find a summation sign, regard it as an instruction to take the sum of what follows.

2 We use the letter X to refer to the scores in the distribution and $\bar{X}$ (read "X bar") to represent their mean. Since there are several such individual scores, we think of them as being numbered, and we put subscripts on the letter X: we write X_1 for the first score, X_2 for the second, and so on. In order to represent any single score without specifying which one, we write X_i. In order to say, "Take the sum of all the values of X," we write ΣX_i, and we then think of the subscript i as taking, successively, each of the numbers from 1 to N. As i goes from 1 to N, X_i stands in turn for each of the 50 scores. Whenever there is any possible doubt about the values i may take, we make a notation about these values below and above the summation sign: $\sum_{i=1}^{N} X_i$ refers explicitly to the sum of all the X's as i goes from 1 to N.

We can now write the rule for finding the arithmetic mean as $\bar{X} = \sum_{i=1}^{N} X_i/N$. Most of the time it is sufficient to write only $\bar{X} = \Sigma X/N$.

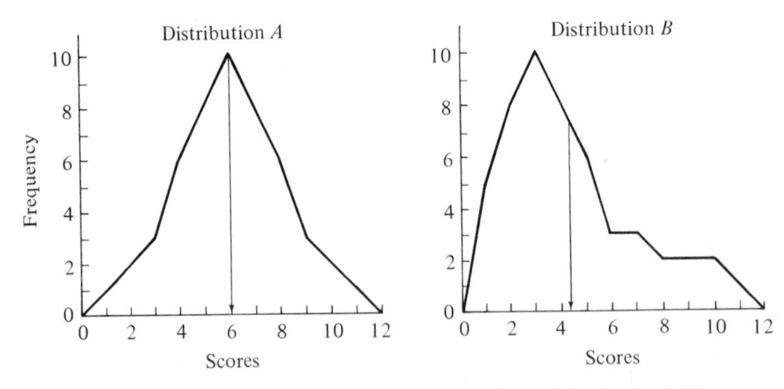

Figure 6-2: Frequency polygons showing two distributions of 50 scores each: distribution A is symmetrical; B is skewed to the right. Arrows indicate positions of the arithmetic means.

Table 6-1 Frequency Distributions from Fig. 6-2

Score X	Distribution B Frequency f	Frequency Times Score $f \cdot X$	Distribution A f	$f \cdot X$
1	5	5	______	______
2	8	16	______	______
3	10	30	______	______
4	8	32	______	______
5	6	30	______	______
6	3	18	______	______
7	3	21	______	______
8	2	16	______	______
9	2	18	______	______
10	2	20	______	______
11	1	11	______	______
	$N = 50$	Sum = 217	______	______
		$\frac{217}{50} = 4.34$		______

6-1 The symbol Σ is called the ______ sign. It is an instruction to take the ______ of that which follows it.

summation
sum

6-2 Examine distribution A in Fig. 6-2. Write down the frequencies of the 11 values of X in the spaces under f in Table 6-1.

1; 2; 3; 6; 8; 10; 8; 6; 3; 2; 1

6-3 How many scores are in this distribution? ______. What is the value of ΣX_i? ______

50
300

6-4 The mean of this distribution is ______.

6

6-5 The mean of distribution B is 4.34. The mean coincides with the peak in distribution ______.

A

6-6 Distribution ______ is symmetrical, while distribution ______ is skewed. Its longer tail is toward the (*left; right*).

A
B; right

6-7 B is skewed to the (*left; right*). Its mean lies to the (*left; right*) of the peak.

right
right

6-8 The distributions in Fig. 6-2 are (*frequency; probability*) distributions. They are drawn as ______ polygons; the distribution in Fig. 6-1 is drawn as a ______.

frequency
frequency
histogram

6-9 The mean is defined by the equation $\bar{X} =$ ______. The subscript i is understood as taking, in turn, each ordinal number from ______ to ______.

$\frac{\Sigma X_i}{N}$
1; N

B. THE MEAN AS A CENTER OF GRAVITY

The mean will play a very large role in the lessons which follow. You will be able to understand it better by thinking of it in terms of a simple analogy. The analogy is a mechanical one, already familiar to you if you have studied elementary physics.

The mean of a set of values is like the center of gravity of a set of weights. Think of a seesaw mounted on a support (a fulcrum) with various numbers of 1-pound weights placed and stacked at different distances from this balance point. You know that the seesaw will balance perfectly only under special conditions which depend upon the weight placed at different points on each side and upon the distance of these points from the fulcrum.

Figure 6-3 is a sketch of such a situation. There are 11 unit weights, and the scores at which these weights are placed range from 4 to 10. There are three scores of 7, two each of 6 and 8, and one each of 4, 5, 9, and 10. Notice that the drawing looks like a histogram of the frequency distribution for a set of 11 scores. The scores, listed in order, are 4, 5, 6, 6, 7, 7, 7, 8, 8, 9, 10.

The mean of these 11 scores is 7, and the supporting fulcrum has been drawn at the center of the score 7. We claim that the set of 11 weights will exactly balance if the fulcrum is in this position, and you are probably not disposed to disagree since the distribution of weights is obviously symmetrical around this point. Let us nevertheless show *why* the mean of 7 is the balance point.

In the mechanical analogy we are considering, the pressure, or downward push, exerted by a weight is the product of its own weight and its distance from the fulcrum. Since these weights are all equal to 1 pound, the product for each of the 11 numbers is simply 1 times its distance from the fulcrum. The distance from the fulcrum is the DEVIATION FROM THE MEAN, and it can be written as $X_i - \bar{X}$, where X_i again represents each member of the set of scores in turn. Subtracting $\bar{X} = 7$ from each of the 11 scores, we obtain the following set of DEVIATION SCORES: $-3, -2, -1, -1, 0, 0, 0, +1, +1, +2, +3$. The sum of the positive deviation scores is exactly equal to the sum of the negative deviation scores; the seesaw, mounted at exactly the center of position 7, will balance.

The sum of the deviations of a set of scores from their arithmetic mean is always equal to zero. The sum of the positive deviations (the weight pushing downward on the right-hand side) will always exactly equal the sum of the negative deviations (the weight pushing downward on the left-hand side). In later lessons we shall often recall the fact that $\Sigma(X_i - \bar{X}) = 0$.

When a distribution is skewed so that its longer tail is to the left, the mean will lie somewhat to the left of the distribution's peak. Such a distribution is skewed to the left or negatively skewed, meaning that its tail is toward the left, where the *negative* deviation scores will be found. A distribution like *B* in Fig. 6-2 (page 84), which is skewed to the right, is said to be positively skewed. Its tail is toward the positive deviation scores at the right, and its mean lies to the right of the peak. The mean is always pulled away from the peak toward the extreme scores in the tail of the distribution.

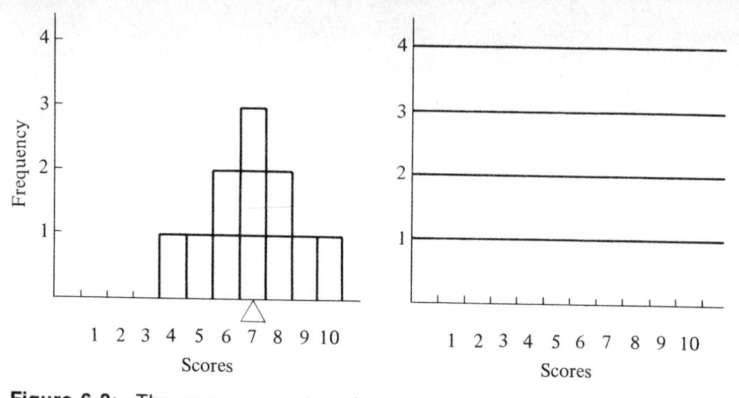

Figure 6-3: The mean as center of gravity of a set of weights.

6-10 The mean of a set of scores is analogous to the ________ of ________ of a set of weights. If the set of weights is supported at the exact point which represents the mean, the weights will ________ perfectly.

center; gravity

balance

6-11 In the space provided at the right of Fig. 6-3, draw a histogram for the following scores: 10, 10, 9, 9, 9, 8, 8, 7, 7, 6, 4. The mean of these scores is ________.

8

6-12 Put a fulcrum in your drawing. It should be placed at the center of the position numbered ________.

8

6-13 This distribution is (*skewed; symmetrical*). Its mean (*does; does not*) coincide with its peak.

skewed
does not

6-14 The tail of the distribution is toward the (*left; right*); it is skewed to the ________.

left
left

6-15 Write down the deviation scores $X_i - \bar{X}$ for this distribution. The sum of the positive deviation scores is ________; the sum of the negative deviations is ________.

+8
−8

6-16 The sum of all the deviation scores in this distribution is ________. The sum $\Sigma(X_i - \bar{X})$ is always equal to ________ for any distribution.

0; 0

6-17 A skewed distribution whose tail is toward the left is (*negatively; positively*) skewed because the tail is toward the (*negative; positive*) deviations.

negatively
negative

6-18 A distribution which is positively skewed has its tail toward the ________ deviations. It is skewed to the ________.

positive
right

Understanding the mean as a center of gravity will enable you to grasp the sense in which the mean $\bar{X}$ is a "best guess" about an unknown score belonging to its distribution. The center of gravity is that point at which the sum of negative deviations (to its left) equals the sum of positive deviations (to its right). If you know nothing about an unknown score X except that it belongs to a distribution with $\bar{X} = 4.34$, your best guess about the value of X is the value 4.34. You will of course not be *exactly* right in any one of a series of such guesses about different scores; none of the scores can be exactly 4.34. But in the long-run series of such guesses your guess will be closer to the true values of the unknown scores than any other value or values you might have chosen.

We shall encounter this concept of a best guess quite often. Whenever we do, the best guess at the value of a variable will be called the **EXPECTED VALUE** or **EXPECTATION** of that variable. The expectation of an unknown score X is written $E(X)$, and we have just defined it: $E(X) = \bar{X}$.

Be careful to avoid letting the word "expectation" suggest to you that $E(X)$ is the value X is literally "expected" to have. $E(X)$ is frequently a value which X can *never* have; 4.34 in our example is just such a value. The expectation of a variable is that value which will, *in a long-run series of guesses,* be closest to the true values of the unknown variable. The long-run amount of error will be less for this value than for any other value that might be chosen.

We can apply these new concepts to binomial probability distributions. The mean of any binomial probability distribution is equal to Np, the product of N and p. For $N = 10$ and $p = 3/4$, the mean is 7.5. Like any other mean, this mean is the balance point of the distribution. It is the expectation of the variable K, $E(K)$.

Let us imagine that we participate in a very large number of 10-trial coin-toss experiments (say, 10,000). Just before each such experiment, we make a guess about the value which K will take. If $p = 3/4$, then the value 7.5 is always the best guess to make. By guessing 7.5 each time, we shall be mistaken by smaller amounts *in the long run* than we would by guessing any other number or numbers. This is true in spite of the fact that 7.5 is a number which can never be exactly right; in no experiment can K turn out to be exactly 7.5. Consult Table 6-2, and you will find that the probability of getting either $K = 7$ or $K = 8$ is .5317. We can literally expect $K = 7$ or $K = 8$ to occur in 5,317 of the 10,000 experiments, and in all those cases our guess will be wrong by only a very small amount (.5). In 8,654 cases, K will be 6, 7, 8, or 9, and we shall not be wrong by more than 1.5. We shall make larger errors much less often than these small errors.

You will see on page 89 that $E(K) = \Sigma K_i p(K_i)$, the sum of all the values of K after each has been multiplied by its own probability. This sum is equal to Np.

Table 6-2 Binomial Probability Distribution for $N = 10$ and $p = 3/4$

K	$p(K)$	$Kp(K)$	K	$p(K)$	$Kp(K)$
10	.0563	______	4	.0162	______
9	.1877	______	3	.0034	______
8	.2815	______	2	.0003	______
7	.2502	______	1	.0000	______
6	.1460	______	0	.0000	______
5	.0584	______			

6-19 For $K = 10$, $p(K) =$ ______, and $Kp(K) =$ ______.

.0563; .5630

6-20 Write the values of $Kp(K)$ in Table 6-2, and find $\Sigma K_i p(K_i)$. Rounded off to two decimal places, $\Sigma K_i p(K_i) =$ ______.

7.50

6-21 7.5 is the ______ of this probability distribution. It can be found more quickly by multiplying ______ times ______.

mean
N
p

6-22 You can understand why $Np = \Sigma K_i p(K_i)$ if you recall that a probability is an *expected relative frequency*. In Lesson 3 (page 42) we used empirical relative frequencies to find empirical ______.

probabilities

6-23 The column $p(K)$ in Table 6-2 is a column of theoretical relative frequencies. Its role is like the role of the column labeled ______ in Table 6-1.

Frequency (f)

6-24 The sum of the values in the column $p(K)$ is ______. The sum of the values in the column Frequency for Table 6-1 is ______.

1

50 (or N)

6-25 The column $Kp(K)$ is like the column ______ in Table 6-1; when we divide $\Sigma K_i p(K_i)$ by 1, it is like dividing ______ by N in a frequency distribution.

Frequency Times Score ($f \cdot X$)
ΣX_i

6-26 The mean Np is the expectation of the variable ______.

K

6-27 The mean $\bar{X}$ is the ______ of the variable X. If a long series of guesses about unknown values of X must be made and these values are known to belong to the distribution whose mean is $\bar{X}$, the best guess is ______.

expectation

$\bar{X}$

6-28 In dealing with a binomial probability distribution with $N = 50$ and $p = 3/4$, the best guess about values K will assume in a long series of experiments is $E(K) =$ ______.

37.5

C. ORDINAL DATA AND THE MEDIAN

Some of the properties which one would like to measure do not have useful yardsticks or other measuring scales. "Beauty" is such a property, and it is not hard to think of countless others. Judges of a beauty contest do not usually attempt to say how beautiful a girl is; instead they compare a group of contestants and rank them in order from most to least beautiful. The resulting series of ordinal numbers, i.e., ranks, is a set of ORDINAL DATA on the relative beauty of the contestants.

In assigning rank numbers, a judge may or may not be allowed to include *tied ranks*. Table 6-3 shows two sets of ordinal data on the same 10 persons; the persons are designated by the letters A through J and arranged in the order of the first judge's ranking. Judge X has assigned no tied ranks; his rank numbers are the whole numbers from 1 to 10. Judge Y found contestants A and D tied for second place, and since the next lower person is G in fourth place, A and D are both assigned the rank 2.5; together they occupy the second and third places in the series. Similarly E and J get the rank 7.5 because they are tied in seventh place and occupy ranks 7 and 8 together.

Table 6-3 Two Sets of Ordinal Data on Ten Persons (*A* through *J*)

Judge	A	B	C	D	E	F	G	H	I	J
X	1	2	3	4	5	6	7	8	9	10
Y	2.5	1	6	2.5	7.5	9	4	5	10	7.5

We can find a midpoint in the set of 10 ranks; it will be the rank 5.5, halfway between the two middle ranks 5 and 6. The technical name for this midpoint is MEDIAN; it divides the set of ranks into two equal subsets. We can compare the rankings of judges X and Y and observe that contestants A, B, and D are rated above the median by both judges, while contestants F, I, and J are rated below the median by both judges.

When we are dealing with ranks instead of scores, we cannot say as much about the relative standings of a group of persons or objects. Ordinal data do not provide information about the relative distances between pairs of observations. Judge X's ratings indicate that he considered A's beauty greater than B's and B's greater than C's, but they do not tell anything about the amount of difference between A and B, between B and C, or between A and C. Technically, we say that ordinal data do not provide information about the INTERVALS between observations. On the other hand, observations such as scores or heights do provide information about these intervals. Therefore we can now begin to designate such observations as INTERVAL DATA, to distinguish them from ordinal observations.

Unless we have interval data, we cannot calculate a mean. The arithmetic mean of a set of *rank numbers* (such as the numbers 1 through 10 in Table 6-3) is not a center of gravity with respect to the variable being studied ("beauty"). With ordinal data we can only determine a median, dividing the ordered series into an upper and a lower half.

6-29 A set of observations is said to be an *ordered set* when the observations can be arranged in a meaningful order, such as from most to least. Ranks in a contest are an __________ set of observations. — *ordered*

6-30 Twelve children in a family can also be arranged in order of their birth. Their names are an __________ set. — *ordered*

6-31 If no information is available on the intervals between members of such an ordered set, the observations are ordinal data. Without information about the ages of the 12 children, observations on their birth order are __________ data. — *ordinal*

6-32 The oldest child is given the rank number 1. The youngest two children are twins; they will receive a tied rank number of ________. — *11.5*

6-33 If the fifth, sixth, and seventh children are triplets, all three will receive the rank number ________. — *6*

6-34 The median for a set of 12 ordinal observations is ________. It divides the set into two equal subsets, a subset with ranks above the __________ and a subset with ranks below it. — *6.5*; *median*

6-35 In Table 6-3, contestant C is rated __________ the median by judge X and __________ the median by judge Y. — *above*; *below*

6.36 A set of height observations does provide information about the __________ between observations. Such data are called __________ data. — *intervals*; *interval*

6-37 Students are sometimes described as being in the upper (or lower) half of their graduating class. This kind of description is based upon the (*mean; median*) as a central point. — *median*

6-38 The mean and the median are both called indicators of ==========. — *central tendency*

6-39 The center of gravity of a set of observations is the __________. It exists only when the data are __________ data. — *mean; interval*

6-40 The median divides a set of observations into ==========. — *two equal subsets*

6-41 Ranks and birth order are examples of __________ data. It is possible to calculate a __________ for such data, but it is not possible to calculate a __________. — *ordinal*; *median*; *mean*

When we have a set of interval observations such as scores, we can find a mean and we can also identify a MEDIAN SCORE. The median score partitions the set of scores into two equal subsets; there must be as many scores above the median score as below it. Unless the distribution is perfectly symmetrical, the median score will not be the same as the mean.

Table 6-4 Determining the Median Score for Distribution *B* from Table 6-1

Score	Exact Limits	Frequency	Ranks Occupied
11	11.5–10.5	1	1
10	10.5–9.5	2	2–3
9	9.5–8.5	2	4–5
8	8.5–7.5	2	6–7
7	7.5–6.5	3	8–10
6	6.5–5.5	3	11–13
5	5.5–4.5	6	14–19
4	4.5–3.5	8	20–27
3	3.5–2.5	10	28–37
2	2.5–1.5	8	38–45
1	1.5–0.5	5	46–50

Rank	Score
	4.5
20	
21	4.25
22	
23	4.0
24	
25	3.75
26	
27	
	3.5

Table 6-4 shows how the median score is determined for distribution *B* in Table 6-1. The two scores of 10 are considered to occupy ranks 2 and 3 together; they are also considered as occupying the space between 10.5 and 9.5 on the scale of scores. The space occupied by scores having a particular value is indicated in the Exact Limits column of the table.

Since there are 50 scores, the middle rank is 25.5 and the median score is a score with that rank. It is one of the eight scores of 4 which occupy ranks 20 through 27. These scores occupy the space from 4.5 to 3.5 on the scale of scores. The space occupied by these 8 scores is enlarged at the right of Table 6-4 to show that 25.5 is three-fourths of the distance from the top of rank 20 to the bottom of rank 27. The median score must also be three-fourths of the distance between 4.5 and 3.5; it is therefore 3.75.

Median scores are not much used in statistical decision making, but they can be important whenever a distribution is extremely skewed. For example, the distribution of personal incomes in the United States is extremely skewed to the right; the mean income is misleadingly high because of a few very large incomes. Therefore, the median income is more often stated; half the population have incomes above the median and half have incomes below it.

Medians are also used instead of means whenever a distribution is *open-ended*. Table 6-5 shows a distribution of time scores; 100 persons were given up to 1 hour to complete a task, and the amount of time required by each has been tabulated. Since four persons did not finish the task, the time category "60 and over" is an inexact category, and the distribution is open-ended because the upper end of the range is unspecified.

Table 6-5 Distribution of Times Required by 100 Persons to Complete a Task

Time t	Frequency	Midpoint	Time t	Frequency	Midpoint
1–5.5	1	3.5	36–40.5	6	38.5
6–10.5	1	8.5	41–45.5	3	43.5
11–15.5	7	13.5	46–50.5	3	48.5
16–20.5	23	18.5	51–55.5	3	53.5
21–25.5	18	23.5	56–59.5	1	58.0
26–30.5	21	28.5	60 and over	4	—
31–35.5	9	33.5			

6-42 The data in Table 6-4 are (*interval; ordinal*) data. The data in Table 6-5 are ________ data. *interval* *interval*

6-43 The distribution in Table 6-5 does not give exact times for each of the ________ persons studied. It groups their time scores into ________ categories to save space. *100* *13*

6-44 Times were measured to the nearest 30 seconds; therefore, no score could fall between 5.5 and 6 minutes. A score of 12.5 would be placed in the category whose midpoint is ________. *13.5*

6-45 If a mean had to be calculated *from this table,* the seven scores in that category would all be treated as if they equaled 13.5, and the product Frequency Times Time for that category would be taken as ________. *94.5*

6-46 However, there is one category which does not have a midpoint; it is the category ________. *60 and over*

6-47 Because of this category, the distribution is described as an ________ distribution. *open-ended*

6-48 No scores are available for the ________ persons in this category because they did not finish the task within the time limit. *4*

6-49 Without scores for these four persons, it is not possible to calculate a ________ for the distribution. However, a ________ can be used to describe its central tendency. *mean* *median*

6-50 The median score for the distribution would be between 25.5 minutes (for rank number ________) and 26 minutes (for rank number ________). The median score is ________ minutes. *50* *51; 25.75*

D. CATEGORICAL DATA AND THE MODE

Figure 6-4 shows a frequency distribution for which neither a mean nor a median can be calculated. Since there is no quantitative difference among the categories of the variable "author's illness," these categories cannot be ordered as a numerical series. Such a variable is said to be QUALITATIVE, whereas ordinal and interval data are QUANTITATIVE.

It may seem that one could not speak at all of a central tendency in the distribution in Fig. 6-4. However, it is of some interest to note that "paranoid schizophrenia" occurs with greater frequency than any other category. The most frequently occurring category is called the MODAL category, and this category is usually mentioned as a kind of central tendency in descriptions of qualitative data distributions.

When we are dealing with interval data, as in Fig. 6-2, we can also designate a most frequently occurring *score*. For distribution B it is the score of 3 with a frequency of 10. We call 3 the modal score or simply the MODE of that distribution. Distribution A in Fig. 6-2 also has a mode; its value is 6.

Occasionally we find distributions which have more than one clear peak. If there are two such peaks, we describe the distribution as BIMODAL; a distribution with only one peak is UNIMODAL. If there are more than two peaks, the distribution is MULTIMODAL.

All three indicators of central tendency—mean, median, and mode—may be used to describe any distribution consisting of interval observations. As Fig. 6-5 shows, the three indicators will differ whenever the distribution is not both unimodal and symmetrical. When you intend to make inferences about a population, you should usually choose the mean because it will permit you to employ the most powerful methods for making such inferences. You may choose the median instead when the distribution is either open-ended or extremely skewed. The mode is used only for statements about the most frequently occurring observation.

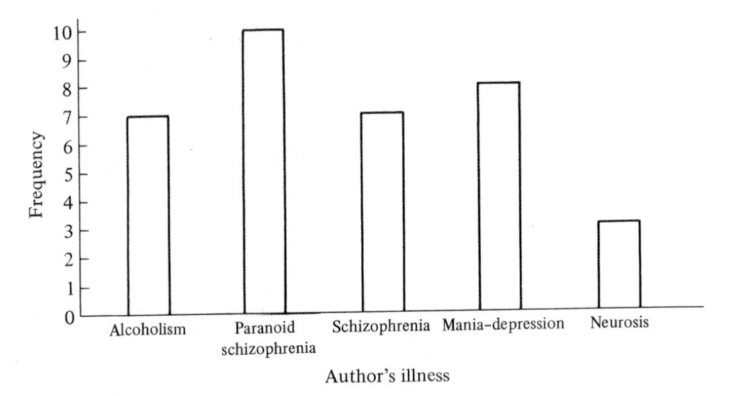

Figure 6-4: Total number of autobiographies of former mental patients available in English in 1959. [Data from R. Sommer and H. Osmond, "Autobiographies of Former Mental Patients," *J. Ment. Sci.*, **106**:653 (1960).]

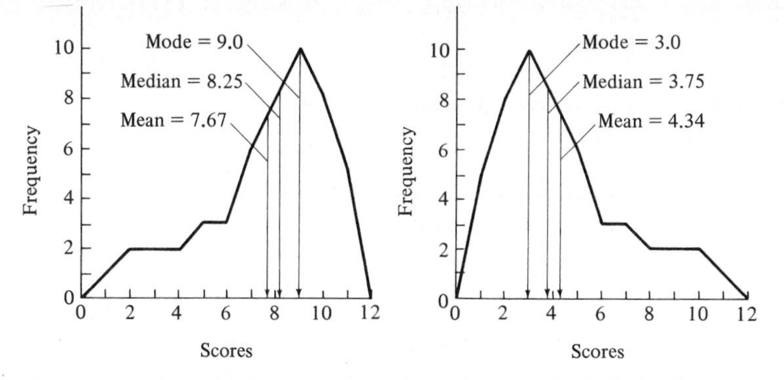

Figure 6-5: Comparative positions of the mean and median relative to the position of the mode in two skewed distributions ($N = 50$, each distribution).

6-51 Categorical data are also called (*qualitative; quantitative*) data because the values of the variable are *qualities*.

qualitative

6-52 In order for data to be quantitative, the values of the variable must be *numbers*. Ordinal and interval data are (*qualitative; quantitative*); categorical data are ________.

quantitative
qualitative

6-53 With qualitative data, the only possible description of central tendency is in terms of the ________ category.

modal

6-54 With interval data, we can describe the most frequently occurring score; it is called the ________ score, or simply the ________.

modal
mode

6-55 Figure 6-5 shows two distributions. The one on the right is from Table 6-1; it is (*negatively; positively*) skewed. The left distribution is very similar, except that it is ________ skewed.

positively
negatively

6-56 In the negatively skewed distribution, the mode is ______, and the median is ______. The mean is ______.

9; 8.25; 7.67

6-57 In a negatively skewed distribution, the extremely (*high; low*) scores in the tail of the distribution affect the mean more than the median, so that the mean is (*larger; smaller*) than the median.

low
smaller

6-58 In the positively skewed distribution, the mode is ______. Both the mean and the median are (*larger; smaller*) than the mode.

3; larger

6-59 When the skew is positive, the tail contains some extremely (*high; low*) scores; these scores affect the (*mean; median*) more than they affect the (*mean; median*).

high
mean; median

REVIEW

6-60 The probability distributions we have been studying have been drawn as histograms. The height of each bar represents the ________ of a particular value of K. *probability*

6-61 When a frequency distribution is drawn as a histogram, the height of each bar represents the ________ of a particular value of X. *frequency*

6-62 The sum of the heights of all the bars in a *frequency* distribution is ________. In a *probability* distribution, the sum is always ________. *N* *1*

6-63 A frequency distribution may also be drawn as a frequency ________. Each frequency is represented by a single point, and lines connecting these points form a many-sided figure, or ________. *polygon* *polygon*

6-64 The symbol Σ indicates the operation of ________. $\Sigma p(K) = 1$ means that the ________ of all the probabilities in a probability distribution is ________. *summation* *sum* *1*

6-65 Let $f(X)$ stand for the frequency of X. $\Sigma f(X) = N$ means that the ________ of the ________ of all the scores in a frequency distribution is ________. *sum; frequencies* *N*

6-66 X_1 refers to the first member of a set of scores, X_2 to the second, and X_N to the last. Any one of these scores can be represented by ________. X_i

6-67 ΣX_i means the ________ of all the values of X. The subscript i is thought of as taking, in turn, all the ordinal numbers from ________ to ________. *sum* *1; N*

6-68 $\Sigma X_i/N$ defines the ________ of a set of scores. It is represented by the symbol ________. *mean* $\bar{X}$

6-69 $\Sigma(X_i - \bar{X})$ refers to the sum of a set of ________ of scores from their own ________. *deviations* *mean*

6-70 $\Sigma(X_i - \bar{X})$ is always equal to ________. *0*

6-71 $E(X)$ refers to the ________ of the variable X. $E(X) =$ ________. *expectation* $\bar{X}$

6-72 $E(X) = \bar{X}$ means that the best long-run guess about the variable ________ is the ________ of its distribution. *X; mean*

6-73 If a long series of such guesses must be made about scores in a distribution whose mean is 50, the long-run error will be less by guessing the value _______ than by guessing any other value or series of values. *50*

6-74 The expectation of the variable K in 10-toss experiments with a fair coin is the __________ of the binomial distribution with $N =$ _______ and $p =$ _______. *mean* *10; 1/2*

6-75 The best long-run guess in this case is that $K =$ _______. *5*

6-76 The mean of a binomial distribution is always equal to the product _______. *Np*

6-77 $E(K) = \Sigma K_i p(K_i)$, the sum of a set of products. Each of these products is obtained through multiplying _______ by its __________. *K* *probability*

6-78 Ranks and birth orders are examples of __________ data. Such data do not provide information about the __________ between pairs of observations. *ordinal* *intervals*

6-79 The central tendency of a set of ordinal observations is described by the __________. *median*

6-80 The median __________ a set of observations into __________. *divides* *two equal subsets*

6-81 A median can be calculated also for interval data. It is used particularly when the distribution is __________ or extremely __________. *open-ended* *skewed*

6-82 Qualitative data have neither mean nor median, but the most frequently occurring category may be described as the __________ category. *modal*

6-83 In a distribution of scores, the modal score is called the __________. *mode*

6-84 A distribution with one mode is called __________; with two modes, it will be called __________. *unimodal* *bimodal*

6-85 When inferences are to be made from a set of scores, the best measure of central tendency to use is the __________. *mean*

6-86 If all three measures are computed for a unimodal negatively skewed distribution, the one with the highest value will usually be the __________. The one with the lowest value will usually be the __________. *mode* *mean*

PROBLEMS

1. Which of the three indicators of central tendency could be used in each of the following situations? If more than one measure is possible, mention each and tell which might be preferred.

a. The frequency of alcoholism among various national groups, such as French, Norwegians, Poles, etc.

b. The frequency of alcoholism among economic classes of Americans divided into various annual income groups, such as "up to $1,999," "$2,000 to $2,999," etc.

c. A group of 20 wine tasters are asked to assign a rank from 1 (highest quality) to 5 (lowest quality) to a particular brand of sherry. A distribution is made up showing the frequency of ratings 1, 2, 3, 4, and 5.

2. Take the following distribution as a set of classroom test scores ($N = 30$). The test had 55 possible points. In assigning letter grades to the test papers, the teacher might follow any one of several procedures. Examine the consequences of each of the following procedures, determining the number of students getting each grade in part *a*, the numbers getting A or B versus C, D, or F in parts *b* and *c*, and the cutoff points for deciding between grades in part *d*.

a. Converting the scores to "percent correct out of 55," and giving A's to those with 90 to 100 percent correct, B's for 80 to 89 percent, C's for 70 to 79 percent, D's for 60 to 69 percent, F's to those below 60 percent.

b. Finding the *mean*, and giving B or A to those above the mean and C, D, or F to those below the mean.

c. Finding the *median*, and giving B or A to those above it and C, D, or F to those below it.

d. Giving A's to the top 10 percent (by raw score), B's to the next 50 percent, C's to the next 25 percent, and D's or F's to the bottom 15 percent.

Score	Frequency	Score	Frequency	Score	Frequency
52	1	43	2	34	0
51	1	42	1	33	1
50	0	41	0	32	1
49	3	40	2	31	0
48	2	39	1	30	1
47	4	38	1	29	0
46	3	37	0	28	1
45	1	36	0	27	2
44	1	35	1		$\Sigma X_i = 1,258$

3. No mean could be computed for the time scores in Table 6-5 because of the open-ended category "60 or over." A mean could be computed, however, for the 96 persons who finished the task. If one is trying to determine the average time that will be required for similar people to do this task, would such a mean be defensible? Why or why not?

LESSON 7
THE VARIANCE AND OTHER MEASURES OF DISPERSION

We entered upon Lesson 6 with a set of binomial distributions in mind. These distributions differed in the locations of their peaks. They also differed in the degree to which they were spread out along the horizontal axis, and we now turn our attention to this difference in dispersion.

The simplest way to describe the dispersion of our four distributions (shown on page 78) is to state the *range* of their nonzero values. We can make the comparisons shown in Table 7-1.

Table 7-1 Comparison of Four Binomial Distributions with $p = 3/4$

Distribution	Mean	Range	General Shape
$N = 10$	7.5	3–10	Unimodal, negatively skewed
$N = 50$	37.5	26–47	Unimodal, slight negative skew
$N = 100$	75	60–89	Unimodal, very slight negative skew
$N = 200$	150	128–170	Unimodal, almost symmetrical

The range of a set of values gives only a very rough indication of its dispersion. The size of the deviation scores $(X_i - \bar{X})$ is a better indicator. Since the sum of the deviation scores, $\Sigma(X_i - \bar{X})$, is always zero, we must use only the size of these deviations, not their sign. We find two ways of doing so.

1 We can *ignore* the algebraic signs and calculate a mean of the ABSOLUTE VALUES of the deviation scores. The absolute value of a quantity such as +5 is written $|5|$. The absolute value of −5 is also $|5|$. The mean of absolute values of deviation scores is $\Sigma|X_i - \bar{X}|/N$, and it is named the AVERAGE ABSOLUTE DEVIATION (AAD) from the mean.† Although this statistic seems reasonable to the beginner, it is not the best way to use deviation scores in statistical work.

2 We can *square* the deviation scores. All the signs then become positive, and a sum as well as an average can be defined for $(X_i - \bar{X})^2$. The only important measures of dispersion are based on $(X_i - \bar{X})^2$.

† If the deviations are taken from the median, we obtain an AAD from the median, $\Sigma|X_i - \text{median}|/N$. In a skewed distribution the value of the AAD is smaller when the deviations are taken from the median than when they are taken from any other value, including the mean. When a distribution is sufficiently skewed to warrant using the median rather than the mean in describing central tendency, the AAD from the median may be used to describe its dispersion.

A. THE SUM OF SQUARED DEVIATIONS FROM THE MEAN

Squared deviation scores give a smaller sum when they are taken from the mean than when they are taken from any other number, including the median. The sum of a set of squared deviation scores, $\Sigma(X_i - \bar{X})^2$, is called the SUM OF SQUARES for the score distribution from which it is drawn. From this point on, we shall refer to the sum $\Sigma(X_i - \bar{X})^2$ simply as the sum of squares (SS).

You will frequently have to compute SS for a sample of N scores. Instead of converting each raw score X into its deviation score $X - \bar{X}$ and then squaring these deviation scores, you can find the value of SS by using this COMPUTING FORMULA:

$$\text{SS} = \Sigma X_i^2 - \frac{T^2}{N}$$

where T (total) is ΣX_i. This computing formula allows you to square the raw scores themselves, take their sum, and then subtract the quantity T^2/N to obtain SS. You will learn in Lesson 13 how to prove that this computing formula is exactly equivalent to the defining equation $\text{SS} = \Sigma(X_i - \bar{X})^2$.

Table 7-2 shows the calculation of SS for the sample of 50 scores used as an example in Lesson 6. From Lesson 6 you already know that $\bar{X} = 4.34$. The value of T is 217; $T^2 = 46{,}989$, and $T^2/N = 939.78$. By the computing formula, $\text{SS} = 1271 - 939.78 = 331.22$.

Table 7-2 Calculation of Sum of Squares for Distribution *B* in Fig. 6-2 (Page 84)

Score (X)	Frequency (f)	fX	X^2	fX^2
1	5	5	1	5
2	8	16	4	32
3	10	30	9	90
4	8	32	16	128
5	6	30	25	150
6	3	18	36	108
7	3	21	49	147
8	2	16	64	128
9	2	18	81	162
10	2	20	100	200
11	1	11	121	121
	$N = 50$	$T = \Sigma X_i = 217$		$\Sigma X_i^2 = 1{,}271$

Whenever you start to work with a set of scores, you should obtain ΣX_i and ΣX_i^2 simultaneously. If you are working at a desk calculator, you will not need to list all the values of X_i^2 as we have done in Table 7-2. Working from a list of the N values of X, you can enter each value in turn and accumulate the two sums ΣX_i and ΣX_i^2 simultaneously. Record and label these sums at the bottom of your list of raw scores, so that you will have them available for use in further statistical analysis.

Table 7-3 Calculation of Sum of Squares for Distribution *A* in Fig. 6-2.

Score (X)	Frequency (f)	$(X-\bar{X})^2$	$f(X-\bar{X})^2$
1	1	______	______
2	2	______	______
3	3	______	______
4	6	______	______
5	8	______	______
6	10	______	______
7	8	______	______
8	6	______	______
9	3	______	______
10	2	______	______
11	1	______	______

7-1 In Lesson 6 (page 85) you found the value of ΣX_i for this distribution of 50 scores to be 300. The mean is ______. *6*

7-2 For these simple whole numbers, it is easy to calculate SS by finding $\Sigma(X_i - \bar{X})^2$. What is $X_i - \bar{X}$ for a score of 1? ______ *−5*

7-3 Write down the value of $(X_i - \bar{X})^2$ for each of the 11 values of X in the table above. Multiply each squared deviation score by its frequency, and write the product in the column $f(X-\bar{X})^2$. The sum of this column is the value of SS; it is ______. *232*

7-4 Now use the computing formula to obtain SS. You must first square the values of ______. *X*

7-5 Next you must multiply each X_i^2 by its ______. The sum of these products will give you the term ______. *frequency* *ΣX_1^2*

7-6 $\Sigma X_i^2 =$ ______ for this example. In order to obtain SS, you now subtract T^2/N, which equals ______. *2,032* *1,800*

7-7 By this computing formula, the value of SS is ______. *232*

7-8 This distribution of 50 scores is symmetrical; the distribution in Table 7-2 is ______. SS is (*larger; smaller*) for the symmetrical distribution. *skewed; smaller*

7-9 When you begin to analyze a distribution of N scores, your first step is to calculate two sums, ______ and ______. *$\Sigma X_i \leftrightarrow \Sigma X_i^2$*

7-10 One of these is called T; it is the sum ______. *ΣX_i*

7-11 You can then obtain SS by subtracting ______ from ______. *T^2/N* *ΣX_i^2*

B. THE SAMPLE VARIANCE AND STANDARD DEVIATION

We were able to compare the two distributions in Tables 7-2 and 7-3 directly on the basis of their sums of squares because both distributions had the same N. Naturally, SS will be greatly affected by changes in N; the addition of more raw scores means that there will be a larger number of squared deviations, making SS larger. To describe the dispersion of any distribution regardless of size, we divide SS by N.

When the sum of squares from a sample of N scores is divided by N, it becomes the SAMPLE VARIANCE. This variance can best be understood as the mean of the squared deviation scores, or the MEAN-SQUARED DEVIATION. The sample variance for Table 7-2 is $331.22/50 = 6.62$; for Table 7-3 the sample variance is $232/50 = 4.64$.

For descriptive purposes it is customary to calculate the sample STANDARD DEVIATION, which is the square root of the sample variance. The sample standard deviation is designated by the lowercase letter s; the sample variance is then designated s^2. The defining equations are

$$s^2 = \frac{\text{SS}}{N} \qquad \text{and} \qquad s = \sqrt{\frac{\text{SS}}{N}}$$

The standard deviation should have the same number of decimal places as $\bar{X}$. For Table 7-2, $s = 2.57$; for Table 7-3, $s = 2.15$.

The standard deviation can be indicated in a graph of the frequency distribution because it has the same units as the raw X scores. Figure 7-1 shows a distribution with $N = 100$, $\bar{X} = 16$, and $s = 2.4$. Vertical arrows have been drawn at the mean and at 1-standard-deviation intervals around the mean. The value of s^2 has no useful meaning in the context of such a graph.

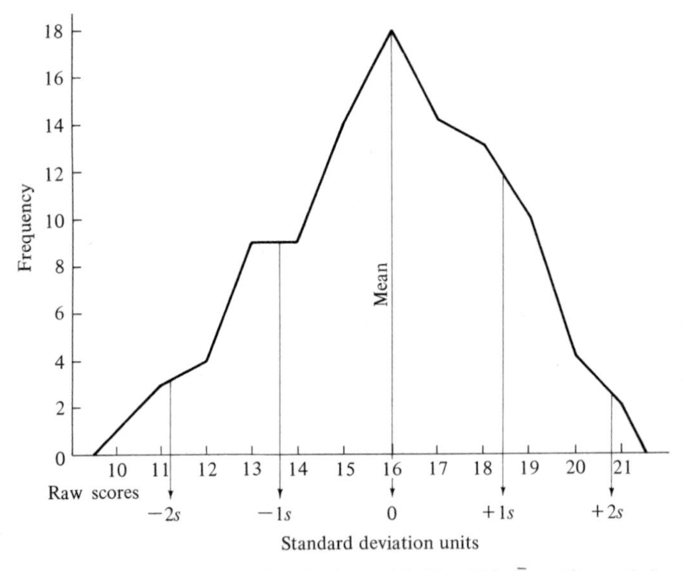

Figure 7-1: A frequency distribution with $N = 100$, $\bar{X} = 16$, $s = 2.4$.

7-12 The mean of the squared deviation scores is SS divided by ______. It is called the sample ______.

N; variance

7-13 The sample variance can be interpreted as the mean ______ deviation of the sample. Its symbol is ______.

squared; s^2

7-14 The symbol s^2 represents the ______ deviation, or variance; the symbol s represents the ______ deviation of the sample.

mean-squared
standard

7-15 In Fig. 7-1, consider that the raw scores are measurements in inches. The standard deviation is ______ inches; the variance is ______ *square* inches.

2.4
5.76

7-16 The standard deviation is represented as a particular distance in Fig. 7-1. The variance cannot be represented as a distance because s^2 is expressed in units of ______.

square inches

7-17 The mean in Fig. 7-1 is at ______ inches. Since $s = 2.4$, 18.4 inches is 1 ______ above the mean.

16
standard deviation

7-18 The value which is 1 standard deviation *below* the mean is ______ inches.

13.6

7-19 The value 20.8 inches is 2 standard deviations ______ the mean.

above

7-20 The distance between 11.2 and 20.8 inches is 9.6 inches. In this distribution, these values are ______ standard deviations apart.

4

7-21 Consider two scores from this distribution, 15 and 17 inches. The distance between them is 2 inches; it can also be described in terms of the standard deviation. It is (what fraction?) ______ of 1 standard deviation.

$^5/_6$

7-22 Three students, A, B, and C, took a college entrance examination in a group of 450 students. For the group, $\bar{X} = 330$ points and $s = 30$ points. A's score of 300 was ______ standard deviation ______ the mean.

1
below

7-23 B scored 315, and C scored 360. B's score is ______ standard deviation below the mean, and it is ______ standard deviations below C's score.

0.5
1.5

7-24 In this example the sample variance is ______. The value of SS is ______.

900
405,000

C. THE POPULATION VARIANCE

With complete information about every element in a population, we can calculate a sum of squares for the population. This SS will be the sum of squared deviations taken from the population mean μ, or $\Sigma(X_i - \mu)^2$. When we divide this population SS by the number of elements in the population, we obtain the population variance. The square root of the population variance is of course the population standard deviation.

The sample standard deviation s is a sample statistic. The population standard deviation is a population parameter; it is therefore given a Greek letter designation. We take the lowercase s from the Greek alphabet, σ (sigma), to designate the population standard deviation; σ^2 then becomes the symbol for the population variance.

The true values of μ and σ^2 are rarely known, for it is rare to have a complete population available for study. It is usually necessary to estimate these parameters from sample statistics. When μ is unknown, we take $\bar{X}$ as an estimate of μ; in this case, the sample statistic $\bar{X}$ itself is an UNBIASED ESTIMATOR of μ. The average of the means of *all possible random samples* from a given population is μ, the population mean; therefore, the expectation of $\bar{X}$ is μ, just as $E(X) = \bar{X}$. Since $E(\bar{X}) = \mu$, $\bar{X}$ is an estimator of μ which will sometimes be too high, sometimes too low, but it will not be consistently too high or too low in a long series of estimates. Because it does not have a consistent bias toward overestimation or underestimation of μ, we call $\bar{X}$ an unbiased estimator of μ.

However, s^2 is not an unbiased estimator of σ^2. The average of the variances of all possible random samples is $[(N-1)/N]\sigma^2$, which is a quantity smaller than σ^2. Since $E(s^2) = [(N-1)/N]\sigma^2$, s^2 as an estimator of σ^2 is biased toward underestimation. We accordingly invent another statistic to use as an estimator. The expectation of $[N/(N-1)]s^2$ is σ^2; the average of the statistic $[N/(N-1)]s^2$ from all possible random samples from the same population is σ^2. Since $s^2 = \text{SS}/N$, we can write our estimator as $\text{SS}/(N-1)$. $E[\text{SS}/(N-1)] = \sigma^2$, and therefore $\text{SS}/(N-1)$ is an unbiased estimator of σ^2. Whenever we must estimate an unknown population variance from sample statistics, we use $\text{SS}/(N-1)$ as our unbiased estimator of σ^2.

Some writers use a special notation for estimates of parameters; the estimated σ^2 is sometimes written $\hat{s}^2$. We shall simply write out estimated σ^2 (or est. σ^2) when we must distinguish between the true but unknown σ^2 and its estimate.

In Lesson 8 we shall need to describe binomial probability distributions in terms of their mean and variance. You already know that the mean of such a distribution is Np. *Its variance is Npq.* For a distribution with $N = 100$ and $p = 1/2$, the mean is 50 and the variance is 25. Since the distribution is a theoretical one (not an empirical sample), the mean and variance are parameters and Npq is designated σ^2.

7-25 To calculate the population sum of squares, deviation scores must be taken from the mean of the ________. *population*

7-26 The sample SS is the sum of squared deviations from the mean of the ________. *sample*

7-27 The population variance is the population sum of squares divided by the number of elements in the ________. Its symbol is ________. *population* σ^2

7-28 If the population is infinite, it (*is; is not*) possible to calculate σ^2. *is not*

7-29 For an infinite population, it is always necessary to ________ the value of σ^2 from sample statistics. *estimate*

7-30 It is also impossible to calculate μ for an infinite population. To estimate μ, we look for a sample statistic which will not be consistently too high or too low in a long series of estimates of population means. We call such a statistic an ________ estimator of μ. *unbiased*

7-31 Any statistic can serve as an unbiased estimator of the *mean* of its distribution. For example, $E(X_i) = \bar{X}$; the mean, or center of gravity, of the X_i distribution is ________. Any single value of X_i can therefore give an unbiased estimate of ________. $\bar{X}$ $\bar{X}$

7-32 $E(\bar{X}) = \mu$. If we take the means of all possible random samples from a population, the center of gravity of the distribution of these means is ________. μ

7-33 Thus, $\bar{X}$ is an unbiased estimator of the population parameter ________. μ

7-34 $E(s^2) = [(N-1)/N]\sigma^2$. s^2 is therefore (*a biased; an unbiased*) estimator of $[(N-1)/N]\sigma^2$; at the same time, s^2 is (*a biased; an unbiased*) estimator of σ^2. *an unbiased* *a biased*

7-35 $E[N/(N-1)s^2] = \sigma^2$. Therefore, $[N/(N-1)]s^2$ is (*a biased; an unbiased*) estimator of σ^2. *an unbiased*

7-36 $s^2 = \mathrm{SS}/N$, so $[N/(N-1)]s^2 =$ ________. In terms of the sample SS, an unbiased estimator of σ^2 is the statistic ________. $SS/N-1$ $SS/N-1$

7-37 The mean of a binomial probability distribution is given by the product ________. Np

7-38 The variance of a binomial probability distribution is given by the product ________. Its standard deviation is ________. Npq; $\sqrt{Npq}$

D. THE STANDARD SCORE (z-SCORE) TRANSFORMATION

The sample variance and standard deviation are descriptive statistics. They are calculated for the purpose of describing a particular set of observations and not for the purpose of making any inferences to a parent population. In this final section we return to the descriptive level, focusing our attention on the standard deviation as a useful tool in comparing two or more sets of observations with each other.

Two distributions with very different means and standard deviations are difficult to compare closely unless both distributions can be put into a standard form. The process of changing a distribution's form is called TRANSFORMATION. When we transform a set of numbers, we do something much like transposing a melody to a different key. We may change the position of the numbers with respect to some absolute reference point. We may change the interval size between the numbers, stretching or shrinking the dimensions of the set. Nevertheless, we retain the structured arrangement of the series; the numbers are still in the same order.

When a transformation does not change the general shape of a graph of these numbers, we say it is a LINEAR TRANSFORMATION. Adding a constant to each member of the set of numbers is a linear transformation; it shifts the whole set farther to the right on the horizontal axis, but it does not change the shape of the distribution. Subtracting a constant is also a linear transformation, since it is the same as adding a negative constant; the distribution shifts to the left on the horizontal axis, but it does not change shape.

Multiplying each element by a constant (or dividing each by a constant) also results in a linear transformation. The operation of multiplication by a constant greater than 1 stretches the dimensions of the distribution, but it does not change the shape because all parts of the distribution are stretched by exactly the same amount. The operation of division (which is multiplication by a constant between 0 and 1) reduces the dimensions of the distribution but again does not change its shape. Both multiplication and division by a constant simply change the *interval size* on the horizontal axis.

Some transformations are NONLINEAR. Squaring a set of deviation scores, for example, produces a nonlinear transformation of these scores. Compare the X and X^2 columns in Table 7-2 (page 100); squaring the scores increases the larger scores more than the smaller ones, and the intervals between neighboring values change by different amounts at the upper and lower ends of the range. Yet the scores remain in the same order. Logarithmic (for example $\log X$) and exponential (for example 2^X) transformations are also nonlinear.

The transformation of raw scores to STANDARD SCORES, called z scores, is a linear transformation. A new set of scores is defined as follows:

$$z_i = \frac{X_i - \bar{X}}{s}$$

A constant, $\bar{X}$, is first subtracted from each score; then the remainder is divided by the standard deviation. When a distribution has been transformed in this way, it is said to be STANDARDIZED.

7-39 A distribution ranges between 10 and 50. If all its scores are raised by 10 points, the distribution has been ________. Its range will now be from ______ to ______; its position has been shifted to the (*left; right*) along the horizontal axis.

transformed
20; 60
right

7-40 If the same scores are now divided by 2, the distribution undergoes another ________. Its range will be from ______ to ______. Its dimensions are (*increased; reduced*), and it shifts to the (*left; right*) on the horizontal axis.

transformation; 10
30; reduced
left

7-41 Both these transformations are linear; in the first transformation, a constant was ________ to all scores; in the second, all scores were divided by a ________.

added
constant

7-42 In a linear transformation, the relative size of intervals between scores is not changed. Suppose $X_1 - X_2 = 1$ and $X_{10} - X_{11} = 2$. After the addition of 10 to all scores, $X_1 - X_2 =$ ______ and $X_{10} - X_{11} =$ ______.

1; 2

7-43 After division by 2, $X_1 - X_2 =$ ______ and $X_{10} - X_{11} =$ ______. The intervals have both been reduced, but $X_1 - X_2$ is still exactly ________ the size of $X_{10} - X_{11}$.

½
1
half

7-44 Subtraction of a constant from all scores and multiplication of all scores by a constant are also ________ transformations.

linear

7-45 Squaring all the scores is also a transformation, but it is ________. The relative sizes of the intervals between scores (*do; do not*) remain the same.

nonlinear
do not

7-46 A z score is a ________ score. It is obtained by subtracting ______ from X and dividing the remainder by ______.

standard
$\bar{X}$; s

7-47 For any distribution, $\bar{X}$ is a constant; s is also a ________. The z-score transformation is a ________ transformation.

constant; linear

7-48 When a distribution has been transformed to standard scores, it has been ________.

standardized

7-49 A score of 38 belongs to a distribution with $\bar{X} = 30$ and $s = 4$. The z-score value for this score is ______.

+2

7-50 In the same distribution, a score of 26 becomes a z score of ______.

−1

We shall now examine two important characteristics of standard scores. First let us look at the matter of *units*. We have already observed that the sample variance s^2 will have units which are the square of the raw-score units; if the raw scores are in inches, s^2 will be in square inches. The standard deviation, on the other hand, will be in the same units as the original raw scores since s is the square root of s^2.

What units does a z score have? The z-score definition tells us when we write with it the units from the height example in Fig. 7-2:

$$z = \frac{X \text{ inches} - \bar{X} \text{ inches}}{s \text{ inches}} = \frac{(X - \bar{X}) \cancel{\text{inches}}}{s \cancel{\text{inches}}} = \frac{X - \bar{X}}{s}$$

The units used in expressing the original raw scores disappear in the transformation to a z score; they cancel out because the same units appear in numerator and denominator. The z score is therefore a dimensionless number. This property of standard scores permits us to compare standardized distributions which initially involved quite different kinds of measurement. For example, a z-score transformation will enable us to compare a set of heights (in inches) with a set of weights (in pounds) or to compare a set of scores from one psychological test with those from an entirely different sort of test.

Second, it is useful to notice that the z-score transformation expresses deviation scores in units equal to 1 standard deviation. Each deviation score is divided by the standard deviation of its own distribution. As a result, the horizontal axis takes on a standard appearance in which the number 0 is at the center, the number +1 represents 1 standard deviation above the mean, the number −1 represents 1 standard deviation below the mean, and so on. Figure 7-2 illustrates this change of scale. Since every set of z scores has the same horizontal axis, any two distributions can be directly compared when each has been standardized.

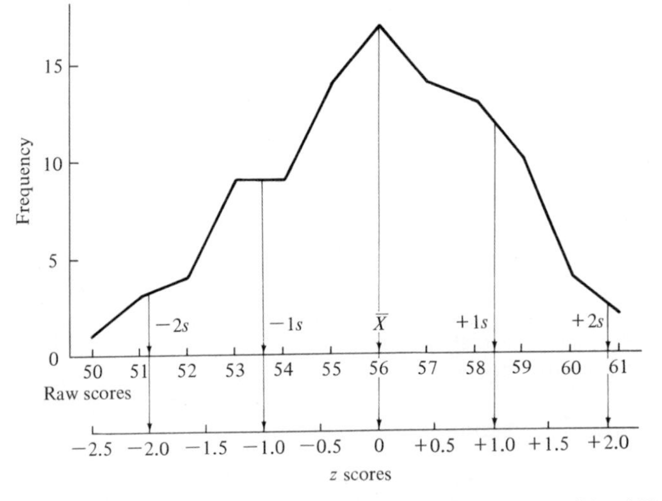

Figure 7-2: Frequency distribution of heights in inches ($N = 100$, $\bar{X} = 56$, $s = 2.4$). The horizontal axis has been given two separate scales, one showing raw scores in inches and the other showing a z-score transformation.

7-51 In Fig. 7-2 a height of 58.4 inches becomes a z score of ______. *+1*

7-52 A z score of +1 always indicates that the original raw score lies exactly ______ standard deviation above the ______ of the distribution to which it belongs. *1* *mean*

7-53 A z score of 0 indicates that the raw score is exactly equal to ______. $\bar{X}$

7-54 Positive z scores indicate that the raw score lies ______ the mean. Negative z scores indicate that the raw score lies ______ the mean. *above* *below*

7-55 The absolute size of the z score describes the size of the interval between the raw score and ______. $\bar{X}$

7-56 It describes this interval in units equal to the statistic ______. s

7-57 The mean of a distribution of z scores is always ______. The standard deviation of a distribution of z scores is always ______. *0* *1*

7-58 A fourth-grade pupil takes two achievement tests. He scores 74 on the arithmetic test, which has a mean of 80 and a standard deviation of 5. His z score on this test is ______. *−1.2*

7-59 The reading test has a mean of 82 and a standard deviation of 4.5; his score of 91 is the equivalent of a z score of ______. *+2*

7-60 This pupil scored ______ standard deviations above the mean on reading, but he scored ______ standard deviations ______ the mean on arithmetic. *2* *1.2* *below*

7-61 Another pupil has the same z score on both tests. His score is 85 on the arithmetic test; his score on the reading test is ______. *86.5*

7-62 In terms of raw scores, it seems that this pupil did better in reading than in arithmetic. Which comparison should be used, the raw-score or the z-score comparison? ______ *z-score*

7-63 The z scores are more meaningful because they take into account the difference between the two tests with respect to the statistics ______ and ______. $\bar{X} \leftrightarrow s$

REVIEW

7-64 This lesson is concerned with measurement of the spread, or __________, of a distribution. — *dispersion*

7-65 A measure of dispersion is a measure of the variability in a distribution. If all the elements are clustered very close to the mean, the variability is (*high; low*). — *low*

7-66 A rough indication of the degree of variability can be obtained by stating the highest and lowest scores in the distribution. This indicator is called the __________ of the distribution. — *range*

7-67 A narrow range of scores indicates a low degree of __________. — *variability (or dispersion)*

7-68 Other measures of dispersion are based on deviation scores. $X_i - \bar{X}$ is a deviation of the score X_i from the __________ of its distribution. — *mean*

7-69 $X_i -$ median is also a __________ score, but the deviation is taken from the __________ instead of from the mean. — *deviation; median*

7-70 $\Sigma(X_i - \bar{X})$ is always equal to _______ because the sum of the positive deviations is always equal to the __________ of the negative deviations. — *0; sum*

7-71 The algebraic signs of these deviation scores can be ignored. The values of the deviations are then called __________ values. — *absolute*

7-72 If $\Sigma|X_i - \bar{X}|$ is divided by _______, we obtain a statistic called the average absolute deviation from the __________. — *N; mean*

7-73 An AAD can also be calculated from the median. It is (*larger; smaller*) than the AAD from the mean. — *smaller*

7-74 Whenever a distribution is skewed enough to warrant using the median as the indicator of central tendency, the dispersion can be measured by the __________ from the __________. — *AAD; median*

7-75 When all the values of $X_i - \bar{X}$ are squared, their algebraic signs become __________. The sum of these squared deviation scores is called the sample __________. — *positive; sum of squares*

7-76 $\Sigma(X_i - \bar{X})^2/N$ defines the __________ of the __________. Its symbol is _______. — *variance; sample; s^2*

7-77 The square root of s^2 is called the ________ standard deviation. Its symbol is ______.

sample
s

7-78 The variance of the *population* is defined by ______. The deviations are taken from ______, and N is the number of elements in the ________.

$\frac{\Sigma(X_i - \mu)^2}{N}$
μ
population

7-79 The abbreviation SS refers to the (*population; sample*) sum of squares in which deviations are taken from ______.

sample
$\bar{X}$

7-80 The population variance is a (*parameter; statistic*), and its symbol is ______.

parameter
σ^2

7-81 The variance of a binomial probability distribution is equal to the product ______.

Npq

7-82 For a binomial distribution with $N = 64$ and $p = 1/2$, the mean is ______ and the variance is ______.

32; 16

7-83 The standard deviation of this binomial distribution is equal to ______.

4

7-84 A z-score transformation can be made for $K = 36$ in this distribution. First the value of the mean is ________ from 36; then the remainder is divided by the value of the ________.

subtracted
standard deviation

7-85 When it is transformed to a z score, $K = 36$ becomes $z =$ ______. It is exactly ______ standard deviation above the ________ of its distribution.

+1; 1
mean

7-86 A transformation which does not change the relative size of ________ within the distribution is called a linear transformation.

intervals

7-87 The z-score transformation is a (*linear; nonlinear*) transformation. Squaring all scores in a distribution is a ________ transformation.

linear
nonlinear

7-88 When μ is unknown, it can be estimated from a sample statistic. $\bar{X}$ is an ________ estimator of μ because $E(\bar{X}) =$ ______.

unbiased
μ

7-89 When σ^2 is unknown, it can be estimated from s^2. Since $E[N/(N-1)s^2] = \sigma^2$, an unbiased estimator of σ^2 is ______ times s^2.

$\frac{N}{N-1}$

7-90 Since $s^2 = SS/N$, σ^2 can also be estimated directly from SS. The unbiased estimator is SS divided by ______.

$N - 1$

PROBLEMS

1. For the distribution of 30 classroom test scores given in Prob. 2 on page 98, you can now consider another procedure which the teacher might follow. Suppose she calculated both $\bar{X}$ and s for the distribution, then determined the letter grades in terms of the following breaking points:

A = z scores above +1.5
B = z scores between 0 and +1.5
C = z scores between 0 and −1.0
D = z scores between −1.0 and −2.0
F = z scores below −2.0

How would this procedure affect the number of students getting A's, B's, etc.?

2. If a distribution has a mean of 118 and s of 11, what z scores correspond to raw scores of 115, 134, and 99?

3. Compute the range, sum of squares, variance, and standard deviation for the distribution of scores given below:

x	f	fX	x	f	fX
20	1	20	14	10	140
19	4	76	13	8	104
18	6	108	12	5	60
17	9	153	11	4	44
16	10	160	10	2	20
15	16	240		$N = 75$	$T = 1125$

4. The distribution described below represents the numbers of students in a sample of 212 who arrived at class at various times after the scheduled class hour.

Frequency	Minutes Late	Frequency	Minutes Late
180	0	2	5
13	1	1	6
6	2	1	7
4	3	1	8
3	4		

Taking the median as 0.57 minutes, consider how the mean and standard deviation, on the one hand, and the median and AAD from the median, on the other, would differ as methods of describing the variability in these data.

LESSON 8 THE NORMAL PROBABILITY DISTRIBUTION IN BINOMIAL TESTS

In Lesson 5 we learned that increasing the number of trials from 10 to 20 would reduce the probability of a type II error. With $N = 20$, however, the probability β is still disconcertingly large—.256 for H_2: $p = 3/4$. Evidently an even larger sample is necessary if β is to be about as low as α.

The calculation of binomial distributions for large N would be laborious. Fortunately there is a good shortcut for making binomial tests when N is large, and Lessons 6 and 7 have now put us in a position to use it.

When N is very large, all binomial probability distributions take a form closely resembling the NORMAL PROBABILITY DISTRIBUTION. The properties of this distribution have been calculated and set down systematically in tables. Whenever we are dealing with a distribution which closely resembles the normal probability distribution, we can simply look up the probabilities we need in a table like Table A-1 (page 351). The normal-curve table is extremely easy to use once you have understood the simple logic on which it is based.

We shall begin by looking again at the four binomial distributions on page 78. For all of them, $p = 3/4$. But as N increases, the shape and position of the distributions change. With $N = 200$ the distribution has become almost symmetrical; it is also much flatter in appearance and more spread out than the other distributions. Using the concepts from Lessons 6 and 7, we can now state the mean, variance, and standard deviation for each of these four distributions. We also know that we can use these parameters to make z-score transformations which will show all these distributions on the same standard scale. Presently we shall make such transformations for $N = 50$ and $N = 200$.

Table 8-1 Parameters of Four Binomial Distributions with $p = 3/4$

N	Mean (Np)	$\sigma^2(Npq)$	σ ($\sqrt{Npq}$)
10	7.5	1.875	1.37
50	37.5	9.375	3.06
100	75	18.75	4.33
200	150	37.5	6.12

Before going on, remind yourself of what you know about the area inside these distributions. Because they are probability distributions, each vertical bar represents the probability of a particular K value, $p(K)$. The width of each bar is also 1, and the area inside a particular bar is $p(K)$. Since $\Sigma p(K) = 1$, the total area inside each of the four distributions is 1.

A. NORMALIZING THE BINOMIAL DISTRIBUTION

Table 8-2 shows a linear transformation of the distribution for $N = 50$ from page 78. The first two columns show the values used in plotting Fig. 5-4 (page 78), where K is the value on the horizontal axis and $p(K)$ is the height of the bar for that value of K. The next column shows the subtraction of $Np = 37.5$ from each value of K. The fourth column completes the transformation of K into z; its values are z scores for the new horizontal axis.

The fifth and last column in Table 8-2 requires careful explanation. The values in this column will be used as the vertical heights of the bars in our transformed distribution. Instead of $p(K)$, the heights will now be $p(K)\sqrt{Npq}$. We change the bar heights in order to keep the bar *area* equal to $p(K)$. When we divide every value of $K - Np$ by $\sqrt{Npq}$ to obtain a z score, we reduce the width of every bar; the width was 1 unit, and it becomes $1/\sqrt{Npq}$ unit. If the area of the bar is to be $p(K)$ in the transformed distribution, its height must be increased by the same amount; we therefore multiply $p(K)$ by $\sqrt{Npq}$ to obtain the new height. By changing the heights of all bars in this way we ensure that the total area inside the probability histogram will still equal 1.

Table 8-2 Normalization of the Binomial Distribution for $p = 3/4$ and $N = 50$

Original Distribution		Transformed Distribution		
K	$p(K)$	$K - Np$	$z = \frac{K - Np}{\sqrt{Npq}}$	$p(K)\sqrt{Npq}$
26	0	−11.5	−3.76	0
27	.001	−10.5	−3.43	.003
28	.002	−9.5	−3.10	.006
29	.004	−8.5	−2.78	.012
30	.008	−7.5	−2.45	.024
31	.015	−6.5	−2.12	.046
32	.026	−5.5	−1.80	.080
33	.043	−4.5	−1.47	.132
34	.064	−3.5	−1.14	.196
35	.089	−2.5	−0.82	.272
36	.111	−1.5	−0.49	.340
37	.126	−0.5	−0.16	.386
38	.129	0.5	0.16	.395
39	.120	1.5	0.49	.367
40	.098	2.5	0.82	.300
41	.072	3.5	1.14	.220
42	.046	4.5	1.47	.141
43	.026	5.5	1.80	.080
44	.012	6.5	2.12	.037
45	.005	7.5	2.45	.015
46	.002	8.5	2.78	.006
47	0	9.5	3.10	0

8-1 To convert a value of K into a z score, we take K minus ______ and divide it by the standard deviation, ______.

Np; $\sqrt{Npq}$

8-2 If $N = 100$ and $p = 1/2$, the mean of the binomial distribution is ______ and its σ^2 is ______. The value of σ is ______.

50; 25; 5

8-3 In such a distribution, $K = 45$ lies exactly ______ standard deviation ______ the mean. Its z score is ______.

1
below; −1

8-4 In the binomial distribution for $N = 100$ and $p = 1/2$, the probability that K will be exactly 50 is given by $\binom{N}{K} p^K q^{N-K}$ for $K = 50$. This probability can be worked out rapidly by logarithms; it is .07959, or about .08. In a probability histogram, the bar for $K = 50$ would have a width equal to ______ and a height equal to ______.

1
.08

8-5 The area of this bar for $K = 50$ would be ______. The area of all the bars in the histogram, added together, would be equal to ______.

.08
1

8-6 When $K = 50$ is transformed into a z score, it becomes $z =$ ______. The width of its bar in the transformed distribution will be $1/\sqrt{Npq}$, which is ______.

0
1/5

8-7 If its height remains at .08, the area of the bar will be reduced to ______.

.016

8-8 In order to keep the area of the bar equal to $p(K)$, the height of the bar must be multiplied by ______.

5

8-9 Instead of having a height of .08, the bar for $K = 50$ must have a new height in the transformed distribution; it is ______.

.4

8-10 We have made a transformation for just one bar in the distribution for $N = 100$ and $p = 1/2$. Because that distribution is symmetrical, its mean coincides with its mode, and the bar for $K = 50$ is the highest bar in the distribution. Its transformed height, which we have just calculated, is ______; the maximum height given in Table 8-2 for $N = 50$ and $p = 3/4$ is ______.

.4
.395

8-11 Remember that we rounded off .07959 to .08 in Frame 8-4. The comparison just made shows that the transformation has given these two distributions approximately the same maximum ______.

height

8-12 In these transformed distributions, the probability $p(K)$ is no longer represented by the ______ of the bar at K; it is represented only by its ______.

height
area

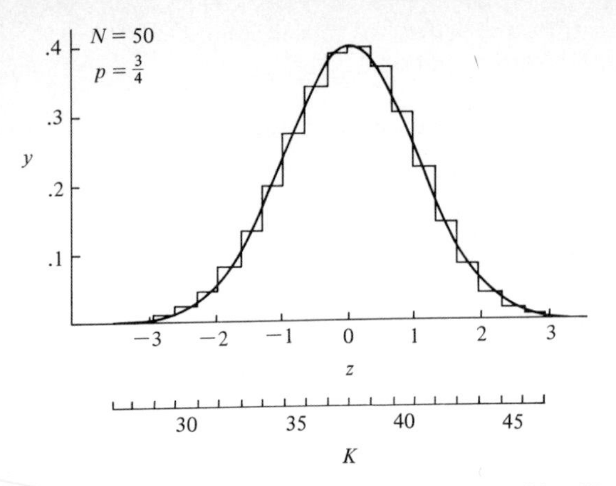

Figure 8-1: Normalized binomial distribution ($N = 50$, $p = 3/4$), with normal probability distribution superimposed.

Figure 8-1 shows the binomial distribution for $N = 50$ and $p = 3/4$, as transformed in Table 8-2. Figure 8-2 shows the same transformation of the distribution for $N = 200$ and $p = 3/4$. Notice the great similarity of these two transformed distributions: their means are of course both 0, both range from about −3 to about +3 on the horizontal axis, and both have a maximum height near .4 on the vertical axis.

These distributions have been *normalized*. They have undergone a z-score transformation which changed the scale of the horizontal axis. They have also had the vertical scale changed so that the area inside the distribution remains equal to 1. As a result, the distributions have come to resemble the normal probability distribution quite closely, and we call the entire transformation a normalization of the original binomial distribution. A smooth curve, representing the normal probability distribution, has been drawn in both figures; the similarity to the normal distribution is particularly marked in Fig. 8-2, where there are more bars and each bar is quite narrow.

You should now look back at page 78 again to satisfy yourself that the normalized distributions are similar to the normal probability distribution because of the size of N, not because of the transformation we have just performed. Figure 8-1 is no more symmetrical than the original distribution for $N = 50$; it is simply drawn to the same scale as the normal curve which is superimposed on it here. The distribution for $N = 100$, if we were to normalize it, would clearly be closer to the normal distribution than the distribution for $N = 50$, but it would not be as close as the one for $N = 200$. Binomial distributions approximate more and more closely to the normal probability distribution as N grows infinitely large. Normalization does not *produce* this increasing similarity; it simply makes the similarity obvious.

Be careful to remember that the normalization has changed the height of every bar in such a way that height no longer represents probability. *Probability is now represented by area alone.* Hence we no longer designate the vertical axis as p; we call it y, the normal-curve ordinate.

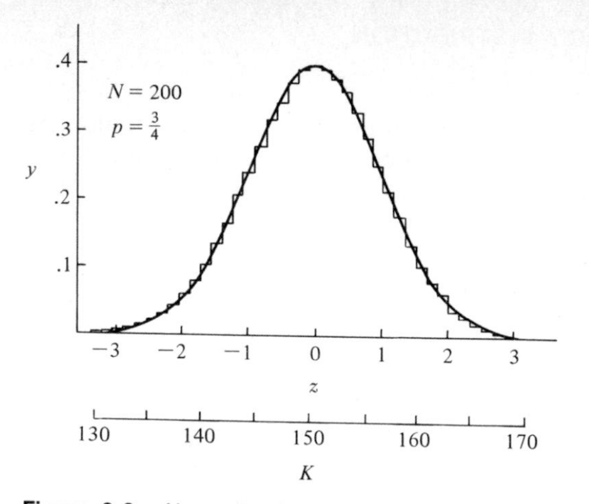

Figure 8-2: Normalized binomial distribution ($N = 200$, $p = 3/4$), with normal probability distribution superimposed.

8-13 Figure 8-1 can be compared directly with the normal curve superimposed on it. In its left half, it is a little (*higher; lower*) than the normal curve; in its right half, it is a little (*higher; lower*) than the normal curve.

lower
higher

8-14 Look again at the binomial distribution for $p = 3/4$, $N = 10$ (page 78). It is skewed to the __________; if it were normalized, its left side would fall much (*above; below*) the normal curve.

left
below

8-15 In Fig. 8-2 you can still detect the skewness; the left side falls a little (*higher; lower*) with respect to the normal curve than the right side does.

lower

8-16 A binomial distribution becomes more like a __________ probability distribution as the parameter N increases.

normal

8-17 If you want to know the probability of $K = 150$ from Fig. 8-2, you can read off the height of its bar (about .4) and divide this height by the value of ______ for this distribution.

σ

8-18 Since $\sigma = 6.12$ in this case, the probability of $K = 150$ in this distribution is about _______.

.065

8-19 You cannot take .4 as the probability of $K = 150$ because bar height does not represent probability in a __________ binomial distribution.

normalized

8-20 Probability, in a normalized distribution, is represented by __________. The probability of $K = 150$ is represented by the __________ of the bar for $K = 150$.

area
area

B. THE NORMAL PROBABILITY DISTRIBUTION

The normal probability distribution is a smooth or *continuous* distribution, as you have already seen in Figs. 8-1 and 8-2. It does not have any vertical bars at all; the jagged outline produced by the bars in binomial distributions has become a smooth, unbroken curve. It is possible to imagine that a binomial probability distribution has been drawn for N so large—infinitely large—that the number of vertical bars needed would be infinite, the width of each bar would be infinitesimal, and the tops of the bars therefore would be no wider than a small dot. In such a case the outline of the distribution would indeed be smooth and continuous.

The curve which characterizes a normal probability distribution can be described mathematically as an equation. You need not memorize this equation, but it is useful to know that it exists. The equation is an exponential one:

$$y = \frac{1}{\sqrt{2\pi}} e^{-z^2/2}$$

where e is the number 2.718 (used as the base for natural logarithms), π (pi) is the familiar number 3.1416, and the fraction $-z^2/2$ is an exponent of the number e. The only variables in this equation are y and z; all the other values are constants. The equation shows how y, the height above the horizontal axis, varies as the value of z changes. We shall not need to use this equation ourselves, but it has been fundamental in those mathematical studies which give us the normal-curve table.

Figure 8-3 illustrates two important facts about the normal distribution:

1 It is symmetrical about its mean. The area in its right half exactly equals the area in its left half; each half has an area of .5.

2 Between the mean and $z = +1$ lies exactly .3413 of the total area under the curve. A like proportion lies between the mean and $z = -1$. Figure 8-3 shows areas lying between other limiting z values. Remember that areas under the normal curve are fully determined by the z-score limits of those areas.

The normal probability distribution is a theoretical distribution. However, many empirical frequency distributions can be approximately described by the normal-curve equation. Whenever a population of observations forms an approximately normal distribution, we say that the population is NORMALLY DISTRIBUTED.

In its raw-score form, an approximately normal frequency distribution may not look exactly like Fig. 8-3. As in the binomial distributions on page 78, a z-score transformation may be necessary to bring out its similarity to the normal curve. But even without this transformation, the probabilities represented by normal-curve areas will apply to its relative frequencies. For example, about 34 percent of its observations will lie between $\bar{X}$ and 1 standard deviation above $\bar{X}$.

Remember that a z-score transformation can be performed for *any* distribution, whether it is normal or not. However, the normal-curve proportions will apply only to variables which are normally distributed.

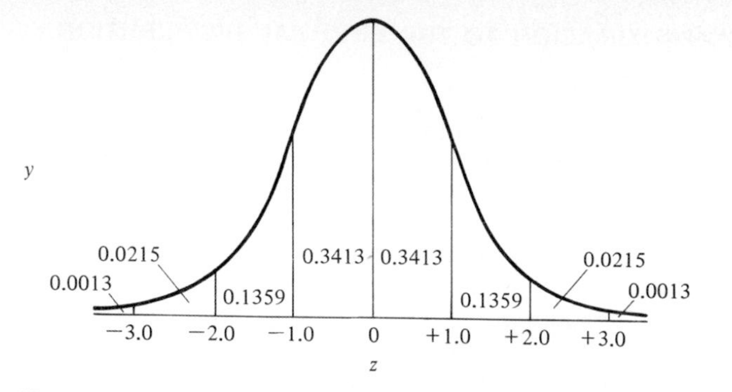

Figure 8-3: A normal probability curve, showing some of its characteristic areas.

8-21 Look at the right half of the normal curve. As z increases from 0 to +1, the steepness of the curve (*decreases; increases*). *increases*

8-22 As z increases from +1 to +2, the steepness reaches a maximum and then starts to decrease. Beyond $z =$ ______, it levels off close to the horizontal axis. *+3*

8-23 The area under the curve beyond $z = +3$ is only ______. This area represents the probability of drawing, at random, an element with a z score of ______ or greater from a normally distributed population. *.0013* *3*

8-24 Values of K in coin-tossing experiments are ______ distributed when N is large. *normally*

8-25 When $N = 100$ and $p = 1/2$, $Np = 50$ and $\sigma = 5$. A K value of 65 is ______ standard deviations above the mean. The probability of getting $K \geqslant 65$ is ______. *3* *.0013*

8-26 The area under the normal curve from $z = -1$ to $z = +1$ is equal to ______. About ______-thirds of the observations fall within 1 standard deviation around the mean. *.6826; two*

8-27 The probability that an observation will fall within 2 standard deviations around the mean in a normal distribution is given by the area from $z =$ ______ to $z =$ ______. This area is ______. *−2; +2* *.9544*

8-28 In a normal distribution, about two-thirds of the area lies within ______ standard deviation around the mean, and about 95 percent lies within ______ standard deviations. *1* *2*

8-29 These facts apply also to populations of observations whenever the population is ______ distributed. *normally*

C. THE NORMAL APPROXIMATION TO THE BINOMIAL DISTRIBUTION

Use of the normal-curve table can replace the actual calculation of probabilities by the binomial expansion when N is large. It is time to find out how this is done.

We can use the normal distribution as an approximation to any binomial probability distribution that is approximately normal in shape. A binomial probability distribution will be normal in shape if p is very close to $1/2$, no matter how large N may be. Figure 8-4 shows that the binomial distribution for $(p+q)^{20}$, when $p = 1/2$, is quite similar to the normal curve superimposed on it—except for the roughness, or *discontinuity*, of its outline. When p is farther from $1/2$, the binomial distribution will be somewhat skewed until N grows rather large; Fig. 5-4 (page 78) has shown this to be the case for $p = 3/4$.

To show how the normal approximation works, let us take a problem which is already familiar. When $N = 20$, $\alpha = .01$, and H_1 is $p = 1/2$, we already know that the critical value of K will be 16. The binomial calculations done in Lesson 5 told us that the area in Fig. 8-4 for $K \geqslant 16$ is not greater than $\alpha = .01$.

We could use the normal approximation to find this same probability, the probability of $K \geqslant 16$. The normal-curve table gives information about areas under the normal curve. Our first problem is to decide what area under the normal curve corresponds to the area inside the histogram bars for $K = 16$, 17, 18, 19, and 20.

Since the normal curve is continuous, we must look for the area under the normal curve which lies *beyond 15.5*, not the area beyond 16. The bar for $K =$ 16, when it is superimposed on a continuous distribution, must be thought of as extending from 15.5 to 16.5. Taking 15.5 instead of 16 as the critical value is called making a CORRECTION FOR CONTINUITY. We must always make such a correction when we go from a discontinuous distribution (like the binomial distribution) to a continuous one.

Notice that the area under the smooth normal curve beyond 15.5 is very similar to the area inside the binomial distribution for $K \geqslant 16$. The curve cuts off a small corner of each bar, but it includes a similar area which is not inside the bar.

Before we can take the number 15.5 to the normal-curve table, we must convert it to a z score. For $N = 20$ and $p = 1/2$, $Np = 10$ and $\sqrt{Npq} = 2.236$. Our z score is $(15.5 - 10)/2.236 = +2.46$. The area beyond $z = 2.46$ under the normal curve is .0069. The area for $K \geqslant 16$ in the binomial probability histogram is .0059. The probability obtained by the normal approximation is thus very nearly the same as that obtained by the more laborious binomial calculation. More important, it leads to the same critical value of K; we would have selected $K = 16$ as the critical value even if we had used the normal approximation in working out our color-touch example.

Observe that it is not necessary to normalize the entire binomial distribution for $N = 20$ and $p = 1/2$ when we use this normal approximation. We can find the probability we require by converting only one value of K to its equivalent z score and looking in the normal-curve table to find its probability.

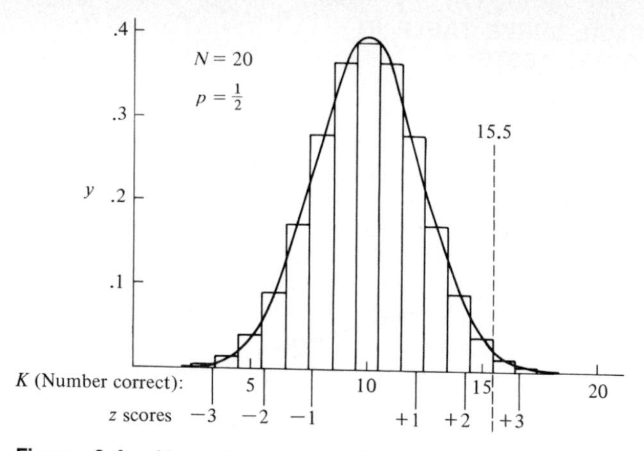

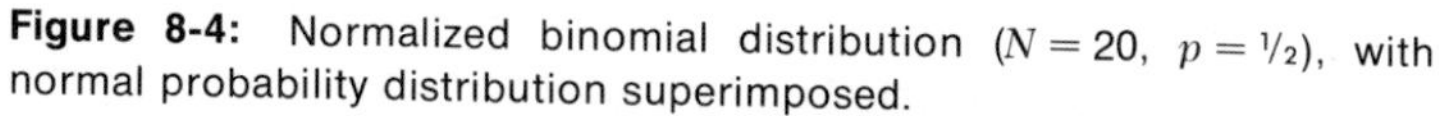

Figure 8-4: Normalized binomial distribution ($N = 20,\ p = 1/2$), with normal probability distribution superimposed.

8-30 The histogram in Fig. 8-4 is (*skewed; symmetrical*); the normal curve fits it (*better; less well*) than the normal curve fits the histograms for $p = 3/4$ (Figs. 8-1 and 8-2).

symmetrical
better

8-31 All binomial distributions tend toward the normal distribution when ______ becomes large enough. This tendency appears at lower values of N when p is close to ______.

N
$1/2$

8-32 The z scale in Fig. 8-4 is a (*continuous; discrete*) scale. There are no gaps in the values which z may take.

continuous

8-33 The scale of K values is a ______ scale since K can take no values except whole numbers.

discrete

8-34 In order to compare the histogram to the normal curve, we must look at the K values *as if* they were continuous, and we must think of the bar for $K = 16$ as extending from $K =$ ______ to $K =$ ______.

15.5
16.5

8-35 The area inside the bars for $K = 16, 17, 18, 19$, and 20 must therefore be considered as the area beyond $K =$ ______.

15.5

8-36 Taking 15.5 as the cutoff point instead of 16 is called making a correction for ______. We are comparing a discontinuous distribution of K values to a ______ distribution of z scores.

continuity
continuous

8-37 In order to find the area beyond $K = 15.5$, we must convert this K value to a ______. We can then look it up in the ______ table.

z score
normal-curve

D. USE OF THE NORMAL-CURVE TABLE IN ONE-TAILED BINOMIAL TESTS

We now turn to the actual normal-curve table which must be consulted in using the normal approximation to a binomial distribution. In this lesson we shall confine our attention to tests which are extensions of the color-touch example. Because H_2 is directional in that example ($p > 1/2$), the rejection region always has just one part and the statistical test is one-tailed. Lesson 10 will extend the normal approximation to cases which require two-tailed tests.

Turn to Table A-1 (page 351), and insert a marker. The entire normal-curve table describes a single probability distribution; it gives the *cumulative* probability for each value of z. Just two columns are needed, one for the value of z and one for the corresponding Area (Mean to z). These pairs of columns are broken into sections and placed side by side (instead of end to end) for printing convenience.

The values of z start near 0 (the mean) and increase to 3.72. The column of Areas starts at .0199 and increases to .4999. Thus the table lists probabilities for only the right half of the normal distribution; the left half, for negative z values, is exactly like the right half, and there is no need to write the same probabilities twice. The area from the mean to $-z$ is exactly equal to the area from the mean to $+z$.

Each Area (Mean to z) is the probability of obtaining an outcome lying *inside* the region between the mean and the value of z stated. Such probabilities increase cumulatively as we move outward from the mean toward $z = 3.72$; each successive area includes all the previously listed ones, plus an increment. To find the probability of obtaining an outcome lying *beyond* the stated value of z, the Area (Mean to z) must be subtracted from .5000.

Now suppose you are just starting to find the critical value of K for $N = 20$, $\alpha = .01$, and $H_1: p = 1/2$. In the right half of the distribution there must be a z value beyond which .01 of the area lies. The area from the mean to that z value will be .49; you therefore look in the Area column of Table A-1 for the value .49. The corresponding value of z is the critical value for $\alpha = .01$. You will find from the normal-curve table that the critical value of z is 2.33.

In order to obtain the critical value of K, write $z = (K - Np)/\sigma = (K - 10)/2.236$. Thus $K = 10 + 2.33\ (2.236) = 15.2$. The critical value in practice must be either 15 or 16; 15 will be too low to meet $\alpha = .01$, and therefore the critical value has to be 16.

The critical value of 16 will have a probability somewhat lower than .01. If you want to find exactly what that probability is, you can follow the steps already outlined on page 120. First you correct for continuity: you take 15.5 as the limiting value instead of 16. Then you convert 15.5 to a z score; it becomes $z = 2.46$. Turning to Table A-1, you look up $z = 2.46$ and find that the Area (Mean to z) is .4931. The area beyond $z = 2.46$ is then $.5 - .4931 = .0069$.

8-38 In a normal distribution, the Area (Mean to z) when $z =$ 1.65 is ______. The area *beyond* $z = 1.65$ is thus ______. *.4505; .0495*

8-39 If $z = -1.65$, the area beyond z in the *left* end of the distribution is ______. *.0495*

8-40 Suppose you must find the z score beyond which .1 of the distribution lies. You look in the table for the area ______. *.4000*

8-41 The area .4000 is not listed in the table; the nearest value listed is ______, and its corresponding z score is ______. *.3997; 1.28*

8-42 You must find the area between $z = +1$ and $z = +1.5$. The area from the mean to $z = 1$ is ______; the area from the mean to $z = 1.5$ is ______. The area from $z = 1$ to $z = 1.5$ is ______. *.3413* *.4332; .0919*

8-43 Find the area from $z = +1$ to $z = -1.5$. To obtain this area you must add ______ and ______. The area is ______. *.3413; .4332; .7745*

8-44 Find the area under the normal curve which lies *to the right of $z = -0.5$.* The area from the mean to $z = -0.5$ is ______; the area to the right of the mean is ______. The area to the right of $z = -0.5$ is ______. *.1915; .5000* *.6915*

8-45 When $N = 100$ and $p = 1/2$, $Np = 50$ and $\sigma = 5$. For this distribution, find the area lying beyond $K = 59.5$. The z score for $K = 59.5$ is ______. *1.9*

8-46 The area lying beyond $z = 1.9$ is ______. This area represents the probability that K will be at least ______. *.0287* *60*

8-47 Find the probability of getting $K = 50$ in this distribution. You must consider the bar for $K = 50$ as extending from $K =$ ______ to $K =$ ______. *49.5; 50.5*

8-48 The z-score equivalent of $K = 49.5$ is ______; the z score for $K = 50.5$ is ______. The probability that $K = 50$ is ______. (Compare the exact value calculated by logarithms in Frame 8-4.) *−0.1* *+.01; .0796*

8-49 Find the critical value of K in this distribution when $\alpha =$ .05. To find the z score beyond which .05 of the area lies, look for the area ______ in the table. *.4500*

8-50 When Area (Mean to z) = .4505, $z =$ ______. *1.65*

8-51 For this distribution, $z = 1.65$ corresponds to $K =$ ______. The whole-number critical value of K for $\alpha = .05$ is ______. *58.25* *59*

After finding the critical value of K for experiments of 20 trials, we went on in Lesson 5 to determine the power of the binomial test for H_2: $p = 3/4$ and H_2: $p = 9/10$. To find $1 - \beta$ for each of these hypotheses, we had to calculate the probability of getting $K \geqslant 16$ when $p = 3/4$ and when $p = 9/10$. We could have used the normal approximation to shorten the labor of these calculations.

However, p deviates rather far from $1/2$ in each of these hypotheses. Will the binomial distributions for $N = 20$ be too skewed to allow use of the normal approximation? We can find out by comparing the results of the normal approximation with the results obtained by binomial expansion. In the process, you will see how the normal approximation simplifies calculation of a power function.

Table 8-3 shows the power calculation for $p = 3/4$. To find z scores for the approximation, we again subtract .5 from each value of K as a correction for continuity. Then we subtract $Np = 15$ and divide by $\sqrt{Npq} = 1.94$ to obtain the series of z scores. The last column tells the probability of getting a z score at least as great as the one listed. Compare this column with the third column: the approximation overestimates the probabilities for the first two K values and underestimates the remaining ones. If we use this approximation to derive $1 - \beta$, we find $1 - \beta = .3974$ instead of .4148, probably a negligible difference.

Table 8-3 Comparison of Binomial Expansion and Normal Approximation for $N = 20$ and $p = 3/4$

By Binomial Expansion			By Normal Approximation			
K	$p(K)$	Cumulative Probability	$K - 0.5$	z	Area Mean to z	Area Beyond z
20	.0032	.0032	19.5	2.32	.4898	.0102
19	.0211	.0243	18.5	1.80	.4641	.0359
18	.0669	.0912	17.5	1.28	.3997	.1003
17	.1339	.2251	16.5	0.77	.2794	.2206
16	.1897	.4148	15.5	0.26	.1026	.3974

Let us turn to the other alternate hypothesis, $p = 9/10$. The binomial distribution for $N = 20$ and $p = 9/10$ is quite skewed (look again at Fig. 5-2, page 69). One would not usually use the normal approximation for such a skewed distribution unless N is at least 30. But a comparison like that in Table 8-3 shows that we would not go very far wrong by taking the normal approximation even in this case. Instead of the value $1 - \beta = .9567$, which we would obtain by the binomial expansion, we would obtain a slightly larger value by the normal approximation, $1 - \beta = .9686$. Even with N as low as 20 and p as extreme as $9/10$, the normal approximation is not bad.

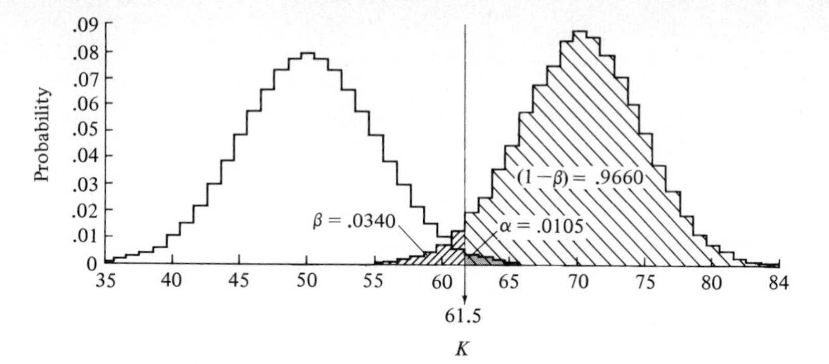

Figure 8-5: Binomial distributions for H_1: $p = 1/2$ (*left*) and H_2: $p = .70$ (*right*) in a color-touch experiment with $N = 100$. β and $1 - \beta$ are shown for a decision based on $\alpha = .01$.

8-52 We now examine a color-touch experiment with $N = 100$. The H_1 distribution is shown at the left in Fig. 8-5; its mean is at $K =$ ______. The figure shows the rejection region for $\alpha = .0105$. The critical value of K is ______.

50
62

8-53 The distribution on the right in Fig. 8-5 is H_2: $p = .70$. If this hypothesis is true, the probability of getting K *less than* 62 is ______. This probability is the probability that H_2 will be ______.

.0340
rejected

8-54 When H_2: $p = .70$ is true, the probability of a type II error is therefore ______. This probability is called (β; $1 - \beta$).

.0340; β

8-55 When H_2: $p = .70$ is true, the probability of a correct decision is the probability of getting K at least ______. This probability, called ______, is equal to ______.

62
1 − β; .9660

8-56 Therefore, the power of our test to discriminate between H_1 and H_2: $p = .70$ is ______.

.9660

8-57 Figure 8-5 is based on binomial expansions of $(1/2 + 1/2)^{100}$ and $(.70 + .30)^{100}$. Now calculate $1 - \beta$ by using the *normal approximation* to the binomial distribution for H_2: $p = .70$. The mean of this distribution is ______; its standard deviation is 4.58.

70

8-58 Since the normal approximation is continuous, we must regard the rejection region as beginning at $K =$ ______.

61.5

8-59 61.5 is below the mean of 70. Its z-score equivalent in the normal approximation is ______.

−1.86

8-60 The area to the right of $z = -1.86$ is ______, according to the normal-curve table.

.9686

REVIEW

8-61 All binomial probability distributions regardless of the value of p become like the __________ probability distribution as ________ becomes very large.

normal
N

8-62 For the *same* value of N a binomial distribution for $p = 1/2$ will be (*less; more*) similar to the normal distribution than a binomial distribution for $p = 3/4$.

more

8-63 When a distribution is similar to the normal distribution, the variable whose values it describes is said to be __________.

normally distributed

8-64 A table giving properties of the normal distribution at different values of z is called the __________ table.

normal-curve

8-65 The normal-curve table we have been using provides the __________ under the normal curve between the __________ and different values of z.

areas; mean

8-66 The total area under the normal curve is ________. The area to the right of the mean is ________.

1
0.5

8-67 Because the normal curve describes a probability distribution, the areas under the curve can be interpreted as __________.

probabilities

8-68 For instance, the area to the right of the mean can be interpreted as the probability of drawing, at __________ and from a __________ distributed population, an element whose z score is greater than ________.

random
normally
0

8-69 The area under the curve and to the right of $z = 2.33$ can be interpreted as the probability of randomly drawing an element whose z score is greater than ________ from such a population.

2.33

8-70 The normal-curve table (page 351) does not give this probability directly. It gives the area between the __________ and $z = 2.33$. To find the area beyond $z = 2.33$, the value in the table must be __________ from ________.

mean

subtracted; .5000

8-71 The probability of drawing, at random from a normally distributed population, an element whose z score is at least 2.33 is ________. (Give the probability to two places.)

.01

8-72 The probability of drawing at random from such a population an element whose z score is *not greater than* -2.33 is ________.

.01

8-73 The binomial distribution required for testing $H_1: p = 1/2$ in a 50-trial experiment can be approximated by a __________ distribution with mean = ________ and $\sigma = 3.54$.

normal
25

8-74 To normalize this distribution, we convert the K values to __________. We subtract ________ from each K and divide the remainder by ________.

z scores; 25
3.54

8-75 Then we change the vertical height of each bar so that its __________ will still equal $p(K)$.

area

8-76 The z-score transformation has reduced the width of each bar from 1 to ________. The original height of the histogram bar, before normalization, is equal to ________.

$\frac{1}{3.54}$
p(K)

8-77 To make the area of the bar again equal $p(K)$, we multiply its original height by ________.

3.54

8-78 The new height, $3.54p(K)$, (*does; does not*) represent the probability of K. This probability is represented by the __________ of the bar in the normalized distribution.

does not

area

8-79 If we want to use the normal curve to find an area inside a normalized binomial distribution, we must remember that the normal distribution is (*continuous; discrete*) while the binomial distribution is __________ even after it has been normalized.

continuous
discrete

8-80 To find the critical value of K for $\alpha = .01$, we first look for the critical value of z in the normalized binomial distribution. This critical value is ________.

2.33

8-81 A z score of 2.33 in the normalized binomial corresponds to a K of ________. Since a 50-trial experiment can give only whole numbers as values of K, the critical value of K for $\alpha = .01$ will be ________.

33.25

34

8-82 The decision rule in terms of K_o is: reject H_1 if ________ is at least ________.

K_o
34

8-83 To find the power of this test to detect $p = 3/4$, we take as our H_2 distribution the normalized binomial with mean = ________ and σ = ________ (see Table 8-1).

37.5; 3.06

8-84 Remembering the correction for continuity, we take as our cutoff point K = ________. This value becomes a z score of ________ in the H_2 distribution.

33.5
−1.31

8-85 The normal-curve table indicates that $1 - \beta$ will be about ________ for this H_2.

.9049

PROBLEMS

1. Assume a distribution of 75 scores which is approximately normal. How many scores would you expect to find between the following limits?

a. Between $+1s$ and $-1s$.

b. Between $+2s$ and $-2s$.

c. Between $+1s$ and $+2s$.

2. If a normal distribution has a mean of 118 and $s = 11$:

a. What percentage of scores will lie between the scores 107 and 118?

b. What percentage will lie below a score of 129?

c. What percentage will lie above 140?

3. For a particular normal distribution of scores, the following facts are known: 16 percent of the scores fall below 57, and a z score of $+0.5$ corresponds to a raw score of 69.

a. What is the mean of the distribution?

b. What is the standard deviation?

c. What is the raw score above which only 2 percent of the scores will be found?

4. In a normal distribution, what is the z-score equivalent of the *median*? What is the z score above which only 16 percent of the distribution lies? What percentage of the scores lie below a z score of $+2.0$?

5. For the color-touch example, assume $\alpha = .05$ and construct a table showing the power of the test to discriminate H_1: $p = 1/2$ from each of the alternate hypotheses listed below when $N = 100$. Compare with the power function for these same conditions when $\alpha = .01$ (page 125).

H_2	z	$1 - \beta$
$p = .55$	______	______
$p = .60$	______	______
$p = .65$	______	______
$p = .70$	______	______
$p = .75$	______	______

6. For the color-touch example, show how the power of a binomial test to discriminate H_1 from H_2: $p = 2/3$ changes as the value of N increases. Take $\alpha = .01$, and find the critical value of K for each of the following values of N. Between what values of N does β first fall as low as .05?

N	Critical K
36	______
64	______
100	______
144	______
196	______

LESSON 9
RANDOM SAMPLING DISTRIBUTIONS

What you know about statistical tests has so far been applied only to categorical data with dichotomous observations. We turned to coin-toss problems at the beginning of Lesson 3 because probability calculation is simplest for cases with dichotomous observations. Now we are in a position to recall those two examples introduced in Lesson 1 to illustrate continuous and discrete variables, the college-height example and the comparison among three vocabulary-learning methods. We cannot consider our survey of basic concepts complete until you see how the logic of statistical tests is applied to such examples.

Interval data present a special problem because the observations can take on such a large number of different values. With a continuous variable like heights, the number of possible values is infinite. Even with discrete scores, the number of values is usually very large. Interval data are best visualized as a frequency distribution, and they can be summarized by stating the mean and standard deviation of the distribution. Our problem is to see how a decision rule can be formulated for a statistical test of such data.

In Lesson 2 we used the college-height example to illustrate the choice of statistical hypotheses. A side-by-side comparison of this example with a coin-toss problem, for which we already understand the solution, will help us identify the exact question we must now answer.

	Coin-Toss Problem	College-Height Example
Assumed H_1	$p = 1/2$	$\mu_A - \mu_O = 0$
Directional H_2	$p > 1/2$	$\mu_A - \mu_O > 0$
Nondirectional H_2	$p \neq 1/2$	$\mu_A - \mu_O \neq 0$
Sample statistic	K_o	$\bar{X}_A - \bar{X}_O$

In our early discussion of the college-height example, we spoke loosely of the "average" height for our samples from Alpha and Omega colleges. These averages can now be more precisely referred to as the sample means $\bar{X}_A$ and $\bar{X}_O$, just as we call the population means μ_A and μ_O. We are interested in a possible difference between the population means, and we must make our inference about this difference on the basis of a sample statistic—the observed difference between the sample means.

Our comparison shows that the statistic $X_A - X_O$ must play the same role in the height example that the observed number of heads K_o has played in the

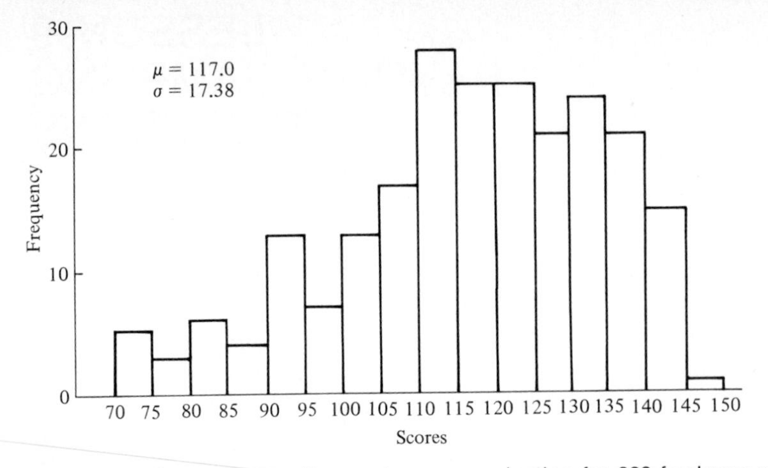

Figure 9-1: Scores on a college entrance examination for 228 freshman men entering Oberlin College in 1956.

coin-toss problem. To test H_1 in a coin-toss problem, we find the probability of a result as extreme as K_o when H_1 is true; we then compare this probability with our α level. Therefore, to test H_1 in the height problem, we need to find the probability of a result as extreme as $\bar{X}_A - \bar{X}_O$, that is, the probability of getting a difference between the sample means which is at least as large as the one actually observed, when H_1 is true. This probability will then be compared with our α level. It is the task of this lesson to find an answer for the question, "Since it will not be appropriate to use a binomial probability distribution for the height problem, where can we find the conditional probability distribution of values of $\bar{X}_A - \bar{X}_O$ which we need for a statistical test?"

A. A SAMPLING DISTRIBUTION OF MEANS

We shall begin by making a comparison between a sample and its parent population. Figure 9-1 shows a population of 228 scores on a college entrance examination. The mean is 117.0, and the standard deviation is 17.38. The distribution is quite negatively skewed.

A random sample of 20 scores drawn from this population will have a mean $\bar{X}$ and a standard deviation s. Another random sample of 20 scores from the same population would most likely have a different mean. How much variability between such sample means should we expect? We can actually estimate the *amount of variability* to be expected.

Figure 9-2 shows a distribution of 50 values of $\bar{X}$, each one derived from an actual random sample of $N = 20$ drawn from this population. Both the horizontal and the vertical scales have been enlarged so that Fig. 9-2 can be compared directly with Fig. 9-1. This distribution is called a RANDOM SAMPLING DISTRIBUTION (RSD) OF MEANS. It is a distribution of the values of means, and these means are all taken from random samples of the same size drawn from the same population.

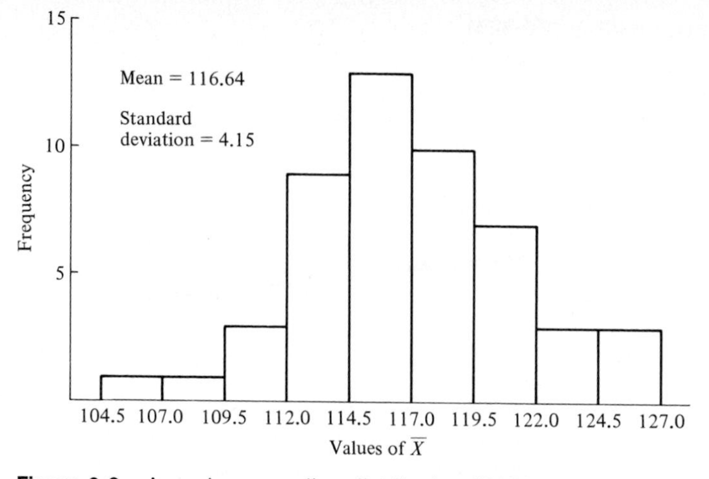

Figure 9-2: A random sampling distribution (RSD) of the means of 50 samples of $N = 20$, each drawn randomly from the population of 228 scores shown in Fig. 9-1.

9-1 Compare the population distribution in Fig. 9-1 with the RSD of means in Fig. 9-2. The mean of the RSD is ______; the population mean is ______.

116.64
117

9-2 The range of scores in the population is ______ score points. The range of $\bar{X}$ values in the RSD is ______ points. Mark the range of the RSD on Fig. 9-1 so that you can see this difference clearly.

80
22.5

9-3 The value of σ is roughly ______ times the value of the standard deviation of the RSD.

4

9-4 The mean and mode are more nearly in the center of the range in the (*population distribution; RSD*) than in the other distribution. This distribution is somewhat more symmetrical.

RSD

9-5 This RSD contains the means of ______ samples of $N =$ ______, drawn randomly from the ______ shown in Fig. 9-1.

50
20; population

9-6 An RSD of means is a random ______ distribution of the ______ of samples.

sampling
means

9-7 All the samples whose means appear in the RSD of means must be of the same ______, must come from the same ______, and must meet the conditions of ______ chance and ______.

size
population; equal
independence

B. THE CENTRAL LIMIT THEOREM

In Fig. 9-2 each value of $\bar{X}$ came from an actual sample. Most frequently, however, RSDs of means are derived theoretically by mathematical analysis. The theoretical derivation relies on a very powerful principle called the CENTRAL LIMIT THEOREM, which we shall discuss in two parts.

PART 1 When samples of size N are drawn randomly from a population with mean μ and variance σ^2, the *form* of the RSD of sample means approaches a normal distribution around a mean $= \mu$ as N becomes large. For very large N the RSD of the means is approximately normal.

You should notice, first, that the theorem does not mention any particular number of samples of size N. It is concerned with a theoretical RSD, one which may be imagined to include an infinitely large number of sample means.

Notice, further, that the theorem does not say that the population must be normally distributed. It simply indicates that we may expect the RSD of means to be normally distributed whenever N is large enough. The population distribution does make some difference; the RSD of means will become normal at smaller values of N when the population itself is closer to normal. But Fig. 9-3 indicates how powerful the tendency toward normality of the RSD really is and how little difference the population distribution really makes. The left-hand drawing shows a negatively skewed population distribution; a normal distribution of the same size is superimposed on it to show the skewness. The right-hand drawing is an RSD of the means of samples of $N = 4$ drawn from that same skewed population; the RSD is enlarged in scale so that it can be compared with the population distribution (just as Fig. 9-2 was enlarged to compare with Fig. 9-1). The figure shows that a population with considerable negative skewness will yield an RSD of means which is close to normal *even when N is only 4.* A sample N of 30 is usually considered large enough to ensure that the RSD of means will be normal regardless of population skewness.

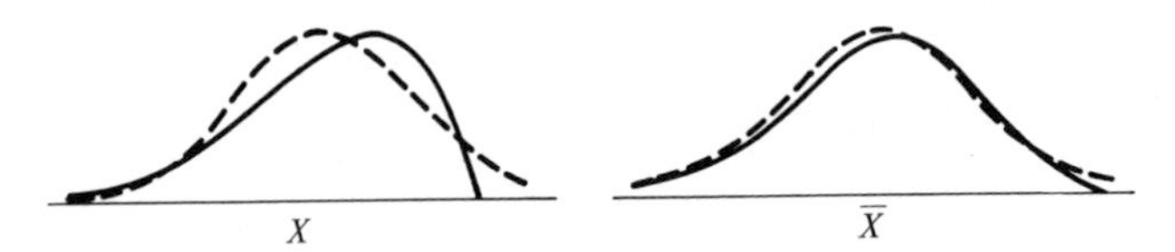

Figure 9-3: A negatively skewed population distribution (*solid line, left figure*) with a comparable normal distribution (*dotted line*), and an RSD of the means of samples of $N = 4$ drawn from that same population (*solid line, right figure*). (*From William L. Hays,* Statistics for Psychologists, *p. 239, Holt, Rinehart and Winston, Inc., New York, 1963.*)

This first part of the central limit theorem also says that the mean of the RSD of means will be μ. You have already learned that $E(\bar{X}) = \mu$ (page 104); now you can see that this statement is derived from the central limit theorem. Just as $E(X) = \bar{X}$, the mean of the X distribution, the expectation of $\bar{X}$ is the mean of the $\bar{X}$ distribution, i.e., the mean of the RSD of means. And because $E(\bar{X}) = \mu$, $\bar{X}$ is an unbiased estimator of μ.

9-8 The RSD of means is a distribution of the __________ of a set of __________ from the same population. *means* *random samples*

9-9 The central limit theorem describes a theoretical RSD in which the number of sample means is imagined to be __________ large. *infinitely*

9-10 In Fig. 9-2 there are only 50 sample means. This RSD is only a small part of the theoretical RSD of means with $N =$ ______ drawn randomly from this same population. *20*

9-11 The theorem states that the shape of the RSD is __________ for samples of very large size. *normal*

9-12 The theorem is true for skewed populations as well as normally distributed ones. If the population is skewed, the RSD for very *small* samples will be somewhat __________. *skewed*

9-13 A "large" sample is any sample of $N = 30$ or greater. Even when the population is skewed, the RSD of means will be nearly normal if the sample N is at least ______. *30*

9-14 The RSD in Fig. 9-3 is for means of samples of $N =$ ______. Even though the population is skewed and the sample size is small, the RSD is almost __________ distributed. *4* *normally*

9-15 According to the central limit theorem, the mean of the RSD of means is the __________ mean. *population*

9-16 In a frequency distribution of X values, X is the variable, and $\bar{X}$ is its mean. $E(X) =$ ______. $\bar{X}$

9-17 Recall that any statistic can serve as an unbiased estimator of the __________ of its own distribution. *mean*

9-18 Since $\bar{X}$ is the mean of the X distribution, any single X_i can serve as an unbiased estimator of ______. $\bar{X}$

9-19 In an RSD of means, $\bar{X}$ is a variable, and ______ is its mean. The expectation of $\bar{X}$ is ______. μ μ

9-20 Since μ is the mean of the $\bar{X}$ distribution, any single $\bar{X}$ can serve as an unbiased estimator of ______. μ

9-21 When we want an unbiased estimator of the parameter μ, we look for a statistic whose *expectation* is μ. $E(\bar{X}) = \mu$; an unbiased estimator of μ is ______. Likewise, since $E[\text{SS}/(N-1)] = \sigma^2$, an unbiased estimator of σ^2 is ______. $\bar{X}$ $\dfrac{SS}{N-1}$

The standard deviation of the RSD of means is commonly called the STANDARD ERROR OF THE MEAN. It is written $\sigma_{\bar{X}}$; the subscript $\bar{X}$ is a reminder that this σ refers to a distribution of values of $\bar{X}$. The second part of the central limit theorem tells us about the size of $\sigma_{\bar{X}}$.

PART 2 If the population is very large and its variance is σ^2, the variance of the RSD of the means of samples of size N is equal to σ^2/N.

Again, we may notice several things about this part of the theorem. If we know the value of the population variance, we can calculate the value of the variance of the RSD of sample means exactly. The RSD variance $\sigma_{\bar{X}}^2$ depends directly on population variance σ^2; the greater the population variance, the greater the value of $\sigma_{\bar{X}}^2$. However, the RSD variance also depends upon sample size; it is larger for small samples than for large ones. This relationship between sample size and the amount of variability among sample means is reasonable. When samples are small, one expects to find a rather large amount of variation among sample means. Larger samples will resemble one another more than small samples do.

Of course, we usually do not know the exact value of σ^2. When we do not know the exact σ^2, we estimate it from the sample SS by taking SS/$(N-1)$. In that case the *estimated* $\sigma_{\bar{X}}^2$ will be SS/$N(N-1)$, which is the same as (est. σ^2)/N.

When we do know the parameters μ and σ for some population, as we do for the population in Fig. 9-1, the central limit theorem enables us to decide whether a particular sample of size N and mean $\bar{X}$ is likely to have come from that population. For example, the population in Fig. 9-1 has $\mu = 117$ and $\sigma = 17.38$. The theoretical RSD of means of samples with $N = 20$ drawn from this population has a mean of 117 and a variance of $17.38^2/20 = 15.10$. According to the central limit theorem, the shape of this RSD is normal, and we can use the normal-curve table as a probability distribution in determining whether some sample comes from this population.

Now suppose we have a sample with $\bar{X} = 105$. If the sample came from this population, it is part of the RSD of means which we have just described. Its mean is 12 points below the mean of that RSD. The standard error of the mean is $\sqrt{15.10} = 3.89$. This $\bar{X}$ is therefore $12/3.89 = 3.09$ standard deviation units below the mean. How often does a z of -3.09 or less occur in a normal distribution?

The normal-curve table (page 351) gives an Area (Mean to z) of .499 for $z = -3.00$. Only .001 of the area in the distribution lies beyond -3.00. The probability of getting an $\bar{X}$ as low as 105, or lower, in random sampling from this population is less than .001. With $\alpha = .001$, it can be concluded that the sample did not come from this population. The probability is less than 1 in 1,000 that this decision is a type I error.

9-22 The second part of the central limit theorem concerns another parameter of the RSD of means. This parameter is its ________.

variance

9-23 The population must be very ________ in order for the statement about RSD variance to apply.

large

9-24 When the population is very large, the variance of the RSD of means is equal to the ________ variance divided by ________.

population
N

9-25 The symbol for the variance of the RSD of means is ________. Its square root is the standard deviation of the RSD of means; it is also called the standard ________ of the ________.

$\sigma_{\bar{X}}^2$
error
mean

9-26 The value of $\sigma_{\bar{X}}$ can be calculated exactly whenever σ^2 is known. It is the square root of ________.

$\frac{\sigma^2}{N}$

9-27 When σ^2 is not known, it must be estimated. Since $E[\text{SS}/(N-1)] = \sigma^2$, we take ________ as the estimated σ^2.

$\frac{SS}{N-1}$

9-28 Sometimes we want to know $\sigma_{\bar{X}}^2$ when σ^2 is not exactly known. We then use the ________ value of σ^2 and divide it by ________ to obtain an estimate of $\sigma_{\bar{X}}^2$.

estimated
N

9-29 The estimated variance of the RSD of means of samples of size N will then be ________.

$\frac{SS}{N(N-1)}$

9-30 A population of 10,000 scores has a μ of 100 and σ^2 of 225. The RSD of means of samples of $N = 50$ from this population has a mean of ________ and a variance of ________.

100; 4.5

9-31 The square root of 4.5 is 2.12. The standard error of the mean for samples of $N = 50$ from this population is ________.

2.12

9-32 Look at page 119, and consider the normal curve drawn there as a picture of this RSD. Two-thirds of all the samples in this RSD will have means between ________ and ________. (Give the limits in score points.)

97.88; 102.12

9-33 About 95 percent of all the means in the RSD will be between ________ and ________ score points.

95.76; 104.24

9-34 The probability of getting a mean which lies more than 3 standard deviations above 100 is shown in the normal-curve drawing. This probability is ________.

.0013

9-35 Therefore, if we have a sample whose mean is 106.36, the probability that it comes from this population is only ________.

.0013

C. RANDOM SAMPLING DISTRIBUTIONS OF OTHER STATISTICS

All probability distributions employed in statistical decision making are random sampling distributions of some statistic. We have used the binomial probability distribution for making statistical decisions in coin-tossing problems. The binomial probability distribution is in fact a random sampling distribution of the statistic K.

You must be sure to understand that the binomial distribution is really an RSD. Imagine carrying out the procedure for calculating an *empirical* RSD in coin-toss problems with $N = 10$ and $p = 1/2$. What would you do? You would obtain a fair coin, perform a large number of 10-trial experiments, and tabulate the values of K obtained. If you then drew a relative frequency distribution for these values of K, you would have an empirical RSD of the statistic K.

We can describe this RSD theoretically, without actually doing any experiments. The central limit theorem, as it applies to coin-toss problems, runs something like this: For coin-tossing samples of N trials each, where the probability of heads is p and the probability of tails is q, the RSD of the statistic K is the binomial expansion of $(p + q)^N$. As N becomes very large, the RSD of K approaches a normal distribution around a mean Np with a variance Npq.

In Lesson 4, when we pointed out that every statistical test is based on a probability distribution, we noted that the statistical test is ordinarily named for its probability distribution. Thus statistical tests employing the binomial distribution are binomial tests. Lesson 9 adds a further detail to this picture: the probability distribution is always a random sampling distribution of some statistic.

With interval data like those in the college-height example, the statistic in which we are interested is usually a mean or a combination of means. $\bar{X}_A - \bar{X}_O$ is a combination of means, a difference between two means. When we need to find $p(\bar{X}_A - \bar{X}_O)$ given H_1, we must think in terms of the RSD of means as it has been described in this lesson. The probability distribution we need for a statistical test of the college-height difference can be derived from this RSD of means.

The RSD of means has many derivatives, each one designed for a special purpose. For example, $\bar{X}_A - \bar{X}_O$ is not a mean but a *difference* between means; it requires an RSD of the difference between means of samples drawn from the same population. This RSD is easily derived from facts about the RSD of means; you will learn the details in Lesson 16. The test you will use is called a t test because it employs a probability distribution of the statistic t. This statistic is very similar to z, but it has special properties which arise whenever population variance σ^2 must be estimated from sample statistics.

The vocabulary-learning example from Lesson 1 involves three means. Comparisons among more than two sample means are often carried out by analysis of variance, using the RSD of the statistic F (another close relative of z and t). You have already learned the basic logic underlying these statistical tests; you have nothing left to learn but the technique of using each RSD.

9-36 When we use the binomial probability distribution to make a statistical test, the test is called a __________ test. *binomial*

9-37 When N is very large, we do not calculate the binomial probability distribution directly. We approximate this distribution by using the __________ distribution. *normal*

9-38 The binomial distribution is a random sampling distribution of the statistic ______. To obtain an empirical RSD of this statistic, we would have to perform many experiments, each with ______ trials, and take the statistic ______ from each experiment. *K* *N; K*

9-39 This procedure is analogous to obtaining an empirical RSD of sample means. For such an RSD, we draw many samples of size ______ and take the statistic ______ from each sample. *N; $\bar{X}$*

9-40 Instead of obtaining an empirical RSD, we can use the __________ theorem to find the theoretical RSD of an infinitely large number of sample means. *central limit*

9-41 This theorem tells us that the RSD of sample means will be __________ distributed when ______ is large enough. *normally; N*

9-42 The same theorem tells us that the RSD of K will be __________ distributed when ______ is large enough. *normally; N*

9-43 By the theorem the RSD of means is known to have a mean = ______ and a variance = ______; the RSD of K is known to have a mean = ______ and a variance = ______. *μ; σ^2/N* *Np; Npq*

9-44 Statistical tests of the difference between just two sample means often use the RSD of the statistic t. The tests are called ______ tests. *t*

9-45 The t distribution is similar to the normal curve (or z distribution), but it applies when the value of ______ must be estimated from sample statistics. *σ^2*

9-46 The basic logic is the same for all statistical tests. A particular statistic is derived from the sample or samples observed, and the __________ of this statistic is found. With interval data the statistic usually is a __________ or a combination of __________. *RSD* *mean* *means*

9-47 The RSD indicates the __________ of drawing at random a sample (or samples) at least as extreme as the one(s) observed when ______ is true. *probability* *H_1*

D. THE NORMAL DISTRIBUTION INTERPRETED AS AN ERROR FUNCTION

We shall conclude Part One with another look at the normal probability distribution. We know that the binomial distribution tends toward the normal when N is large; we also know that the RSD of sample means tends toward the normal when sample N is large. The normal distribution thus describes the variation among values of K in coin-toss experiments and the variation among sample means for interval data.

It is variation which leads to *error* when we have to make guesses about a future value of K or $\bar{X}$. The normal distribution describes the probability of error in predicting variables like K and $\bar{X}$ which are normally distributed. We are now going to show that it also describes the probability of error in making measurements of variables which cannot be measured precisely. By sketching the theory of measurement error, we shall show why the normal distribution is frequently called the NORMAL CURVE OF ERROR.

Figure 9-4 shows a normal curve superimposed on a binomial distribution for $N = 100$ and $p = 1/2$. We are accustomed to thinking of this binomial distribution in relation to a lot of coin-toss experiments in each of which a single fair coin is tossed 100 times. Let us now think of it in relation to an experiment in which each of 100 fair coins is tossed one time. Logically these experiments amount to the same thing, and the distribution applies equally well to both. But the one-toss experiment suggests what may be happening when we make just one measurement which can be influenced by 100 *independent, random,* and *additive* influences.

The quantity we are measuring has some true but unknown value; let us call it X_T. Each of the 100 influences can produce a tiny error in the measurement. When an influence is present on the occasion of a particular measurement, it adds an amount C to the resulting measurement. When an influence is absent, an amount C is subtracted from the resulting measurement. Each influence is random, and the probability that it will be present on a given occasion is $1/2$.

Figure 9-4 describes the relative frequency with which the different values of K will occur, where K is now the *number of influences present* during a single measurement attempt. Because the influences are independent, random, and additive, they are like tosses of a fair coin: having an influence present is like getting a head; having it absent is like getting a tail. Figure 9-4 also describes the probability distribution of actual measurements obtained. When K influences are present, the measurement obtained will be $X_T + KC - (N - K)C$. When $K = N - K = 50$, the measurement will be exactly X_T. Since the mean of the distribution in Fig. 9-4 is 50, the mean of the probability distribution of measurements actually obtained will be X_T.

We can conclude that when a measurement is affected by 100 independent, random, and additive influences, the resulting series of repeated measurements will have a probability distribution like Fig. 9-4 with a mean at the true value of the quantity being measured.

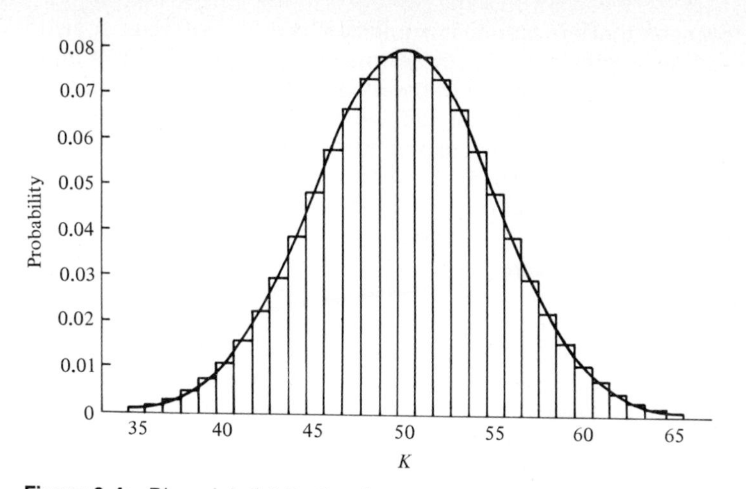

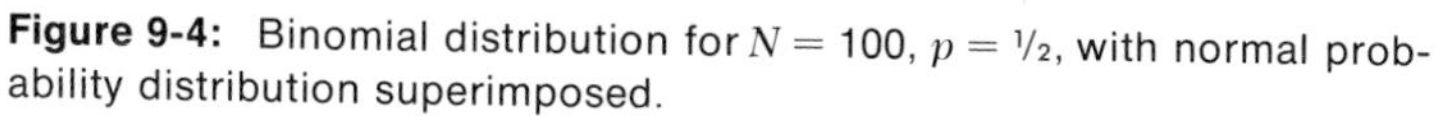

Figure 9-4: Binomial distribution for $N = 100$, $p = 1/2$, with normal probability distribution superimposed.

9-48 The binomial distribution in Fig. 9-4 is for coin-toss experiments with $N =$ ______ and $p =$ ______.

100; 1/2

9-49 It will also apply to measurement of any variable which cannot be measured precisely, whenever there are exactly ______ influences making for error in this measurement.

100

9-50 These influences will have to operate like tosses of a fair coin. No toss is affected by the outcome of any other toss; the influences, like the tosses, must be ______ of each other.

independent

9-51 Each toss of the coin can have either of two outcomes, heads or tails. Likewise, each influence making for error may be either ______ or ______ on the occasion of a particular measurement.

present ↔ absent

9-52 The probability of heads is ______ for a fair coin; the probability that an error influence will be present is ______, when the influences are random.

1/2
1/2

9-53 Each head that occurs increases the statistic K by 1; it affects K in an additive way. Similarly, each error influence must operate in an ______ way, increasing the measurement by C when it is ______ and decreasing it by C when it is ______.

additive
present
absent

9-54 Figure 9-4 applies only if the 100 error influences are ______, ______, and ______.

independent ↔ random ↔ additive

We have now seen that the binomial expansion of $(\frac{1}{2} + \frac{1}{2})^{100}$ can apply to errors of measurement, provided that the errors are produced by 100 influences each of which is (1) independent of the others, (2) randomly varying, and (3) able to affect the measurement in an additive way. These three conditions are also the hallmarks of a coin-toss experiment. Since the distribution of measurement error is like the RSD of K, it will become a normal distribution as the number of independent, random, and additive influences becomes very large. The mean of the distribution will be X_T.

This line of reasoning is the basis for arguing that errors of measurement will be normally distributed, whether they arise through insufficiently precise measuring instruments, relatively sloppy measurement conditions ("experimental error"), or sampling variability. It may seem artificial to think of a lot of independent, random, and additive influences acting simultaneously to produce error. Nevertheless it is empirically true that a normal distribution of values will be obtained when a difficult measurement is repeated many times. The concept of independent additive influences, randomly varying from one measurement to another, does help to explain how such a normal distribution arises.

Remember, however, that the notions of independence and additivity of the influences are crucial to the theory of a normal distribution of measurement error. Whenever there is any reason to believe that errors arise through influences which are not independent of each other or which are not additive in their effects, the situation will no longer be like a coin-toss experiment. The expectation of a normal distribution of error is not justified in such cases. This point will become extremely important in Part Three, where we shall introduce analysis of variance and related methods. These methods cannot properly be used unless the error due to sampling variability is normally distributed. Thus you will find that we have to assume independence and additivity of all influences producing measurement error in order to use these statistical methods.

We should mention, finally, that a considerable number of biological variables also have a normal distribution within a defined population of persons or animals. Height is such a variable; the heights of men descended from the same racial stock and reared in the same geographical area will form a normal distribution. Because of this fact, it is thought reasonable to assume that height is determined by a very large number of randomly varying influences, some of them genetic and some environmental, which operate independently and additively. If each influence increases height by a certain amount when it is present and decreases it by the same amount when it is absent, the height of any given individual will depend upon the number of such influences present in his case. The mean height for this population will occur when just half the influences are present.

Thus whenever a set of observations is found to be approximately normally distributed, we entertain the possibility that the population distribution is normal and that the variable is determined by many independent, random, and additive influences. Evidence for a normal distribution is presumptive evidence of the operation of many independent, random, and additive influences. On the other hand, when a variable may be influenced by factors which do not operate both independently and additively, we do not expect it to be normally distributed.

9-55 The binomial distribution can be used to describe the number of measurement influences present to produce error, provided that these influences are ________, ________, and ________.

independent ↔ random ↔ additive

9-56 When there are 100 possible influences, the true measurement X_T will be obtained on those occasions with ______ influences present and ______ influences absent.

50
50

9-57 When the number of influences is much greater than 100, the distribution describing the number of influences present will be ________.

normal

9-58 The probability distribution of actual measurements obtained under these conditions will also be normal, and its mean will be ______.

X_T

9-59 Each actual measurement obtained can be called X. Since $E(X) = \bar{X}$ for any distribution, the expectation of X in this particular case is ______.

X_T

9-60 When we repeat the measurement several times, we can expect the mean of our actual measurements to get closer and closer to ______ as the number of repetitions increases.

X_T

9-61 We say that errors of measurement are normally distributed because the values of the measurements actually made will tend to be ________ distributed around X_T.

normally

9-62 Each value that is not equal to X_T is in error by some amount. In two-thirds of the measurements, the amount of error is expected to be no more than ______ standard deviation of the measurement distribution.

1

9-63 In 95 percent of the measurements, the error is expected to be no greater than ______ standard deviations.

2

9-64 Some biological variables, like height, have a normal distribution. When they do, it is reasonable to think they may be determined by a very large number of random influences acting ________ and ________.

independently ↔ additively

9-65 Conversely, any variable which is known to be controlled by many independent, random, and additive influences will be expected to have a ________ frequency distribution.

normal

9-66 But if a variable or a measurement is influenced by factors which are not independent or not additive, its distribution should not be expected to be ________.

normal

REVIEW

9-67 This lesson introduces the concept of the RSD, which is a ________ distribution of some statistic.

random sampling

9-68 If the statistic is $\bar{X}$, its RSD is a distribution of values of ________ taken from a large number of ________ samples, all drawn from the same ________ and all having the same ________.

$\bar{X}$; random
population
size

9-69 The term "RSD" means nothing by itself until you know the ________ whose RSD is being discussed.

statistic

9-70 The central limit theorem applies to the RSD of means. It concerns the theoretical RSD, for which the number of sample means is considered to be ________.

infinite

9-71 The theorem states that the shape of this RSD will be ________ when N is large enough. N refers to the sample ________.

normal
size

9-72 A sample N of at least ________ is sufficiently large to ensure that the RSD of means will be normal, even when the population is skewed.

30

9-73 The theorem also states that the mean of the RSD of means will equal ________, the ________ mean. For this reason, we can say that $E(\bar{X}) =$ ________.

μ; population
μ

9-74 The theorem gives information about the population variance for cases in which the population is ________. In such cases, the RSD variance is ________.

large
$\frac{\sigma^2}{N}$

9-75 The variance of the RSD of means is represented by the symbol ________. The standard deviation of this RSD is called the ________ of the mean.

$\sigma_{\bar{X}}^2$
standard error

9-76 When the population variance is not known, it can be estimated from the sample. The estimated population variance is ________.

$\frac{SS}{N-1}$

9-77 When σ^2 is not known, $\sigma_{\bar{X}}^2$ cannot be calculated precisely. However, it too can be estimated from SS; the estimated $\sigma_{\bar{X}}^2$ is ________.

$\frac{SS}{N(N-1)}$

9-78 The normal distribution is also known as the normal curve of error because the distribution of ________ of measurement is usually found to be normal.

errors

9-79 When measurement errors are normally distributed, it is reasonable to assume that they arise through a very __________ number of influences. *large*

9-80 In order for the distribution of error to be normal, these influences must also be __________ of each other, must operate in an __________ way, and must be random. *independent* *additive*

9-81 The mean of the distribution of measurement error is the __________ value of the variable being measured. We have been calling this value X_T. *true*

9-82 Whenever a difficult measurement is being made, the measurement is repeated several times, and a mean is calculated. Each measurement is X; the mean of the sample of measurements is ______. The population mean μ is the true value, ______. $\bar{X}$ X_T

9-83 If we want to estimate X_T, the best estimate available will be ______. $\bar{X}$

9-84 Whenever a variable is found to have a normal distribution, it is reasonable to believe that it is determined by a __________ number of randomly varying influences, each acting in an __________ and __________ way. *large* *independent* ↔ *additive*

9-85 If a variable is known to be determined by influences which do not act independently, its population distribution (*should; should not*) be expected to be normal. *should not*

9-86 If the influences determining a variable are known to act in a way that is not additive, the population distribution should not be expected to be __________. *normal*

9-87 All statistical tests are based on the use of probability distributions which are __________ of some statistic. The name of the statistical test is usually taken from the name of its __________. *RSDs* *RSD*

9-88 With interval data the statistic used is ordinarily a __________ or a combination of __________. *mean; means*

9-89 When two or more means are being compared, the RSD used will be a derivative of the RSD of __________. *means*

9-90 If the influences producing measurement error are not independent and additive, the __________ due to sampling variability may not be normally distributed. *error(s)*

PROBLEMS

1. The standard deviation of the population of Stanford-Binet IQ scores of a very large sample of white American-born citizens of the United States is 16. What is the standard error of the mean for samples of $N = 144$?

2. We have determined the mean number of hours spent in sleep during a certain 3-day period for each of a randomly selected group of 100 students at several different colleges. We find a mean of 7.15 and s of 1.10. What is the sample sum of squares? What is the estimated population variance? What is the estimated $\sigma_{\bar{X}}^2$? What is the estimated standard error of the mean?

3. A measurement is repeated 30 times; $\bar{X}$ is 365, and SS is 87,000. Estimate the standard error of the mean. Since the RSD of the means of samples of $N = 30$ will be normal around X_T as its mean, what is the probability that this $\bar{X}$ lies within 1 standard error of X_T? At this probability, $\bar{X}$ might be too *low* by 1 standard error; what, then, is the highest value X_T could have at this probability level? $\bar{X}$ might, on the other hand, be too *high* by 1 standard error; what is the lowest value X_T could have at this probability level?

PART TWO
Methods for Qualitative Data

LESSON 10
TESTS FOR DICHOTOMIZED DATA

This lesson is a direct continuation of Lesson 8, in which you first learned to use the normal-curve approximation in making one-tailed binomial tests. We must now adapt this method for use in cases which require two-tailed tests. This lesson also has a further aim: to identify the kinds of cases in which a binomial test should be considered.

Before planning to use a binomial test, you should always recall the characteristics of a coin-tossing experiment and ask yourself whether your data have these characteristics. There must be some number N of independent trials, each of which has only two possible outcomes. There must also be some way of stating H_1 as an exact hypothesis about the value of p. Let us distinguish three kinds of cases which meet these requirements.

Case 1: One-sample cases in which *p* is determined by theory Our color-touch example is such a case; we compare our one sample of N trials with a theoretical population in which $p = q = 1/2$ because we assume that the parent population of our sample arises through guessing.

A similar situation is found in multiple-choice tests. Our color-touch trials had only 2 alternatives, red and blue; multiple-choice tests customarily have 5 alternatives from which the correct answer is to be chosen. The probability of a correct choice by chance alone is $1/5$ in such a five-alternative case. We still compare a single sample with a theoretical population based on the guessing hypothesis, but the population is now assumed to contain $1/5$ correct and $4/5$ incorrect choices.

Case 2: One-sample cases in which *p* is determined from *empirical* relative frequencies In such cases we are usually interested in determining whether our sample can be considered to be a random sample from some empirically known population. Here is an example.

An introductory zoology class at a certain college has 90 students, of whom 50 are women. Does this indicate that women elect zoology more frequently than men at this college? To answer this question, we must consider the relative numbers of men and women in this college population. With 1,500 students, 600 of whom are women, the empirical relative frequency of women in the population is .40. To test the hypothesis that men and women are equally likely to elect zoology, we assume that the class of 90 is a random sample from the college population. H_1 becomes $p = .40$, and H_2 is $p > .40$. A binomial test with $N = 90$ and $p = 2/5$ can be used.

Case 3: Two-sample cases in which two related samples are to be compared with each other Related samples can arise in many ways. When N persons are observed under each of two conditions, there will be a sample of N observations for condition A and a related sample of N observations for condition B. The observations obtained may be a set of scores (interval data) or a set of ranks (ordinal data). In either case the observation for condition A can be subtracted from that for condition B for each person, and these rank or score differences (some positive and some negative) will form a single sample which can be dichotomized according to algebraic sign. If the two conditions are not different, the number of positive differences will be about the same as the number of negative differences. The assumed hypothesis becomes $H_1: p = 1/2$, where p is the probability of a difference with a positive sign. A binomial test will show whether H_1 is supported by the data.

This method of dichotomizing data according to algebraic sign is so common that it has acquired a special name, the SIGN TEST. The raw data are quantitative (interval or ordinal data), not qualitative. They are converted into categories in order to carry out the test. But the sign test is usually not the only test which could be used in such cases. It is merely the simplest to do, and it is often used to give a quick estimate of significance before another more elaborate test is undertaken.

A. REVIEW OF ONE-TAILED BINOMIAL TESTS

In a typical extrasensory-perception (ESP) experiment, a perceiver (percipient) attempts to tell correctly the pattern displayed on a card which is visible only to another person, the sender or agent. Proper conditions prevent the transmission of any sensory information between the two persons, directly or indirectly. ESP cards have 5 patterns, making the experiment a 5-alternative multiple-choice test. The usual experiment consists of 25 trials, and trials on which the percipient's choice is correct are called hits. H_1 is the guessing hypothesis, $p = 1/5$. Since the experimenter hopes to find more than 5 correct choices in 25, the proper alternative hypothesis is $H_2: p > 1/5$. This directional H_2 requires a one-tailed test.

The binomial distribution for $p = 1/5$ is skewed when $N = 25$; we shall therefore consider experiments of $N = 100$ so that we can use the normal approximation. We can carry out a binomial test in either of two ways:

1 Keeping the decision rule in terms of α, we may elect to calculate the exact probability of K_o, the number of hits observed in 100 trials. We take K_o from the observed sample, convert it to a z score in the normalized binomial for $N = 100$ and $p = 1/5$, and consult the normal-curve table for the probability of a z at least as large as this one. If this probability is not greater than α, H_1 is rejected.

2 We may choose to find a critical value of K and state the decision rule in terms of K. In this case we start with α, using the normal-curve table to find the z score beyond which the area under the curve equals α. We then convert this critical value of z into a critical value of K. When K_o is at least as large as the critical value of K, H_1 is rejected.

10-1 State the hypotheses for the ESP experiment. H_1 is ______. H_2 is ______.

$p = 1/5$; $p > 1/5$

10-2 The decision-making distribution for this 100-card experiment will be a binomial distribution with $N =$ ______ and $p =$ ______.

100; 1/5

10-3 When this binomial distribution is normalized, its mean is ______ and its standard deviation is ______.

20; 4

10-4 Take $\alpha = .01$, and state the decision rule in terms of α: reject ______ if K_o has a probability no ______ than .01 when H_1 is ______.

H_1; greater true

10-5 Since H_2: $p > 1/5$ is a ______ hypothesis, H_1 will be rejected only when K_o is (*larger; smaller*) than $E(K) = 20$.

directional larger

10-6 We may keep the decision rule in terms of α and calculate the probability of getting a value of K which is at least as ______ as K_o when ______ is true.

large; H_1

10-7 If the number of hits obtained is 26, can H_1 be rejected? To find the probability by the normal approximation to the binomial, you must make a correction for ______. You calculate the probability of $K_o \geq$ ______.

continuity 25.5

10-8 $K = 25.5$ has a z-score equivalent in the normalized binomial distribution; its z-score equivalent is ______.

+1.375

10-9 In a normal distribution, the Area (Mean to z) for $z = 1.375$ is ______. The probability of getting z at least this large is ______. H_1 must be ______.

.415 .085; accepted

10-10 On the other hand, we may prefer to find a critical value of K. With H_2:$p > 1/5$, the rejection region will have (*1 part; 2 parts*), and there will be (*1; 2*) critical value(s) of K.

1 part; 1

10-11 To find this critical value of K, we must know the critical value of z beyond which ______ of the area under the normal curve lies.

.01

10-12 We look up the area ______ in the normal-curve table, and we find that the critical value of z is ______.

.49 2.33

10-13 This z score is equivalent to $K =$ ______ in our normalized binomial distribution. The critical value of K is ______. We can now state a decision rule in terms of K_o: reject ______ if K_o is ______.

29.32 30 H_1 ≥ 30

B. TWO-TAILED TESTS USING THE NORMAL-CURVE TABLE

When H_2 is nondirectional, a two-tailed test is required. Figure 4-4 (page 63) will remind you of the way in which the rejection region (area $= \alpha$) is divided into two equal parts for such a test, each part with area $= \alpha/2$. Keeping this diagram in mind, you will find it easy to modify the two procedures already discussed for one-tailed tests.

Procedure 1 When the exact probability of an observed result must be found, remember that you are looking for the probability of a result which is at least this extreme in either direction. Thus, if $K_o = 26$ in the ESP experiment, the set of results which are at least this extreme includes not only $K \geq 26$ but also $K \leq 14$. When $N = 100$, the distribution is symmetrical around $E(K) = 20$; the probability of $K \geq 26$ is equal to the probability of $K \leq 14$. Your procedure is to calculate one of these probabilities and double it; in this way you obtain the probability of a result at least as extreme in either direction.

When N is small, the binomial distribution for $p = 1/5$ is not symmetrical, and the exact probabilities must be calculated separately for each tail of the distribution. This situation does not arise very often; two-tailed tests are not usually required unless H_1 assumes $p = 1/2$, and the binomial distribution is then symmetrical for even very small N.

Table 10-1 presents some data for a sign test; 25 persons took a reading test under two conditions, designated here as conditions A and B. The persons and their scores have been arranged in the order of their performance under condition A.

Table 10-1 Reading Performances under Two Conditions

Person	A	B	$A-B$	Person	A	B	$A-B$
1	15	13.5		14	10	11.5	
2	14	12.5		15	9.5	8	
3	13.5	12.5		16	9.5	9	
4	13	13.5		17	9	8.5	
5	13	12		18	9	9.5	
6	12.5	11.5		19	9	7.5	
7	12	11.5		20	8.5	9.5	
8	11.5	12		21	8.5	8	
9	11	9.5		22	8	9	
10	11	10		23	7.5	8	
11	11	12		24	7.5	6	
12	10.5	9		25	6	5	
13	10	9					

10-14 Write the difference score $A - B$ for each person in Table 10-1, taking care to include the algebraic sign of each difference. Count the number of positive signs in your list. There are ______ positive signs.

17

10-15 If conditions A and B are not different in their effect upon reading performance, the probability that a difference will be positive is ______.

$1/2$

10-16 Under H_1: $p = 1/2$, the expectation $E(K)$ of the number of positive signs is ______. The actually observed number K_o is ______.

12.5
17

10-17 When H_2 is $p \neq 1/2$, the decision rule is: reject H_1 if the probability of a result at least as ______ as 17, in either direction, is not greater than α.

extreme

10-18 In other words, the probability which must be compared to α is the probability of getting $K \geqslant$ ______ plus the probability of getting $K \leqslant$ ______.

17
8

10-19 The binomial distribution for H_1 is the distribution with $N =$ ______ and $p =$ ______. Its normal approximation has a mean of ______ and σ of ______.

25; 1/2
12.5; 2.5

10-20 Because the normal distribution is continuous, the probability of $K \geqslant 17$ is found by converting $K =$ ______ to a z score and consulting the normal-curve table.

16.5

10-21 The z-score equivalent of 16.5 is ______. The probability of a z score at least this large is ______.

−1.6
.0548

10-22 The z-score equivalent of 8.5 is ______. The probability of a z score at least this small is ______.

−1.6
.0548

10-23 The probability of a z score at least as *extreme* as $|1.6|$ is .0548 times ______.

2

10-24 The probability given H_1 of getting *either* $K \geqslant 17$ or $K \leqslant 8$ is therefore ______.

.1096

10-25 Take $\alpha = .05$. The probability of a result at least as extreme as 17 is ______ when H_1 is true. H_1 will be ______.

.1096; accepted

10-26 The probability of the observed result ($K_o = 17$) is ______ by a two-tailed test. The probability of the same result is ______ by a one-tailed test.

.1096
.0548

Procedure 2 When the decision rule is to be stated in terms of a critical value of K, a two-tailed test requires you to find two such critical values, one at each end of the probability distribution. Since the area of each rejection region must be only $\alpha/2$, you begin by finding the value of z beyond which $\alpha/2$ of the normal-curve area lies. The upper rejection region is bounded by the positive critical value of z; the lower rejection region is bounded by the same critical value with a negative sign. Each of the critical z's must be converted to a critical value of K.

In the reading test with $\alpha = .05$, we look in the normal-curve table (Table A-1) for a z score beyond which .025 of the area will lie. Accordingly, we look for an Area (Mean to z) of .475. The corresponding z score is 1.96. We can reject H_1 if we get a z score either as large as 1.96 or as small as -1.96.

The z-score equivalent of 1.96 is $12.5 + 1.96\ (2.5) = 17.4$. The critical value of K at the upper end of the distribution is therefore 18. Since the z-score equivalent of -1.96 is $12.5 - 1.96\ (2.5) = 7.6$, the critical value of K at the lower end of the distribution is 7. The probability of K at least as low as 8 would be a little greater than .125. The decision rule now becomes: reject H_1 if K_o is either at least as large as 18 or at least as small as 7.

It is worth pausing to compare the results of one- and two-tailed tests on the same data. Suppose that conditions A and B were not just two potentially equivalent conditions but were two different times of testing—B "before" and A "after" a period of training in reading efficiency. Scores after training will be expected to show superiority over scores before training, and $A - B$ should be positive if the training has any effect.

The experimenter, hoping to demonstrate that the training was indeed effective, would still assume $H_1: p = 1/2$, but his alternate hypothesis would now be $H_2: p > 1/2$. With the same α level (.05), his one-tailed test would differ from the two-tailed test we have just discussed. Whereas H_1 would have to be accepted by a two-tailed test, H_1 could be rejected by the one-tailed test.

Table 10-2 Comparison of One- and Two-Tailed Binomial Tests on Data from Table 10-1 ($\alpha = .05$)

H_2	Probability of Result as Extreme as $K_o = 17$	Critical Values of K	Decision When $K_o = 17$
$p \neq 1/2$	.1096	18 and 7	Accept H_1
$p > 1/2$	.0548	17 (exactly 16.6)	Reject H_1

It should be evident from these comparisons that a one-tailed test is more likely to permit rejection of H_1 than a two-tailed test. The probability of an obtained result is less under the one-tailed test; the critical value of K is also less extreme when the test is one-tailed. But this does not mean that one-tailed tests can be used indiscriminately in order to obtain this advantage. Unless there is a clear reason to expect a difference in one direction rather than the other *before the data are seen*, H_2 must be stated as a nondirectional hypothesis and the statistical test must be two-tailed.

10-27 Take $\alpha = .01$. For a two-tailed test of the data in Table 10-1, you must find (*one; two*) critical values of K. *two*

10-28 With $\alpha = .01$, each of the two rejection regions must have an area equal to ______. *.005*

10-29 At the upper end of the normal curve there is a z value beyond which .005 of the area lies. The area from the mean to this z value is ______. *.495*

10-30 Look up this area in the normal-curve table. The critical values of z for $\alpha = .01$ are ______ at the upper end of the distribution and ______ at the lower end. *2.58* *−2.58*

10-31 The mean of the normalized binomial for $N = 25$ and $p = \frac{1}{2}$ is ______. Its standard deviation is ______. *12.5; 2.5*

10-32 To find the critical value of K at the upper end of the distribution, we add ______ times 2.5 to ______. *2.58; 12.5*

10-33 The upper critical value of K is ______. *18.95*

10-34 We can reject H_1 by a two-tailed test at $\alpha = .01$ if we get at least ______ positive signs. *19*

10-35 The lower critical value of K is ______. *6.05*

10-36 We can reject H_1 by a two-tailed test at $\alpha = .01$ if we get no more than ______ positive signs. *6*

10-37 In other words, H_1 can be rejected by a two-tailed test at $\alpha = .01$ provided that the number of positive signs is as great as ______ or as low as ______. *19; 6*

10-38 If we get 6 or fewer positive signs, we shall get ______ or more negative signs. Thus we can reject H_1 by this test if we get at least ______ signs of one kind, either positive or negative. *19* *19*

10-39 For a one-tailed test at $\alpha = .01$, we would seek the critical value of z beyond which ______ of the normal-curve area lies. *.01*

10-40 From Table A-1 you can determine the critical value of z for such a test. It is ______. This critical value is (*less; more*) extreme than the critical value at this end of the distribution when the test is *two*-tailed. *2.33; less*

C. THE RELATIVE POWER OF ONE- AND TWO-TAILED TESTS

From the comparisons we have just made, can we conclude that a one-tailed test always has greater power than a two-tailed test? The answer is a qualified yes. To understand the qualification, let us look at the power functions for the data from Table 10-1 which we have been considering.

Figure 10-1 shows three normal distributions. The distribution in the center, with mean at 12.5, represents H_1. The right-hand distribution represents $H_2: p = .70$; its mean is at 17.5. The binomial distribution for $p = .70$ is of course somewhat skewed when N is only 25; it is represented here by a normal curve because we know, from Lesson 8 (page 124), that the normal approximation will give a very close estimate of $1 - \beta$ even for $N = 25$.

Recall from Table 10-2 that the critical value of K at $\alpha = .05$ is 17 for a one-tailed test and 18 for a two-tailed test. With the correction for continuity, we draw the boundary of the rejection region at 16.5 for the one-tailed test. All the area to the right of the dotted line at 16.5 represents $1 - \beta$ when H_2 is $p > 1/2$, that is, when a one-tailed test is appropriate; the probability $1 - \beta$ is .67 for $H_2: p = .70$. However, when H_2 is $p \neq 1/2$, the boundary of the upper rejection region is exactly at the mean of this H_2 distribution, i.e., at 17.5. The power of a two-tailed test for detecting $H_2: p = .70$ is exactly .5.

But we cannot stop here; we must consider the left-hand distribution in Fig. 10-1. It represents $H_2: p = .30$, with mean at 7.5. Because the dotted line at 16.5 does not touch even the extreme edge of this curve, the power of a one-tailed test to detect such an H_2 would be infinitesimally small. The two-tailed test, however, is just as able to detect $H_2: p = .30$ as to detect $H_2: p = .70$; the power is again .5.

A more complete comparison of the power functions for one- and two-tailed tests is shown in Fig. 10-2.

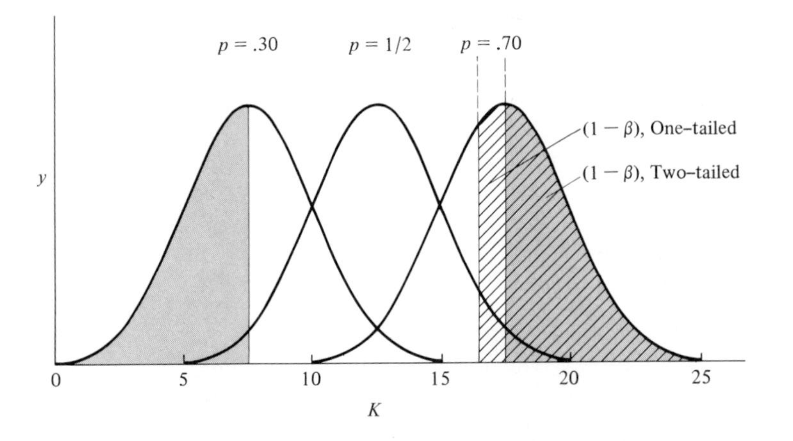

Figure 10-1. Normal curves representing the probability distributions with $N = 25$ for $H_1: p = 1/2$ (middle curve), $H_2: p = .30$ (left curve), and $H_2: p = .70$ (right curve). Rejection boundaries are shown for one- and two-tailed tests at $\alpha = .05$. For each test, the area shown as $1 - \beta$ applies to $H_2: p = .70$.

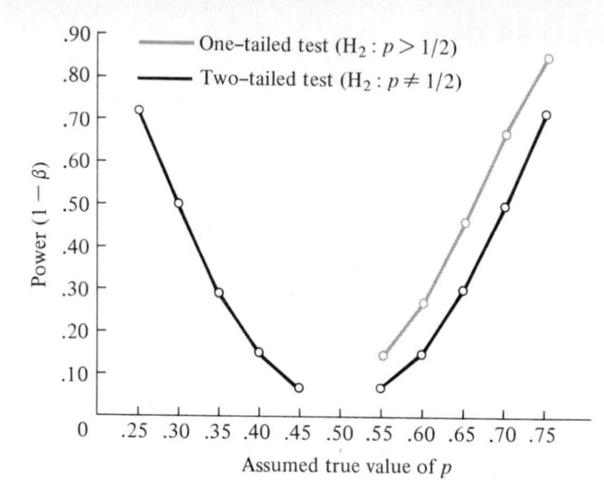

Figure 10-2: Power functions for one- and two-tailed tests of $H_1: p = 1/2$ with $N = 25$ and $\alpha = .05$.

10-41 Figure 10-2 shows the power functions for one- and two-tailed tests at H_2's ranging from $p =$ ________ to $p =$ ________. *.25; .75*

10-42 If H_2 is $p > 1/2$, the experimenter does not expect to have to detect H_2's in which p is ________ than $1/2$. *less*

10-43 In this case he uses a ________-tailed test. Such a test is designed to detect H_2's in which p is ________ than $1/2$. *one* *greater*

10-44 If H_2 is $p \neq 1/2$, the experimenter may have to detect H_2's which deviate from $1/2$ in either direction. He uses a ________-tailed test; its power to detect H_2: $p = .30$ is exactly equal to its power to detect H_2: $p =$ ________. *two* *.70*

10-45 If $p = .70$ is really true, the ________-tailed test has greater power than the ________-tailed test to detect it. *one* *two*

10-46 The power function for the one-tailed test lies (*above; below*) that for the two-tailed test for all H_2's with $p > 1/2$. *above*

10-47 A one-tailed test is always more powerful than a two-tailed test on the same data, *provided* that the H_2's to be detected all deviate from H_1 in the same ________. *direction*

10-48 When a two-tailed test is used, we sacrifice some of the power for detecting $p > 1/2$ in order to be able to detect ________. $p < 1/2$

10-49 In other words, a one-tailed test is always more powerful provided that a ________ H_2 is really justified in the particular case being tested. *directional*

D. A TABLE OF CRITICAL VALUES OF z

Calculating a power function, like the ones in Fig. 10-2, requires finding exact probabilities of a given result under several different H_2's. In such calculations one must use the procedure for determining the exact probability of an observed result.

In most of the practical uses of a binomial test, however, we do not need to find the exact probability of K_o. Such a probability is seldom stated in reports of investigations. For the data in Table 10-1 it is sufficient to state that $K = 17$, with $p > .10$ by a two-tailed test or $p = .05$ by a one-tailed test. Almost no one would bother to state that $p = .1096$ by a two-tailed test or that $p = .0548$ by a one-tailed test.

Conventional usage makes it possible to get along very well with only a few values from the normal-curve table as long as no power functions have to be calculated. The values most often used are those which mark the probabilities .10, .05, .02, .01, and .001. The values of z which are required to reach these conventional α levels can be called CRITICAL VALUES of z. You will eventually memorize some of them. Meanwhile, it will be convenient for you to have a table which gives only these values of z for quick reference.

Table 10-3 provides spaces in which you can now write these critical values of z. You already know that they will be more extreme for a two-tailed test than for a one-tailed test. Since you have recently used the critical values of z for $\alpha = .05$ and $\alpha = .01$, write these values now in Table 10-3. (For $\alpha = .05$, see page 124, Frame 8-49, and page 152. For $\alpha = .01$, see page 153, Frames 10-30 and 10-40.) The frames on page 157 will direct you in completing the table. Hereafter, whenever you need one of these critical values of z, you will be able to refer to your own table of critical values instead of looking in the more complicated normal-curve table.

Tables of critical values are widely used for distributions other than the normal distribution. You will meet another such table in Lesson 11. Such tables will be easy for you to comprehend after you have generated a critical-value table of your own.

Table 10-3 Critical Values of z from the Normal-Curve Table

One-tailed test ($+z$)		Two-tailed test ($-z$, $+z$)	
Probability α	z	Probability α	z
.10	______	.10	______
.05	______	.05	______
.02	______	.02	______
.01	______	.01	______
.001	______	.001	______

10-50 The small drawing in Table 10-3 indicates that the rejection region for a one-tailed test has (*1 part; 2 parts*). *1 part*

10-51 To find the z for a one-tailed test at $\alpha = .10$, you must find the z beyond which exactly ______ of the area under the normal curve will lie. You therefore look in the normal-curve table for an Area (Mean to z) which is ______. *.10* *.40*

10-52 This area corresponds to a z value of ______. Enter this critical value of z in your table. *1.28*

10-53 The small drawing for a two-tailed test reminds you that the rejection region for such a test has ______ part(s). *2*

10-54 When $\alpha = .10$, each of these parts has an area of ______. To find the critical value of z for a two-tailed test at $\alpha = .10$, you must look in the normal-curve table for an Area (Mean to z) which is ______. *.05* *.45*

10-55 This area corresponds to a z value of ______. Enter this critical value of z in your table. *1.65*

10-56 Notice that the critical value of z for a *two*-tailed test at $\alpha = .10$ is exactly the same as the critical value for a *one*-tailed test at $\alpha =$ ______. *.05*

10-57 The critical z for a two-tailed test at $\alpha = .02$ is the same as the critical z for a one-tailed test at $\alpha =$ ______. *.01*

10-58 You already know the critical z for a one-tailed test at $\alpha = .01$. Enter this value for a two-tailed test at $\alpha =$ ______. *.02*

10-59 The small drawings in Table 10-3 will help to remind you that the rejection region for a one-tailed test at α must have the same area as a *single part* of the rejection region for a two-tailed test at (2α; $\alpha/2$). *2α*

10-60 Now look up all the z values needed to complete Table 10-3. After you have done so, use the principle in Frame 10-59 to fill the table below. See the top of page 158 to check your results.

	Probability α						
One-tailed test	.10	.05	.025	.01	.005	.001	.0005
Two-tailed test	.20	.10	.05	.02	.01	.002	.001
z	______	______	______	______	______	______	______

Table 10-4 Critical Values of z from the Normal-Curve Table

	Probability α						
One-tailed test	.10	.05	.025	.01	.005	.001	.0005
Two-tailed test	.20	.10	.05	.02	.01	.002	.001
z	1.28	1.65	1.96	2.33	2.58	3.09	3.29

REVIEW

10-61 A binomial test may be used whenever an experiment has the characteristics of a coin-toss experiment. Such an experiment consists of ________ trials with only ______ outcomes per trial. *independent; 2*

10-62 It is not necessary, however, that the experiment have trials in a literal sense. A multiple-choice test consists of questions called "items"; the independent trials in such an experiment are the individual ________, each of which is answered either correctly or incorrectly. *items*

10-63 In the reading experiment, 25 persons took a reading test under two different conditions, A and B. The difference $A - B$ represents the difference in a person's performance under these two ________. If $A \neq B$, the difference must have either a positive or a negative ________. *conditions* *sign*

10-64 If conditions A and B do not make any difference in performance, the probability that $A - B$ is positive should be ______. *1/2*

10-65 None of the 25 differences in Table 10-1 is 0. Therefore, we can treat the 25 difference scores as a set of 25 independent ________, each of which has only two possible ________. *trials; outcomes*

10-66 Since these outcomes are algebraic signs, this use of the binomial test is commonly called a ________ test. *sign*

10-67 We say that a sign test is appropriate for comparing two related samples. In this case we actually have two sets of 25 scores, one from condition A and one from condition B. These two samples are related because the 25 scores in both sets were contributed by the ________ group of 25 people. *same*

10-68 Because the two sets of scores are related in this way, we are able to transform them into one set of scores which we call ________ scores. We can make this transformation only because the two samples are ________ to each other. *difference* *related*

10-69 A binomial test will be one-tailed when H_2 is a ________ hypothesis. *directional*

10-70 When H_2 is $p > 1/2$, we do *not* expect to have to discriminate between H_1: $p = 1/2$ and any H_2's in which p is ________ than $1/2$. *less*

10-71 In such a case, our test will be more powerful if it is ________-tailed than if it is ________-tailed. *one; two*

10-72 When H_2 is $p \neq 1/2$, we may have to discriminate between H_1 and some H_2's which lie in either direction. In such a case our test should be ________-tailed. *two*

10-73 A student gets only 20 correct on a 75-item multiple-choice test. We use a binomial test to decide whether he could have been merely guessing. Since each item has 5 alternatives, the probability of a correct answer by guessing is ________. *$1/5$*

10-74 We use a normalized binomial distribution with mean of ________ and $\sigma^2 =$ ________. *15; 12*

10-75 Since $\sqrt{12} = 3.46$, the score of 20 correct corresponds to a z of ________. *1.45*

10-76 We are not interested in detecting any H_2 with $p < 1/5$, and so our binomial test will be ________-tailed. *one*

10-77 The probability of such a high score by guessing alone is ________. We (*can; cannot*) reject the guessing hypothesis at the .05 level. *.0735; cannot*

10-78 An experimenter has data for a sign test. He has 104 difference scores, 4 of which are equal to 0. He should therefore use the normalized binomial with $N =$ ________ and $p =$ ________. *100; $1/2$*

10-79 His binomial distribution has a mean of ________ and a standard deviation of ________. *50* *5*

10-80 If p is the probability of a positive sign and H_2 is $p > 1/2$, he can reject H_1 if he obtains as many as ________ positive signs. *59*

10-81 If H_2 is $p \neq 1/2$, he can reject H_1 if he obtains as many as ________ positive signs or as few as ________ positive signs. *60; 40*

PROBLEMS

1. In a college of 1,000 students, 400 came from large high schools and 600 came from small high schools. The honor roll listed 50 students in a particular year, of whom 25 came from large high schools. Use a normal-curve test to determine whether the number of honor-roll students coming from large high schools is significantly larger than the number coming from small high schools.

How would the result of this test differ if the hypothesis being tested were, "The number of honor-roll students coming from large high schools is significantly *different* from the number coming from small high schools"?

2. A coin is tossed 900 times. The owner of the coin alleges that the coin is biased. How many heads must turn up in 900 tosses for $H_1: p = 1/2$ to be rejected at the .05 level?

3. Out of 57 men at a weekly college dance, 36 had been at the dance the week before, and of these, 23 had brought the same date on both occasions, and 13 had brought a different date or had come alone. Test whether the number of men who came both weeks with the same date is significantly different from the number who came both weeks but not with the same date.

4. A new therapy for schizophrenia has been tried for 6 months with 54 randomly chosen patients in three different hospitals. At the end of this time, 25 patients are recommended for release from the hospital; the usual proportion released in 6 months is one-third. Use a binomial test to determine whether the new therapy has produced significantly more improvement than the usual therapy.

LESSON 11
THE PEARSON χ^2 TEST

The binomial test is suitable for experiments in which the outcomes of independent trials can be classified into just two categories. When the outcomes fall into more than two categories, a new statistic can be calculated to replace K; it is called the PEARSON χ^2 (small Greek chi, squared) STATISTIC, and tests based on this statistic are called χ^2 (read "chi-square") tests.

You have already seen in Lesson 10 that the concept of trials can be very broadly interpreted to cover N items on a test or N different individuals. Since you understand why these trials, items, or persons must be independent in the binomial case, keep in mind that the same requirement of independence applies to the χ^2 test. Think of the χ^2 test as an extension of the binomial test to problems involving more than two outcomes per trial.

The binomial test enables us to decide whether some hypothesis about the value of p is acceptable in the light of our evidence. We do not have to say anything specific about q in the hypothesis because $q = 1 - p$. As we go to independent trials with three outcomes per trial, our assumed hypothesis must cover at least two of these three outcomes. It is conventional to put the hypothesis in terms of EXPECTED FREQUENCIES rather than probabilities, but this is merely a change of wording. When we hypothesize that $p = 1/2$ in a 100-trial coin-toss experiment, our hypothesis implies that the expected frequency of "heads" is 50.

We shall use C to represent the number of categories into which outcomes are classified. In a binomial test, $C = 2$. We state an explicit hypothesis about the probability (or expected frequency) of only $C - 1 = 1$ of these categories, and we say that our test has 1 DEGREE OF FREEDOM (df). When $C = 3$, we must state an explicit hypothesis about the expected frequency of $C - 1 = 2$ of these categories; the last expected frequency will be determined by subtracting the sum of all the others from N, the number of independent trials. The test thus has $C - 1 = 2$ degrees of freedom.

The Pearson χ^2 statistic is based on the difference between *expected* frequency f_e and *observed* frequency f_o in each of the C categories. As the difference $f_e - f_o$ grows large, it becomes difficult to believe the hypothesis on which f_e is based. Consider that the N trials are a random sample from a population which contains elements belonging to each of C categories; the proportion of C_1 in this population is the same as f_e/N for C_1, the proportion of C_2 in the population equals f_e/N for C_2, and so on. Any random sample from such a population will depart in some measure from the expected frequencies; it will not mirror perfectly the proportions in the population from which it is drawn. But of course it will usually not deviate too far from the population proportions. In order to make a statistical test, we need a probability distribution conditional upon the expected (or population) frequencies. This distribution will show how often we can expect the observed frequencies in a random sample to deviate from expectation by a particular amount.

A. THE PEARSON χ^2 STATISTIC

At the time of the 1960 census, 33 percent of the families in California had one child, 33 percent had two children, 20 percent had three, and 14 percent had four or more. A study of 2,290 juveniles referred to probation officers for various offenses during a single year in one county showed that their families had a very different distribution on this variable. Of these families, 5 percent had one child, 14 percent had two, 19 percent had three, and 62 percent had four or more. If this sample is randomly drawn from the general population of the state, we would expect to find that the relative frequency of families falling into each of these four categories would not differ greatly from the relative frequency for the state as a whole.

H_1 is the hypothesis that the sample is indeed randomly drawn from the state population. H_2 is the hypothesis that the juvenile offenders come from a population which differs significantly from the state population with regard to family size. To find the probability of drawing such a sample at random from the state population, we determine the statistic

$$\chi^2 = \sum \frac{(f_e - f_o)^2}{f_e}$$

Table 11-1 shows the calculation of χ^2 for this example. The first two columns show the percentages already quoted, stated as proportions. To obtain f_e the proportions which characterize the state (the hypothetical parent population) are multiplied by $N = 2{,}290$; this is the number of independent trials (families of juvenile offenders) in our sample. To obtain f_o the proportions which characterize the sample are multiplied by N. The further steps are easy to follow. Notice that the quantity $(f_e - f_o)^2/f_e$ is calculated for each category; this quantity is the CONTRIBUTION of that category to χ^2. The category "three children" makes a negligible contribution, while "four or more" makes an enormous contribution to χ^2.

Table 11-1 Calculation of χ^2 for Comparison on Family Size†

State	Sample	f_e	f_o	$f_e - f_o$	$(f_e - f_o)^2$	$\frac{(f_e - f_o)^2}{f_e}$
.33	.05	.33(2,290) = 755	115	640	409,600	54,252
.33	.14	.33(2,290) = 755	320	435	189,225	25,063
.20	.19	.20(2,290) = 458	435	23	529	1
.14	.62	.14(2,290) = 322	1,420	−1,098	1,205,604	374,411
						$\Sigma = 453{,}727 = \chi^2$

† Data from A. W. McEachern et al., The Juvenile Probation System, *Am. Behav. Sci.*, **12:**10 (1968)

11-1 The Pearson χ^2 test is an extension of the binomial test to problems with ________ outcomes per trial. *more than two*

11-2 Just as in the binomial test, it is essential that all the N trials should be ________ of each other. *independent*

11-3 In the binomial test H_1 tells the probability of one of the two outcomes. In the χ^2 test H_1 indicates the probabilities of the C outcomes, but these probabilities are expressed as ________ frequencies. *expected*

11-4 In Table 11-1 the probability under H_1 of drawing a family with just one child is ________. This probability is taken from the empirical relative frequency of such families in the assumed parent ________. *.33* *population*

11-5 For a χ^2 test this probability is converted to an ________ through multiplying it by ________. *expected frequency; N*

11-6 In the assumed parent population, 33 percent of the families are one-child families. If we were to draw a large sample of 2,290 families at random from this population, the expected number of one-child families in the sample would be ________. *755*

11-7 This expected frequency is like the expectation $E(K)$ in a binomial test. We (*do; do not*) actually expect to get $E(K) = Np$ heads on every N-trial coin-toss experiment. *do not*

11-8 Similarly, we do not expect to get $f_e = 755$ one-child families in every random sample of 2,290 families. But f_e, like $E(K)$, would be our ________ guess in a long series of guesses about the number of one-child families in such samples. *best*

11-9 Our long-run error will be less if we always guess ________ than if we guess any other number. f_e

11-10 The observed frequency of one-child families is 2,290 times ________, or 115. The difference $f_e - f_o$ for this category is ________. *.05* *640*

11-11 We expect 755 such families, but we observe ________ in our actual sample. We want to know the probability that a sample from this population would show such a ________ deviation from expectation. *115* *large*

Let us now look more closely at the χ^2 statistic,

$$\chi^2 = \sum \frac{(f_e - f_o)^2}{f_e}$$

The term $f_e - f_o$ reflects the size of the difference, for a given category, between an "average" random sample from the population and our observed sample. Since we want a numerical index of the amount of deviation from expectation, this term is not a bad beginning. Note carefully the following characteristics of the χ^2 statistic.

1 χ^2 is a *nondirectional* index. The deviation $f_e - f_o$ is directional; some of these deviations are positive and some are negative. In Table 11-1 you can verify that the sum of the negative deviations equals the sum of the positive deviations. Both Σf_e and Σf_o must equal N; therefore, $\Sigma(f_e - f_o)$ must equal zero. You will learn in Lesson 13 that $\Sigma(f_e - f_o) = \Sigma f_e - \Sigma f_o$.

As soon as we square this term, we remove all evidence about the direction of the deviation. Positive deviations will count just the same as negative deviations. The statistic is clearly going to be a nondirectional index of deviation from expectation. Because χ^2 is a nondirectional index, it is suitable only for nondirectional tests. The χ^2 statistic itself cannot be used to make a directional (one-tailed) test of H_1.

2 χ^2 *emphasizes large deviations* from expectation. By squaring $f_e - f_o$ we exaggerate the effect of large deviations. Table 11-1 illustrates how the large deviations receive extra weight through this squaring operation.

3 χ^2 is a sum of *mean squared deviations* from expectation. When $(f_e - f_o)^2$ is divided by f_e, the squared deviation for a particular category is converted into a mean-squared deviation for that category. In this way the χ^2 statistic takes into account the difference in f_e between categories and the difference in N between experiments. For example, in Table 11-1 f_e for one-child families is 755; dividing 409,000 by 755 gives the average contribution made by each of the families expected to fall into that category. With $f_e = 75$ instead of $f_e = 755$, a much smaller squared deviation would produce the same average contribution per family and the same contribution of the category to χ^2.

When you memorize the χ^2 definition, don't worry about the order of terms in the numerator. A deviation $(f_o - f_e)^2$ will be numerically exactly the same as $(f_e - f_o)^2$. *Do* worry about getting the right term (f_e) in the denominator. Remember that you are testing a hypothesis and that f_e is the frequency which represents that hypothesis when it is applied to a particular category.

Finally, let us notice that the size of χ^2 will depend directly on the number of categories C. The more categories there are, the more contributions there will be to the value of χ^2 and the larger the value of χ^2 is likely to become. We need a way to take the number of categories into account when we evaluate the probability of χ^2. This is done in a very simple way. We have to consult a random sampling distribution, found in a table, in order to find the probability of a certain χ^2 value. If we increase the number of categories, we merely consult a *different* RSD.

11-12 Compare the categories in Table 11-1 which make the largest and the smallest contributions to χ^2. The deviation $f_e - f_o$ for category 3 is only ________; the deviation for category 4 is ________.

23
−1,098

11-13 When $f_e - f_o$ is squared, the contrast between categories 3 and 4 becomes (*less; more*) striking. Squaring exaggerates the influence of relatively (*large; small*) deviations.

more
large

11-14 Category 3 has a (*larger; smaller*) expected frequency than category 4. When $(f_e - f_o)^2$ is divided by f_e, the contribution of category (*3; 4*) to χ^2 becomes negligible.

larger
3

11-15 The number of families expected to be in category 4 is ________. The squared deviation for this category is ________. The contribution to χ^2 is the ____________ squared deviation for each family expected to be in the category.

322; 1,205,604
mean

11-16 The size of a category's contribution to χ^2 depends on two factors. Increases in the absolute size of $f_e - f_o$ will make the contribution ____________, while increases in the size of f_e will make its contribution ____________.

larger
smaller

11-17 Some values of $f_e - f_o$ are positive, and some are negative. $\Sigma(f_e - f_o) =$ ________, because $\Sigma f_e = \Sigma f_o =$ ________.

0; N

11-18 The values of $f_e - f_o$ contain information about the direction of the deviation from expectation. This information is lost when $f_e - f_o$ is ____________.

squared

11-19 The statistic χ^2 is a ____________ of mean squared deviations. It lumps together deviations which were originally positive with deviations which were originally ____________.

sum
negative

11-20 χ^2 is an index of *amount* of deviation from expectation, but it does not contain any information about the ____________ of particular deviations from expectation.

direction

11-21 The statistic z (*does; does not*) contain information about the direction of deviation from expectation.

does

11-22 z is a directional index of amount of deviation from expectation. χ^2 is a ____________ index of ____________ of deviation from expectation.

nondirectional; amount

11-23 The χ^2 statistic is a sum of ________ terms, one for each of the possible ____________ of a single "trial." Each term is a ____________-squared deviation from ____________.

C
outcomes
mean; expectation

B. DISTRIBUTIONS OF CHI SQUARE

The random sampling distributions for the Pearson χ^2 statistic are complex, and in practice they are never calculated exactly. Instead, an approximation is used. When N is large, the RSD of χ^2 is very similar to the RSD of a variable whose properties are easier to study mathematically. This variable is also called CHI SQUARE; we shall spell out its name in order to distinguish it clearly from the Pearson χ^2 *statistic*.

The chi-square *variable* has a special relation to the normal distribution. Imagine that you draw z scores, one at a time, from a very large normally distributed population. You will expect about two-thirds of the z scores you draw to have values between -1 and $+1$, about 95 percent to have values between -2 and $+2$, and so on. If you square each randomly drawn z score and form a new distribution, you will obtain a distribution of z^2 values whose probabilities are described by the first row of Table 11-2.

Look at the value listed in the column 0.10; 10 percent of the values in your z^2 distribution will be as large as 2.71 or larger, according to this table. Now in order to get a z^2 of 2.71 or greater, you will have to draw a z of $|1.65|$ or greater ($\sqrt{2.71} = 1.65$). A quick check of the normal-curve table (Table A-1) will remind you that the probability of getting z as large as $+1.65$ is .05; the probability of getting z as small as -1.65 is .05. Therefore, the probability of a z^2 as great as 2.71 is indeed .10. Every value in the first row of this table can be obtained in this way from the normal-curve table. This line of the table describes critical values in the probability distribution of randomly selected z^2 values.

The second row of Table 11-2 also describes a probability distribution. This one could be obtained by drawing two z scores randomly, squaring each of them, and taking the sum. The table indicates that 10 percent of the values in such a distribution will equal or exceed 4.60. Put another way, the probability is .10 that two z scores drawn randomly and squared will add up to a value at least as great as 4.60. The other rows of the table have a similar interpretation; they represent distributions of the sums of three and four squared z scores, respectively.

A variable which is the sum of some number J of squared z scores is called a CHI-SQUARE VARIABLE. The first row of Table 11-2 describes the probability distribution for a chi-square variable with $J = 1$, the second row describes the distribution for a chi-square variable with $J = 2$, and so on.

When N is large, a Pearson χ^2 statistic with J degrees of freedom has an RSD approximately like the probability distribution of the chi-square variable with that value of J. The χ^2 statistic for our example in Table 11-1 has 3 degrees of freedom, and N is large. Therefore, the RSD of this χ^2 statistic is approximately like the distribution in row 3 of Table 11-2, the distribution for a chi-square variable with $J = 3$.

We could use the normal-curve approximation for binomial tests only when N is large. Remember that we use the chi-square approximation for χ^2 tests, as well, only when N is large.

Table 11-2 Critical Values of Chi Square

df	Probability of Obtaining Chi Square Greater than or Equal to the Value Entered in the Table									
	0.99	0.95	0.90	0.70	0.50	0.30	0.10	0.05	0.01	0.001
1	0.00016	0.0039	0.016	0.15	0.46	1.07	2.71	3.84	6.64	10.83
2	0.02	0.10	0.21	0.71	1.39	2.41	4.60	5.99	9.21	13.82
3	0.12	0.35	0.58	1.42	2.37	3.66	6.25	7.82	11.34	16.27
4	0.30	0.71	1.06	2.20	3.36	4.88	7.78	9.49	13.28	18.46

11-24 The first row in Table 11-2 is listed as df = ______. It represents the probability distribution for the squares of single z scores randomly drawn from a ______ distribution.

1

normal

11-25 The row for df = 2 represents the probability distribution for the sum of ______ squared ______, each drawn randomly and independently from a normal distribution.

2; z scores

11-26 The sum of 2 squared z scores, randomly drawn, is called a chi-square ______ with J = ______. Because the 2 z scores are drawn randomly and *independently,* there are J = ______ degrees of freedom (df).

variable; 2

2

11-27 When N is ______, a Pearson χ^2 statistic with 3 degrees of freedom will have an RSD approximately like the distribution of a ______ variable with $J = 3$.

large

chi-square

11-28 When $J = 2$, the probability of obtaining a value as great as 5.99 for a chi-square variable is ______. The probability is .001 for obtaining a value as great as ______.

.05

13.82

11-29 If a Pearson χ^2 statistic has 4 degrees of freedom, the critical value of χ^2 for $\alpha = .05$ is ______. We can reject H_1 at this level if χ^2 is at least as ______ as this value.

9.49

large

11-30 A χ^2 statistic with C categories has $C - 1$ degrees of freedom. Our example has 4 categories; therefore, its df = ______.

3

11-31 By consulting the table for df = 3, we find that χ^2 must be at least ______ in order to meet $\alpha = .01$.

11.34

11-32 χ^2 for our example is 453,727. The largest χ^2 in the table for df = 3 is ______, with p = ______. Our χ^2 is therefore significant far beyond the ______ level.

16.27; .001

.001

C. χ^2 IN OTHER ONE-SAMPLE PROBLEMS

The analogy between the binomial test and χ^2 is closest in the one-sample case, where the observed frequencies in a single sample are compared with the frequencies in some known population. We have already worked out an example of a one-sample problem in Secs. A and B.

Here is another one-sample problem which illustrates an important restriction on the use of χ^2. The χ^2 test should not be used when any one of the expected frequencies is less than 5; this is a kind of rule of thumb, intended to guard against use of the χ^2 test when N is not large enough to permit the use of the chi-square approximation.

Table 11-3 Observed Frequencies of IQ Scores in a Tenth-Grade Sample

	Below 68.1	68.1–84	84.1–100	100.1–116	116.1–132	Above 132	Total
f_o	0	8	25	45	17	5	100
f_e	2.3	13.6	34.1	34.1	13.6	2.3	100

A sample of IQ scores from 100 tenth-grade students shows the distribution of observed frequencies in Table 11-3. The hypothesis to be tested is that this sample is a random sample from the general population, in which IQ is distributed normally around a mean of 100 with a standard deviation of 16. The expected and observed frequencies are grouped into 6 categories which represent intervals of 1 standard deviation. On H_1 we expect 34.1 scores to fall between 100.1 and 116 because this is the region between the mean and $+1\sigma$; we expect 13.6 to fall between 116.1 and 132; and so on.

Two of these f_e's are below 5, and χ^2 cannot be used to test H_1 as this table now stands. However, the low f_e's do not result from a low N in this case; they result from the way the 6 categories have been chosen. These categories are in standard deviation units; the proportion of a normal distribution which lies beyond $z = +3$ or $z = -3$ is very small. Table 11-4 shows the frequencies rearranged into 6 categories which have been chosen so that f_e will be 100/6 for each category. By restating the problem in this way, the RSD of our χ^2 statistic (with 5 df) will be satisfactorily approximated by the probability distribution of a chi-square variable for $J = 5$. This approximation is best when all the f_e's are equal.

Sometimes the problem of f_e's below 5 is solved by combining 2 small categories into 1. For example, we might combine the 2 lowest categories in Table 11-3 into one and the 2 highest categories into another; we would then calculate a χ^2 statistic with df $= 3$. This way of avoiding f_e's below 5 is not recommended. Since we have to see the data *before* choosing such categories, we can no longer claim that only chance has determined the way the observations fall into categories (assuming that H_1 is true). Categories should always be chosen in advance, and the Pearson χ^2 test is most satisfactory when (1) all f_e's are approximately equal and (2) N is at least large enough to ensure that no f_e will be less than 5.

Table 11-4 Data From Table 11-3 Rearranged to Give Categories with Equal Expected Frequencies

	Below 84.4	84.5–93	93.1–100	100.1–107	107.1–115.6	Above 115.6	Total
f_o	8	10	15	26	19	22	100
f_e	$\frac{100}{6}$	$\frac{100}{6}$	$\frac{100}{6}$	$\frac{100}{6}$	$\frac{100}{6}$	$\frac{100}{6}$	100

11-33 In Table 11-3 there are two f_e's of 2.3; χ^2 cannot be used when there is any f_e with a value less than ________. *5*

11-34 This rule of thumb helps to prevent use of χ^2 when ________ is not large enough to make the chi-square distribution a good approximation to the RSD of χ^2. *N*

11-35 Furthermore, this approximation is best when all the ________ frequencies are approximately equal. *expected*

11-36 An investigator planning to use χ^2 should therefore set up his categories, if he can, so that the f_e's in his data table will be as nearly ________ as possible. *equal*

11-37 In the IQ problem, he could have chosen categories in this way at the outset, before seeing his data. He could have set the IQ limits of his 6 categories so that each category would have $f_e =$ ________, as in Table 11-4. $\frac{100}{6}$

11-38 For a proper test of H_1, the choice of categories must always be made (*after; before*) the data are seen. *before*

11-39 Work out χ^2 according to Table 11-4. Keep all calculations in fractions; for "Below 84.4" $f_e - f_o = {}^{100}/_6 - {}^{48}/_6 =$ ________. Then $(f_e - f_o)^2 =$ ________, and $(f_e - f_o)^2/f_e =$ ________. *52/6; 2704/36* *27.04/6*

11-40 Find the contributions of the remaining 5 categories. The value of χ^2 is ________. It has ________ df. *14.6; 5*

11-41 Turn to Table A-2 (page 352). In the row for df = 5, the critical value just *smaller* than 14.6 is ________, with a probability of ________. *12.832* *.025*

11-42 The probability of such a large value of χ^2 as 14.6, when df = 5, is ________ than .025. *less*

D. χ^2 WITH MORE THAN ONE SAMPLE

You have already been introduced to a two-sample problem. The college-height case is such a problem; two samples, one from each of two colleges, are compared with each other in order to decide whether a difference between them is significant. The case of vocabulary-learning methods (Lesson 1) is a *three*-sample problem in which the decision again concerns the differences among actually observed samples.

Both these examples involve quantitative data and are best handled in terms of sample means. Here is a two-sample problem similar to the college-height case, but it involves data which will be treated as qualitative. When decisions must be made about the differences between two or more samples of categorical data, χ^2 may be a suitable test.

A sociologist has observed with some interest that there seem to be periodic changes in the attitudes of psychologists with respect to the "nature-nurture" controversy. In the recent past, most psychologists argued that human personality is almost entirely determined by environmental rather than by genetic factors, but he notes that many of the younger psychologists, perhaps especially the women, now argue for a major influence from biological inheritance. To test his idea, he devises a 100-point questionnaire; persons scoring above 50 are categorized automatically as "environmentalist," and persons scoring 50 or below are categorized as "nativist." He gives this questionnaire to samples of 24 young women and 30 young men, all with doctorates in psychology. Notice that these are independent (not related) samples.

Table 11-5 shows the frequencies of men and women found in each of the two categories. We wish to hypothesize that the men's and women's samples really come from the same population with respect to the variable "environmentalism-nativism." If so, what proportion of nativists does that population contain? Our best estimate comes from the column totals; 27 of the 54 persons tested fall into the "nativist" category. We get f_e for "nativist men" by applying this proportion, 27/54, to the 30 men studied; $f_e = {}^{27}/_{54}(30) = 15$. Since the number of men studied is a row total, we can write a rule for finding f_e for any cell:

$$f_e = \frac{\text{column total times row total}}{N}$$

Since the row and column totals are called the MARGINAL TOTALS, we multiply together the two marginal totals for that cell and divide by N.

Let us look at this rule in another way. When we apply such a rule, we are regarding the marginal totals as *fixed*. The set of expected frequencies must have the same marginal totals as the set of observed frequencies. The rule f_e = (marginal totals product)/N gives a way of stating the hypothesis "Men and women do not differ on this variable," and the statement applies to an experiment with exactly 30 men and 24 women, 27 nativists and 27 environmentalists.

This table is a 2×2 table. There are only two samples, each of which has been dichotomized into only 2 categories. The rule for finding f_e applies to any $R \times C$ table, that is, any table with R rows and C columns.

Table 11-5 Artificial Data For the Environmentalism-Nativism Example

	Nativists	Environmentalists	Total
Men	11	19	30
Women	16	8	24
Total	27	27	54

11-43 To find f_e for "nativist men," we first found the row and column totals for that cell. These totals are called its ________ totals. *marginal*

11-44 We then took the ________ of these two marginal totals divided by ______. *product* *N*

11-45 Since f_e for "nativist men" is 15, f_e for "environmentalist men" must be ________ minus 15, or ______, and f_e for "nativist women" must be ______ minus 15, or ______. *30; 15* *27; 12*

11-46 Therefore, f_e for "environmentalist women" is 24 minus ______, or ______. *12; 12*

11-47 In this 2×2 table, the rule $f_e =$ (marginal totals product)/N needs to be applied to only ______ of the 4 cells. The other f_e's can be determined by subtraction from the appropriate ________ total. *1* *marginal*

11-48 Since we need only one f_e to determine all four f_e's, this 2×2 table has only ______ degree of freedom. *1*

11-49 For the upper left cell, $f_e - f_o$ is equal to ______. In fact $|f_e - f_o|$ is ______ for every cell in this 2×2 table. *4* *4*

11-50 Before squaring $f_e - f_o$, we must subtract 0.5; we do this for any χ^2 test with only 1 df. (The reason will be given on page 172.) Each $|f_e - f_o|$ becomes ______ after this correction. *3.5*

11-51 Write down the contributions of the four cells to χ^2, and find χ^2 for this problem.

Nativist men	______	*.817*
Environmentalist men	______	*.817*
Nativist women	______	*1.025*
Environmentalist women	______	*1.025*
$\chi^2 =$	______	*3.684*

11-52 The critical value of χ^2 for 1 df at $\alpha = .05$ is ______. We (*can; cannot*) reject H_1 at this α level. *3.84* *cannot*

Table 11-6 χ^2 Applied to a Coin-Toss Problem

Category	f_e	f_o	$(f_e - f_o) - 0.5$	$\frac{[(f_e - f_o) - 0.5]^2}{f_e}$
Heads	18	23	4.5	1.125
Tails	18	13	4.5	1.125
				$\chi^2 = 2.25$

A χ^2 with only 1 degree of freedom can be directly compared with a binomial test. The comparison will help you understand better the close relationship between χ^2 and z.

In a coin-toss experiment with $N = 36$, let K_o be 23. By the binomial test for $N = 36$ and $p = 1/2$, we would find $E(K) = 18$, $\sigma = 3$, and $z = 4.5/3 = 1.5$, with the usual correction for continuity. The square of this z score is 2.25. Table 11-6 shows that this z^2 is exactly equal to the Pearson χ^2 statistic for these same data *provided that* the correction for continuity is also applied in calculating χ^2. Whenever df = 1, this equality between χ^2 and z^2 is present, and one can use the normal-curve table (looking up $\sqrt{\chi^2} = z$) instead of the χ^2 table for df = 1. This relationship has been pointed out for two reasons.

1 It will help you understand why the correction for continuity must always be made in calculating the χ^2 statistic when df = 1. You understand the reason for this correction in the binomial case. Because χ^2 with df = 1 is mathematically the same as a binomial test, the correction must be applied in calculating χ^2; each $f_e - f_o$ must be reduced by 0.5 before it is squared.

2 It will enable you to use a one-tailed test when you have a χ^2 statistic with df = 1. Since the sociologist in our example *expected* women to turn out more nativist than men, a directional H_2 would be defensible in his study. He can take $z = \sqrt{\chi^2} = 1.92$ to the normal-curve table for such a test. Since $p = .0274$ for this one-tailed test, H_1 can now be rejected at $\alpha = .05$; H_1 could not be rejected by the nondirectional χ^2 test at this level.

χ^2 for df = 1 is thus a very special case. While you are learning these unusual features, add one more: when χ^2 has df = 1, the minimum f_e permissible is *10*; with df > 1, an f_e as low as 5 is permissible.

Not all problems involving two independent samples have df = 1, only those in which each sample is dichotomized into just 2 categories. Table 11-7 presents a two-sample problem with 2 df. The data are from a hypothetical experiment in which a new drug treatment is tried on randomly selected hospital patients. Some patients are randomly selected to receive the drug; the remainder receive a neutral agent (a sugar pill) instead of the new drug. Following the period of drug administration, doctors give each of the patients a designation "improved," "slightly improved," or "not improved." The two groups are similar in severity and duration of illness, and neither the patients nor the doctors know which persons are receiving the drug and which are not.

The χ^2 statistic, when it is based on a data table with R rows and C columns, will have df = $(R - 1)(C - 1)$. Thus χ^2 for Table 11-7 has 2 df.

Table 11-7 Frequencies of Patients Improving with and without a New Drug Treatment ($N = 108$)

Group	Number Improved		Number Slightly Improved		Number Not Improved		Total
Drug	20	______	25	______	8	______	53
Control	6	______	12	______	37	______	55
Total	26		37		45		108

11-53 Enter the expected frequencies for Table 11-7 in the spaces. Only the first (*1; 2; 3*) f_e's have to be determined as proportions. The remainder can be obtained by subtraction from the ________ totals.

2

marginal

11-54 The number of df is therefore ______. Does this fit the rule $(R-1)(C-1)$? ________, since $R-1=$ ______ and $C-1=$ ______.

2
Yes; 1
2

11-55 In Table 11-5 (page 171) $R-1=$ ______ and $C-1=$ ______. The number of df in that case is ______.

1
1; 1

11-56 When df = 1, the χ^2 test is equivalent to a ________ test, and the correction for ________ must be used.

binomial
continuity

11-57 To make a correction for continuity, the quantity ______ is subtracted from each value of ______.

0.5; $f_e - f_o$

11-58 In Table 11-7, above, there are ______ contributions to χ^2 which must be calculated. The correction for continuity (*is; is not*) applied because df = ______.

6

is not; 2

11-59 Check the values of f_e you have obtained for Table 11-7 by comparing them with the following list. Then add up the contributions (calculation of the contributions has been done for you), finding χ^2 and its probability.

	f_e	f_o	$\frac{(f_e - f_o)^2}{f_e}$
Row 1	12.7	20	4.11
	18.2	25	2.58
	22.1	8	8.98
Row 2	13.3	6	3.96
	18.8	12	2.48
	22.9	37	9.65

$\chi^2 =$ ______ ; df = ______ ; $p <$ ______

30.76; 2; .001

REVIEW

11-60 In a χ^2 test f_e represents __________, and f_o represents __________.

expected frequency
observed frequency

11-61 f_e for a given category is the number of cases expected to fall into that category when ________ is true. It is an expectation, and like $E(K)$ in the binomial test it is a __________ at the number expected when a long series of guesses must be made.

H_1
best guess

11-62 χ^2 is an index of amount of deviation from expectation. Its calculation begins with the deviation ________.

$f_e - f_o$

11-63 We have written $f_e - f_o$, but the deviation $f_o - f_e$ would serve just as well because the deviation is __________ in calculating χ^2.

squared

11-64 The squared deviation for a category is divided by ________ for that category, giving an average squared deviation per case (*expected; observed*) to fall in that category.

f_e
expected

11-65 χ^2 is a __________ of these mean __________ deviations. It thus contains information about the __________ of deviation from expectation but not about its __________.

sum; squared
amount
direction

11-66 The χ^2 test is an extension of the __________ test for use when the outcomes of independent trials fall into more than ________ categories.

binomial
2

11-67 The relation of χ^2 to the binomial test should be remembered because it will help you recall that in both tests the N trials must be __________ of each other.

independent

11-68 The relation is also important because χ^2 with df = ________ is mathematically equivalent to a binomial test.

1

11-69 χ^2 with df = 1 has two peculiarities. It requires a correction for __________, and it requires a minimum f_e of ________. With df > 1 the minimum permissible f_e is __________.

continuity; 10
5

11-70 The correction for continuity is made by subtracting ________ from every value of ________.

0.5; $f_e - f_o$

11-71 χ^2 with df = 1 arises when ________ samples are compared and when each is divided into just ________ categories.

2
2

11-72 The probability distribution for χ^2, when ________ is large, is approximately the same as the probability distribution of a __________ variable.

N
chi-square

11-73 A Pearson χ^2 with df = 1 has an RSD approximately the same as a probability distribution of the squares of randomly drawn ________. *z scores*

11-74 When N is large, a Pearson χ^2 with J degrees of freedom distributes like the sum of ______ randomly drawn squared z scores. *J*

11-75 In a one-sample problem with C categories, the number of df is ______. *$C-1$*

11-76 With R independent samples and C categories, the number of df is ______. *$(R-1)(C-1)$*

11-77 χ^2 can be used to test the difference among samples only when the samples are ________. *independent*

11-78 This requirement arises because χ^2, like the binomial test, applies only when the N trials are all ________. *independent*

11-79 With R independent samples and C categories, we shall have a χ^2 table with ______ row totals and ______ column totals. *R; C*

11-80 These row and column totals are called ________ totals. They are the same for expected frequencies as for ________ frequencies. *marginal* *observed*

11-81 To find f_e for any cell, we take the product of its ________, then divide that product by ______. *marginal totals; N*

11-82 However, we do not need to use this rule in finding every f_e in a table. In a 2×2 table we need to use it for only ______ f_e. The remaining f_e's can be found by ________ from the marginal totals. *1* *subtraction*

11-83 In a 3×5 table, the number of df is ______, and we must use the rule for finding only the first ______ of the 15 f_e's. *8* *8*

11-84 The table used in evaluating the Pearson χ^2 statistic is a table of the distribution of a ________ variable with J degrees of freedom. *chi-square*

11-85 This table is a good approximation to the distribution of the Pearson χ^2 statistic when ______ is large and when all the values of $(f_e; f_o)$ are approximately equal. *N* *f_e*

PROBLEMS

1. A college suspects that its rising tuition rate has changed the composition of its student body with respect to economic level. It classifies the students of the class of 1972 into three groups according to their fathers' incomes (high, middle, low), and, taking into account the nationwide changes in salary levels during the intervening 10 years, it classifies the students of the class of 1962 into three comparable groups. Using χ^2, determine whether it is possible to reject at the .05 level the hypothesis that these two classes came from the same population with respect to relative economic level.

	Fathers' Incomes			
Class	High	Middle	Low	Total
1962	450	750	400	1,600
1972	1,050	1,650	500	3,200
Total	1,500	2,400	900	4,800

2. Calculate χ^2 for the following 2×2 table. How many degrees of freedom does this table have? Does any problem arise in the use of χ^2 because of the low frequencies in two of the cells?

	A_1	A_2	Total
B_1	3	17	20
B_2	22	8	30
Total	25	25	50

3. Out of 120 children treated in a mental hygiene clinic, 32 were the oldest children in a family, 29 were youngest children, 27 were middle children, and 32 were only children. Compute χ^2 to test whether these four categories can be considered to occur equally often among the clinic's patients. How many df does this χ^2 have? Is it significant at the .05 level?

4. A test designed to measure "awareness of social science principles" is administered to the students of a college, and the frequencies of high, middle, and low scores in the four college classes are shown below. On the hypothesis that attendance through 4 years at that college makes no difference in scores on this test, the proportions of high, middle, and low scorers should be the same for all four classes. Find and evaluate χ^2 with $\alpha = .05$.

Score	Freshmen	Sophomores	Juniors	Seniors	Total
High	60	120	200	220	600
Middle	100	180	110	110	500
Low	240	90	50	20	400
Total	400	390	360	350	1,500

LESSON 12 MEASURING STRENGTH OF ASSOCIATION

When we conclude that a difference is statistically significant, we have simply ruled out the possibility that an observed difference between samples is really due to sampling variability. A statistically significant difference is evidence for the *existence* of an association between variables. We shall now see what can be done to measure the *strength* of that association.

The measures available for qualitative data are based on the idea of CONTINGENCY between two variables. Two variables are contingent upon each other whenever they are not fully independent. We can measure the degree of contingency by measuring the degree of departure from complete independence of the two variables. The calculation of this measure begins with a CONTINGENCY TABLE, such as Table 11-5 (page 171). The rows in that table are the two values of the variable "sex"; the columns are the two values of the variable "environmentalism-nativism." From the distribution of frequencies in the four cells of this 2×2 contingency table, we have already derived a χ^2 test of significance. We shall presently use the same frequencies to measure the degree of departure from independence of these two variables.

We can also measure degree of contingency in another way. When two variables are not wholly independent, it is often possible to predict standings on one variable from a knowledge of standings on the other variable. In the second part of this lesson, we shall show how strength of association can be measured in terms of predictive power. This measure also requires data in the form of a contingency table.

A contingency table can be drawn up only when the observations form at least two samples. Binomial tests involve either single samples or two related samples which have been converted into one sample for a sign test. Therefore the measures of strength of association we are about to discuss will not apply to data in the form suitable for binomial tests. Neither do they apply to one-sample χ^2 problems such as the one discussed in Lesson 11C.

We bother to look into these measures of association strength because a difference can be statistically significant without being worthy of further attention. There are times when proof of any degree of association, however weak, is exceedingly important. This is true of our color-touch example and of ESP experiments; any significant difference at all is unexpected and important because of its effect on scientific theory. Most of the time, though, it is fair to say that strong associations have much greater scientific interest than weak ones. Every investigator should feel obligated to report on the strength of whatever associations he has been able to find.

Table 12-1 Contingency Table with *A* and *B* Completely Independent

	A. Frequency Distribution†			B. Products of Conditional Probabilities $p(A \mid B)p(B \mid A)$		
	A_1	A_2	Total	A_1	A_2	Total
B_1	6 (.10)	18 (.30)	24 (.4)	.25(.4) = .10	.75(.4) = .30	.4
B_2	9 (.15)	27 (.45)	36 (.6)	.25(.6) = .15	.75(.6) = .45	.6
Total	15 (.25)	45 (.75)	60 (1.0)	.25	.75	1.0

† Joint and marginal probabilities in parentheses.

A. MEASURING DEPARTURE FROM INDEPENDENCE

We shall first take up a measure which is based upon the concept of conditional probability. We are interested in measuring strength of association between two variables, A, the column variable in a contingency table, and B, the row variable. If there is no association at all between A and B, we say that these variables are independent of each other; changes in one do not affect the values of the other. When some association exists, there is some degree of DEPARTURE FROM INDEPENDENCE. We would like to find an index of the degree of this departure from independence; the index should equal 0 when A and B are fully independent, and it should equal 1 when they are completely dependent.

Observe that we are now talking about independence of two *variables*. In Lessons 10 and 11 we spoke a great deal about the independence of *samples* and of the observations they contain.

Recall that $p(A \mid B)$ is the conditional probability of A, given B. If $p(A \mid B)$ is the same as $p(A)$, then B makes no difference in A, and A is independent of B. Similarly, if $p(B \mid A)$ is the same as $p(B)$, A makes no difference in B, and B is independent of A. We can say that A and B are independent variables only if *both* these conditions are true, i.e., if $p(A \mid B) = p(A)$ and $p(B \mid A) = p(B)$.

We want to apply these conditions to a contingency table. Let us begin with the simplest kind, a 2×2 table with only four cells. Part A of Table 12-1 is a 2×2 table in which the variables A and B are fully independent. The cell frequencies show how the 60 observations are distributed; joint and marginal probabilities, drawn from these frequencies, are given in parentheses. Notice that the distribution of frequencies within each row is exactly like the distribution of column totals (1/4 in A_1 and 3/4 in A_2); furthermore, the distribution of frequencies within each column is like that of the row totals (.4 and .6).

Part B of Table 12-1 shows the products $p(A \mid B)p(B \mid A)$ for each cell. In each case, $p(A \mid B)p(B \mid A) = p(A)p(B)$, the product of the two marginal probabilities for that cell. The *sum* of these four products is 1.0. Whenever A and B are independent, the sum of the products $p(A \mid B)p(B \mid A)$ will be 1. This sum will be 1 for *any* contingency table with A and B independent, whatever the number of cells in the table.

12-1 The contingency table in part A of Table 12-1 has 2 variables designated as ______ and ______. Each variable has ______ values.

A; B
2

12-2 Two variables are contingent when they are not independent of each other. We are going to measure degree of contingency between A and B by finding the degree of departure from complete ______.

independence

12-3 Since the variables in part A of Table 12-1 are designed to be completely independent, the degree of departure from independence should equal ______ in this case.

0

12-4 Cell A_1B_1 contains ______ of the 60 observations. The joint probability for this cell is therefore 6/60 = ______. This figure appears in parentheses in part A of Table 12-1.

6
.10

12-5 To find the conditional probability $p(A \mid B)$ for cell A_1B_1, we divide the cell frequency 6 by the number of observations in that *row,* which is ______. $p(A \mid B)$ for this cell is ______.

24; .25

12-6 What is the probability of drawing at random, from these 60 observations, an observation in cell A_1B_1? The probability is ______; this is the *joint* probability for cell A_1B_1.

6/60 = .10

12-7 What is the probability of drawing at random, from the 24 observations in row B_1, an observation in cell A_1B_1? The probability is ______; this is the ______ probability of A given B for that cell.

.25; conditional

12-8 What is the probability of drawing at random, from the 15 observations in column A_1, an observation in cell A_1B_1? The probability is ______; this is the conditional probability of ______ given ______ for that cell.

.4
B; A

12-9 Part B of Table 12-1 shows these two conditional probabilities in cell A_1B_1. Their product is ______.

.10

12-10 From part B of 12-1, find $p(A \mid B)$ for cell A_2B_2. $p(A \mid B)$ = ______; $p(B \mid A)$ for that same cell equals ______.

.75; .6

12-11 The product $p(A \mid B)p(B \mid A)$ for cell A_2B_2 equals ______. The sum of the products $p(A \mid B)p(B \mid A)$ for all four cells, in this case, is the same as the sum of the four joint probabilities shown in part A of Table 12-1. This sum is ______.

.45
1.0

Table 12-2 Contingency Table with *A* and *B* Completely Dependent

	A. Frequency Distribution†			B. Products of Conditional Probabilities $p(A \mid B)p(B \mid A)$		
	A_1	A_2	Total	A_1	A_2	Total
B_1	24 (.4)	0	24 (.4)	1.0 (1.0) = 1.0	0	1.0
B_2	0	36 (.6)	36 (.6)	0	1.0 (1.0) = 1.0	1.0
Total	24 (.4)	36 (.6)	60 (1.0)	1.0	1.0	2.0

† Joint and marginal probabilities in parentheses.

We have just seen that $\Sigma p(A \mid B)p(B \mid A)$ for all four cells in a 2×2 contingency table will equal 1 whenever the variables A and B are independent. Table 12-2 shows how these products will work out for a 2×2 table in which A and B are completely *dependent*.

In part A of Table 12-2, the frequencies pile up in just two of the four cells, cells A_1B_1 and A_2B_2. The joint probabilities for these cells match the marginal probabilities. Part B of Table 12-2 shows that the product $p(A \mid B)p(B \mid A)$ for each of these two cells is 1.0, while the product for each of the remaining two cells is 0. The sum $\Sigma p(A \mid B)p(B \mid A)$ for this table is therefore 2. This is the maximum value which this sum can have in a 2×2 table.

The minimum value of $\Sigma p(A \mid B)p(B \mid A)$ occurs when A and B are completely independent, and this minimum value is always 1 no matter how many rows and columns are in the table. The maximum value of $\Sigma p(A \mid B)p(B \mid A)$ occurs when A and B are completely dependent. It varies with the size of the contingency table; it is equal to the number of rows R or the number of columns C, whichever is *smaller*. Thus, in a 3×3 table the maximum value of $\Sigma p(A \mid B)p(B \mid A)$ is 3; in a 3×2 or 2×3 table, the maximum value is 2. Since this maximum value is always equal to either R or C, whichever is smaller, let us call the maximum S (for "smaller"). S is the value of $\Sigma p(A \mid B)p(B \mid A)$ when A and B are completely dependent.

Now recall that we want an index which will be 0 when A and B are independent and which will be 1 when A and B are completely dependent. Let us first define a quantity φ^2 (small Greek phi, squared):

$$\varphi^2 = \Sigma p(A \mid B)p(B \mid A) - 1$$

Since $\Sigma p(A \mid B)p(B \mid A)$ can vary between 1 and S, φ^2 can vary between 0 and $S - 1$. If we now divide φ^2 by $S - 1$, we can express φ^2 as a fraction of its maximum possible value. In a 2×2 table, $S = 2$; when $\varphi^2 = .81$, $\varphi^2/(S - 1) = .81$. In a 3×3 table, $S = 3$; when $\varphi^2 = 1.62$, $\varphi^2/(S - 1)$ again equals .81. The index $\varphi^2/(S - 1)$ will always vary between 0 (for complete independence) and 1 (for complete dependence).

For mathematical reasons we actually measure departure from independence as the *square root* of $\varphi^2/(S - 1)$ and call this index φ' (read "phi prime") the PHI COEFFICIENT. For a contingency table with only two rows or only two columns, $S - 1 = 1$, and $\varphi' = \sqrt{\varphi^2} = \varphi$.

Table 12-3 shows how φ' will vary in relation to φ^2. To obtain $\varphi' = .9$ in a 2×2 table, φ^2 must be .81. To obtain $\varphi' = .9$ in a 4×3 table, φ^2 must be $2(.81) = 1.62$. A φ^2 of only .04 will yield $\varphi' = .2$ in a 2×2 table.

Table 12-3 Relation between φ^2 and the Phi Coefficient φ'

φ'	φ^2	$\frac{\varphi^2}{S-1}$	φ'	φ^2	$\frac{\varphi^2}{S-1}$
.9	$.81(S-1)$	.81	.5	$.25(S-1)$	.25
.8	$.64(S-1)$	.64	.4	$.16(S-1)$	.16
.7	$.49(S-1)$	.49	.3	$.09(S-1)$	.09
.6	$.36(S-1)$	.36	.2	$.04(S-1)$	.04

12-12 In Table 12-2 the probability of A_1 is 24/60 = ______. The conditional probability $p(A \mid B)$ for cell A_1B_1 is 1.0 because ______ of the 24 observations in row B_1 are in cell A_1B_1.

.4

24

12-13 The sum $\Sigma p(A \mid B)p(B \mid A)$ for this table is ______. This is the maximum value for this sum in a 2 × 2 table.

2

12-14 Here is a 2×3 contingency table. Write in $p(A \mid B)p(B \mid A)$ for each cell.

	A_1		A_2		A_3		Total
B_1	6	______	18	______	0	______	24
B_2	0	______	0	______	36	______	36
Total	6		18		36		60

.25; .75; 0

0; 0; 1.0

12-15 The sum $\Sigma p(A \mid B)p(B \mid A)$ for this 2×3 table is ______. Since A and B are completely dependent in this table, this is the ______ value which the sum can have in a 2×3 table.

2

maximum

12-16 The maximum value of $\Sigma p(A \mid B)p(B \mid A)$ is always equal to either R or C, whichever is ______. The minimum value of the same sum is always equal to ______.

smaller

1

12-17 We define φ^2 as $\Sigma p(A \mid B)p(B \mid A)$ minus ______.

1

12-18 If S is the maximum value of $\Sigma p(A \mid B)p(B \mid A)$, then ______ is the maximum value of φ^2. The minimum value of φ^2 is ______.

$S-1$

0

12-19 $\varphi^2/(S-1)$ has a maximum value of ______ and a minimum value of ______.

1

0

12-20 The phi coefficient is the square root of ______. It is called ______.

$\varphi^2/(S-1)$

φ'

B. THE PHI COEFFICIENT IN RELATION TO χ^2

Calculating $\Sigma p(A \mid B)p(B \mid A)$ is not difficult, but it does take time and care. There is an alternative way of finding φ'. Because $\varphi^2 = \chi^2/N$, the phi coefficient can be found directly from χ^2. In most cases when χ^2 has already been used to determine the significance of a difference, this measure of strength of association is very convenient.

We shall examine the relation between φ^2 and χ^2 in our examples. For Table 12-1, where A and B are completely independent, the observed frequencies are exactly equal to the expected frequencies; $f_e - f_o$ is 0 for every cell, and $\chi^2 = 0$. We already know that φ^2 and φ' will be 0 for this table.

Table 12-4 shows the calculation of χ^2 for the example from Table 12-2 in which A and B are completely dependent. The calculation of χ^2 is done here *without* the correction for continuity. Since the phi coefficient does not require comparison to any continuous probability distribution, the correction for continuity is not relevant to φ'.

Table 12-4 Calculation of χ^2 for Data in Table 12-2

f_e	f_o	$f_e - f_o$	$\frac{(f_e - f_o)^2}{f_e}$
9.6	24	14.4	21.6
14.4	0	14.4	14.4
14.4	0	14.4	14.4
21.6	36	14.4	9.6
			$\chi^2 = 60$
			$\varphi^2 = \frac{\chi^2}{N} = 1$

The correction for continuity reduces the size of χ^2. If χ^2 has already been calculated with the correction for continuity, χ^2/N will be less than φ^2. Thus, when χ^2 with df $= 1$ has already been calculated with the correction for continuity, φ^2 cannot be obtained from that value of χ^2. In other cases where χ^2 has more than 1 degree of freedom, φ' can be readily obtained, since $\varphi' = \sqrt{\chi^2/N(S-1)}$.

We have been discussing the phi coefficient as a measure of the departure from independence of two variables, A and B. We now find that this measure is very closely related to the Pearson χ^2 statistic. Since that statistic requires a set of N independent *observations*, it will be easy for you to remember that the phi coefficient is based on the same requirement. φ' can be used to measure strength of association only when the contingency table contains N independent observations. In practice this requirement rules out the use of φ' when the data arise from related samples.

Another measure of strength of association is sometimes calculated from χ^2. It is the CONTINGENCY COEFFICIENT $\chi^2/(N + \chi^2)$. The interpretation of this statistic cannot be made in terms comparable to those used for interpreting any other measure of association, and it will not be used in this book.

Table 12-5 χ^2 and φ^2 for Environmentalist-Nativist Example

	χ^2	$\frac{\chi^2}{N}$	$\sqrt{\frac{\chi^2}{N(S-1)}}$
With correction for continuity	3.684	.068	.26
Without correction	______	______ $= \varphi^2$	______ $= \varphi'$

12-21 The use of χ^2 requires a set of N ________ observations. This requirement also applies to the ________ coefficient.

independence
phi

12-22 When this requirement is fulfilled, strength of association can be measured by the phi coefficient. φ' is an index of departure from ________ of the two ________ in the contingency table.

independence; variables

12-23 With df > 1, χ^2 is normally calculated (*with; without*) the correction for continuity. In such cases, φ^2 is equal to χ^2 divided by ______.

without
N

12-24 The drug example in Table 11-7 (page 173) gave $\chi^2 = 30.76$ with $N = 108$. φ^2 for this example is ______.

0.28

12-25 In that example, $R = 2$ and $C = 3$. Therefore $S =$ ______, and φ' is the square root of ______.

2; 0.28

12-26 With df $= 1$, χ^2 should always be calculated (*with; without*) the correction for continuity, and φ^2 (*will; will not*) equal χ^2/N.

with; will not

12-27 Look at Table 12-5, above. When we calculated χ^2 for the environmentalist-nativist example (page 171), we used the correction for continuity. We obtained a χ^2 of ______.

3.684

12-28 For the same data, $\varphi^2 = 0.089$. $\varphi^2 = \chi^2/N$ when χ^2 is calculated (*with; without*) correction for continuity.

without

12-29 Enter φ^2 in Table 12-5, and find the value of χ^2 without the correction. Since $N = 54$, $\chi^2 =$ ______.

4.8

12-30 $S =$ ______ for this example; therefore, φ' equals the square root of the quantity ______, which is approximately .3.

2
.089

12-31 χ^2 for the drug example reached $\alpha = .001$, and $\varphi' = \sqrt{.28} = .53$. χ^2 for the nativism example did not quite reach $\alpha = .05$, and its φ' is (*greater; less*) than the phi coefficient for the more significant drug example.

less

C. THE CONCEPT OF RELATIVE ERROR

If A and B are associated, it may be possible to use knowledge of variable A to predict variable B or to use knowledge of variable B to predict variable A. Weak associations are less useful in such predictions than strong associations.

Part A of Table 12-6 provides an example. Variable A is freshman performance at a certain college, dichotomized into two categories; over the past 2 years, 15 percent of students admitted have gone on probation during their freshman year, and the remaining 85 percent have made satisfactory grades ("success"). All these students also took a college entrance examination; variable B is performance on this test, again dichotomized into two categories, "pass" and "fail." Part A of Table 12-6 gives the empirical probabilities for the combinations of test and freshman performance; part B of Table 12-6 is calculated from part A to show the conditional probabilities for predicting A (freshman performance) when B (test performance) is known.

Of course, no true "prediction" is yet involved; we are talking only about the two freshman classes already observed, whose performance is summarized in the table. We have the advantage of hindsight, therefore, when we say that we *could* have improved our prediction of A by some particular amount if, at the beginning of these students' freshman year, we had used our knowledge of B. To measure the amount of improvement in prediction, we reason as follows.

If we had had no information about B, we would have had to predict "success" (the category with the largest *marginal* probability) in every case. Because 85 percent succeeded, the prediction would have been correct in .85 of the cases. But with information about each student's category on the entrance exam, we would have predicted the category on A which had the largest *conditional* probability for his category on B. For students in category B_1 we would have predicted A_1, and we would have been correct in .9395 of those cases. For students in category B_2 we would have predicted A_2 and been correct in .70 of those cases.

It is conventional, however, to state the proportion of cases in which we would have made *incorrect* predictions and to call these empirical proportions the PROBABILITY OF PREDICTIVE ERROR. We talk as if we intended to make predictions about future classes, taking the observed proportions as empirical probabilities of error in those predictions. Using this language, we say that the probability of error is .15 without knowledge of B; with knowledge of B, the probability of error is .0605 for students in B_1 and .30 for students in B_2. The *average* probability of error with knowledge of B is obtained by weighting these two probabilities according to the proportions of students found in each category. Since .86 are in B_1 and .14 are in B_2, the average probability of error with knowledge of B is $.0605(.86) + .30(.14) = .094$.

Thus knowledge of B reduces error probability from .15 to .094. The remaining (residual) error is $.094/.15 = .63$ of the error without knowledge of B. The RELATIVE ERROR remaining in this case is .63, or 63 percent of the error which would be made without knowledge of B.

Table 12-6 Predicting Freshman Performance from Entrance Examination Scores

Test Performance	Freshman Performance A: A. Joint and Marginal Probabilities			B. Conditional Probabilities	
	Success A_1	Probation A_2	Total	A_1	A_2
Pass B_1	.808	.052	.86	$p(A_1 \mid B_1) = .9395$	$p(A_2 \mid B_1) = .0605$
Fail B_2	.042	.098	.14	$p(A_1 \mid B_2) = .3000$	$p(A_2 \mid B_2) = .7000$
Total	.85	.15	1.00		

12-32 The probabilities in the four cells of part A in Table 12-6 are (*conditional; joint*) probabilities. The sum of these four probabilities is ______.

joint
1.00

12-33 The conditional probabilities in part B of Table 12-6 are to be used in predicting (A; B) from knowledge of (A; B).

A; B

12-34 For cell A_1B_1, $p(A \mid B)$ is found by dividing ______ by ______, the marginal probability of B_1.

.808
.86

12-35 This conditional probability tells us that when a student has passed the examination, the probability of his success in freshman year is ______.

.9395

12-36 When you must predict a student's standing on A without knowing his standing on B, you should predict (A_1; A_2), and the probability that you will be in error is ______.

A_1
.15

12-37 When you know his standing on B, you should predict (A_1; A_2) when his standing is B_1, and you should predict (A_1; A_2) when his standing is B_2.

A_1; A_2

12-38 When his standing is B_1, your probability of error is ______. Since .86 of the students are in B_1, the overall probability of error in predictions about those students is .86 times ______, or ______.

.0605

.0605; .0520

12-39 When his standing is B_2, your probability of error is ______. The overall probability of error in predictions about students in B_2 is ______ times ______, or ______.

.3
.3; .14; .042

12-40 The overall probability of error *with* knowledge of B is therefore ______.

.094

12-41 The probability of error *with* knowledge of B is ______ percent of the probability of error *without* it.

63

D. THE INDEX OF PREDICTIVE ASSOCIATION λ

The concept of relative error provides one of the most easily interpreted measures of association. When this concept is applied to qualitative data, the measure of association is called λ (small Greek lambda), the index of predictive association. We shall later meet other statistics which are based on relative predictive error but which require quantitative data.

All such measures have a common pattern. There is a term describing the amount of predictive error without use of the association; this term is the measure of TOTAL ERROR. Another term describes the amount of error still remaining when the association is used, or RESIDUAL ERROR. The difference between total and residual error is the amount of error *removed* by using the association to aid prediction This amount is expressed as a proportion:

$$\text{Proportion of error removed} = \frac{\text{total} - \text{residual error}}{\text{total error}} = 1 - \frac{\text{residual error}}{\text{total error}}$$

With qualitative data the measure of total error is the probability of error without use of the association. We can state this probability for Table 12-7 without calculating the joint and conditional probabilities. When we are predicting A without knowledge of B, we look for the largest column total; in this case, our best prediction will be A_2. We shall be right for 12 of the 40 cases and wrong for the other 28. The probability of error in predicting A is 28/40; let us call it $p(E_A)$. The numerator of $p(E_A)$ is always equal to N minus the maximum column total; this gives the number of cases wrongly predicted.

The measure of residual error is the probability of error when B is known. We shall call this probability $p(E_A \mid B)$, the conditional probability of error given B. We look in each row for the largest cell frequency in order to identify our predictions; for B_1 we predict A_3, for B_2 we predict A_1, and for B_3 we predict A_2. We will be right about $6 + 5 + 7$ of the 40 cases and wrong about the other 22. $p(E_A \mid B)$ is 22/40. Its numerator is always equal to N minus the sum of the maximum cell frequencies from the rows.

When we are predicting A, we call the index of predictive association λ_A. λ_A is the proportion of $p(E_A)$ which is removed by knowing B. It is defined as

$$\lambda_A = 1 - \frac{p(E_A \mid B)}{p(E_A)}$$

Since both $p(E_A \mid B)$ and $p(E_A)$ have the same denominator (N), we can write a very simple rule:

$$\lambda_A = 1 - \frac{N - \text{sum of maximum cell frequencies from rows}}{N - \text{maximum column total}}$$

For Table 12-7, $\lambda_A = 1 - 22/28 = 3/14 = .21$.

The same association may also be used to help in predicting B from knowledge of A. When we are predicting B, we call the index of predictive association λ_B and define it as

$$\lambda_B = 1 - \frac{p(E_B \mid A)}{p(E_B)} = 1 - \frac{N - \text{sum of maximum cell frequencies from columns}}{N - \text{maximum row total}}$$

Table 12-7 Contingency Table for Calculation of λ_A and λ_B

	A_1	A_2	A_3	A_4	Total
B_1	1	2	6	3	12
B_2	5	3	0	2	10
B_3	4	7	3	4	18
Total	10	12	9	9	40

12-42 The index of predictive association is called λ. When it measures the reduction of error in predicting A, it is called ______.

λ_A

12-43 λ_B measures error reduction in predicting ______.

B

12-44 The proportion of error removed by an association is 1 − residual error/total error. With categorical data, $p(E_A)$ is the measure of ______ error when ______ is to be predicted.

total; A

12-45 $p(E_A | B)$ is the probability of error given knowledge of the standing on ______. $p(E_A | B)$ is the measure of (*residual; total*) error when A is to be predicted.

B
residual

12-46 When B is to be predicted, the measure of total error is the probability ______.

$p(E_B)$

12-47 To find the numerator for $p(E_B)$, we look for the largest (*column; row*) total in Table 12-7 and subtract it from ______.

row; N

12-48 Without knowledge of A, we would predict that the category on B is (B_1; B_2; B_3). We would be right about ______ cases and wrong about ______ cases. $p(E_B)$ is ______.

B_3; 18
22; 22/40

12-49 The measure of residual error in predicting B from A is the conditional probability ______.

$p(E_B | A)$

12-50 To find the numerator of $p(E_B | A)$, we add the largest cell frequencies from each (*column; row*) and subtract the sum from ______.

column
N

12-51 In Table 12-7 the maximum cell frequencies from the columns are ______, ______, ______, and ______.

5; 7; 6; 4

12-52 $p(E_B | A) =$ ______ for this table, and $p(E_B | A)/p(E_B) =$ ______. λ_B is therefore ______.

18/40
18/22; 4/22(*or* .18)

Table 12-8 Contingency Table with $\lambda_B = 0$

	A_1	A_2	A_3	Total
B_1	1	1	3	5
B_2	5	7	6	18
B_3	3	1	0	4
Total	9	9	9	27

When we state that $\lambda_A = .37$ for our prediction of freshman performance from examination scores, we are saying that our information about the score reduces predictive error by 37 percent on the average. Bear in mind, however, that we are talking only about predictions *within* the sample which has already been observed. λ is a descriptive statistic, not a population parameter. It does not tell anything precise about the error we shall make when we try to predict the performance of next year's freshmen before classes begin. Unless the distribution of frequencies for next year's class is exactly like the distribution we have already studied, λ_A will not be precisely .37 for that class. We may expect it to be fairly close to .37 only if the data in our contingency table are representative of the proportions in the population from which all entering freshman classes are drawn.

In our examination of Table 12-7 we have seen that λ_A need not equal λ_B. For that table, the association reduces predictive error by 21 percent when we are predicting A from B but by only 18 percent when we are predicting B from A. Table 12-8 shows another interesting case. λ_B will be 0 for this table because knowledge of A will lead to the same prediction (B_2) in every case, and this prediction would also be made without knowledge of A. λ_A, on the other hand, is not 0; knowledge of B will improve prediction of A by a measurable amount.

Unlike the phi coefficient, the index of predictive association can be used for related samples. Table 12-9 shows how the same 25 persons reported their feelings about the Vietnam war before (B_1) and after (B_2) the Cambodian invasion. $\lambda_A = .15$, showing that the standings of these persons on the war could be predicted with 15 percent less error with knowledge of the position in time (B).

Whenever the N observations meet the requirement of independence, both φ' and the λ indexes can be calculated for the same data. The values obtained will not usually be the same since these two kinds of measures have different meanings. A phi coefficient of 0 means complete independence of A and B, and it will always lead to a value of 0 for both λ_A and λ_B; no prediction is possible when A and B are independent. But the converse is not true. One or both λ indexes may be 0, yet φ' may still show some degree of association between A and B. In such a case, the association which is present does not provide any error reduction in making predictions. On the other hand, if $\varphi' = 1$, at least one of the λ indexes must also be 1.

Table 12-9 Attitudes of 25 Persons on the Vietnam War before and after the Cambodian Invasion

	Highly Favorable A_1	Favorable A_2	Unfavorable A_3	Highly Unfavorable A_4	Total
Before					
B_1	8	10	5	2	25
After					
B_2	2	5	11	7	25
Total	10	15	16	9	50

12-53 Show that $\lambda_B = 0$ for Table 12-8. The largest row total is ________. The sum of the largest cell frequencies from the columns is ________. $\lambda_B = 1 -$ ________ $=$ ________.

18
18; 1; 0

12-54 Find λ_A for Table 12-8. In this table all 3 column totals equal ________. Without knowledge of B, you would have to pick one of the 3 values of A at random, and you would be wrong about the other ________ cases.

9
18

12-55 The sum of the largest cell frequencies from the rows is ________. $\lambda_A = 1 =$ ________.

13; 14/18; 4/18(or .22)

12-56 A phi coefficient can be calculated for Table 12-8 because all 27 observations are ____________. The easiest way to obtain φ' is by calculating the ________ statistic for this table.

independent
χ^2

12-57 χ^2 for Table 12-8 is 5.4, with ________ degrees of freedom. Since the chi-square approximation does not permit f_e's smaller than ________, the χ^2 test itself (*can; cannot*) be used.

4
5; cannot

12-58 φ^2 for this table is ________, and $S =$ ________. The phi coefficient is therefore equal to the square root of ________.

.2; 3
.1

12-59 φ' for Table 12-8 is .32. There is a considerable departure from ____________ of the two variables.

independence

12-60 When $\varphi' = 0$, λ_A and λ_B (*can; cannot*) be greater than 0. When λ_A and λ_B equal 0, φ' (*can; cannot*) be greater than 0.

cannot
can

12-61 Table 12-9 is a contingency table for two (*independent; related*) samples. φ' (*can; cannot*) be calculated for this table.

related; cannot

12-62 $\lambda_B = .44$ for Table 12-9. When we try to guess the time at which the attitude was measured, we shall make ________ percent less error if we know the attitude held.

44

REVIEW

12-63 The measures of strength of association discussed in this lesson are suitable for ________ data. They require data in the form of a ________ table. — *categorical (or qualitative)*; *contingency*

12-64 One of the two measures can be calculated for two or more *related* samples. Its Greek letter name is ______. — λ

12-65 The other measure is like χ^2 in that it requires the entire set of N observations to be ________ of each other. This measure has the Greek letter name ______. — *independent*; φ'

12-66 φ', or the ________ coefficient, is a measure of departure from ________ of two ________. — *phi*; *independence*; *variables*

12-67 λ is an index of ________ association. It represents the percent of predictive ________ removed by the association. — *predictive*; *error*

12-68 When standings on variable A are being predicted from standings on variable B, the index of predictive association is called ______. — λ_A

12-69 If $\lambda_A = .25$, the error in making predictions about A is reduced by ______ percent through knowledge of ______. — *25*; B

12-70 When A and B are independent variables, $p(A \mid B)$ will equal ______ and $p(B \mid A)$ will equal ______ for every cell in the contingency table. — $p(A)$; $p(B)$

12-71 In this case, $\Sigma p(A \mid B)p(B \mid A)$ will equal $\Sigma p(A)p(B)$, and this sum will always be equal to ______, no matter what the size of the table. — *1*

12-72 When the variables A and B are completely dependent, $\Sigma p(A \mid B)p(B \mid A)$ will equal a quantity which we designate by the letter ______. — S

12-73 The letter S represents either the number of ________ or the number of ________ in the table, whichever is ________. — *rows*; *columns*; *smaller*

12-74 $\Sigma p(A \mid B)p(B \mid A)$ can vary between ______ for complete independence and ______ for complete dependence. — *1*; S

12-75 φ^2 is defined as ______ minus 1. It is also equal to χ^2 divided by ______, when χ^2 has been calculated (*with; without*) the correction for continuity. — $\Sigma p(A \mid B)p(B \mid A)$; N; *without*

12-76 φ^2 can vary between ______ for complete independence and ______ for complete dependence. | *0* *S − 1*

12-77 $\sqrt{\varphi^2/(S-1)}$ defines the quantity ______. This coefficient varies between ______ and ______. | φ' *0; 1*

12-78 The phi coefficient is called φ' rather than simply φ because it is equal to $\sqrt{\varphi^2}$ only in a special case, i.e., when $S =$ ______. | *2*

12-79 Go back to Table 12-1 on page 178. φ' for this table is ______. Therefore, you know that λ_A and λ_B must equal ______. | *0* *0*

12-80 We cannot improve our prediction of either variable from knowledge of the other when A and B are completely ______. | *independent*

12-81 Look again at Table 12-2 on page 180. φ' for this table is ______. What is λ_A for this table? ______ What is λ_B? ______ | *1; 1* *1*

12-82 Compare the table in Frame 12-14, page 181. $\varphi' = 1$ for this 2×3 table, and one of the λ indexes equals 1. Knowledge of (A; B) permits perfect prediction of (A; B), but the reverse is not true. | *A; B*

12-83 For this 2×3 table, $\lambda_B = 1$, but $\lambda_A = 8/9$. When $\varphi' = 1$, at least one of the λ indexes must equal ______. | *1*

12-84 When λ_A is greater than 0, λ_B (*can; cannot*) be 0, and φ' (*can; cannot*) be 0. | *can* *cannot*

12-85 The λ indexes are descriptive statistics. This means that they refer only to predictions within the sample already ______, not to predictions for other samples from the same ______. | *observed* *population*

12-86 φ' is also a descriptive statistic. It refers to the degree of ______ from independence of A and B in the contingency table itself. | *departure*

12-87 Neither φ' nor λ tells us the exact strength of association of A and B in the ______ from which such samples are drawn. | *population*

12-88 If we apply φ' or λ to the population, we are making a further assumption. We are assuming that the proportions in the population are exactly like those found in our ______ table. | *contingency*

PROBLEMS

1. A sociologist thinks that later-born children maintain closer relationships with their families than their older brothers and sisters do. To check his hypothesis, he selects a random sample of 20 college students from each of three groups; group 1 consists of students who are oldest children, group 2 of students who have both older and younger siblings, and group 3 of students who are youngest children. He then obtains information about the number of letters written home by each student over a period of 6 months at college.

Letter Frequency	Family Position			
	Oldest	Middle	Youngest	Total
At least 2 per month	12	14	7	33
Less than 2 per month	8	6	13	27
Total	20	20	20	60

a. Calculate χ^2 for these data. Taking $\alpha = .05$, interpret this χ^2 statistic.

b. Calculate the phi coefficient. What can you conclude from the value of this statistic?

c. Calculate the two λ indexes, λ_L for predicting letter frequency from family position and λ_P for guessing family position from letter frequency. What do these indexes tell you about the data?

2. Calculate (*a*) χ^2, (*b*) φ', and (*c*) λ_C and λ_R for the following data:

	C_1	C_2	Total
R_1	16	14	30
R_2	24	6	30
Total	40	20	60

3. Calculate (*a*) χ^2 and (*b*) φ' for the following data:

	C_1	C_2	C_3	Total
R_1	6	9	15	30
R_2	3	3	12	18
R_3	3	6	15	24
Total	12	18	42	72

4. Take $\chi^2 = 3.85$ (without correction for continuity), df $= 1$. Calculate the phi coefficient assuming (*a*) $N = 30$, (*b*) $N = 100$, and (*c*) $N = 19$.

PART THREE
Significance Tests for Ordinal and Interval Data

LESSON 13
FACTORIAL EXPERIMENTS

This lesson prepares you to study statistical tests of the difference between two or more sample means. Before beginning that study, you must become familiar with the terms, symbols, and operations for analyzing data in the form of numerical scores (interval data).

First we introduce some terms used to describe different kinds of investigations. In Lesson 1 we discussed two examples: (1) a comparison of two samples of continuous data (heights) from two colleges, and (2) a comparison among three samples of discrete data (vocabulary scores) from three classes taught in different ways. Both of these examples are SINGLE-FACTOR studies because each bears upon only one of the influences which may affect the observations. In (1) the influence, or FACTOR, is the college in which a student is enrolled; in (2) the single factor is the learning method.

Some investigations are designed to permit decisions about more than one factor at a time. Another factor which might influence French vocabulary learning is the previous study of Latin. The French teacher may arrange his experiment so that the influence of both these factors can be evaluated at the same time. To do so, he must divide his students into 2 groups, those who have previously studied Latin and those who have not; he can then assign one-third of each group *at random* to each of the 3 learning methods. Factor A (learning method) has 3 levels; factor B (Latin) has only 2. There will be 6 experimental subgroups, one for each of the 6 combinations of factor levels.

Studies in which individuals are randomly assigned to different experimental groups are called CONTROLLED experiments. The vocabulary study is a controlled experiment. Learning method is actually controlled or manipulated by the investigator; he has taken a group of students who might all have been treated alike, and he has chosen by a random process the particular individuals who are to use different learning methods. It is natural in such a case to speak of the factor levels as TREATMENTS.

Such a study differs from the college-height example in an important way. The height observations are already in existence before the study begins; they are already distributed in some unknown way between the two possibly different groups (Alpha and Omega colleges). While the French teacher does something which may influence the size and distribution of his observations, the height investigator simply takes his observations as he finds them.

The presence of experimental control affects the interpretation of a significant difference, but it does not affect the statistical methods for decision making. In both cases we look for a difference which can be attributed to factor A. Statistically we speak of the possible EFFECTS of the factor "college" just as we speak of the possible effects of the factor "learning method," even though the word "effects" may seem odd in the height example. "Effects" in its statistical sense does not refer to a product of mechanical causation; it refers only to the evidence of a statistical association between two variables, namely the *factor* and the *observations* which are under study.

Both our examples will therefore be called single-factor experiments, even though the height example does not seem to be what one might usually call an experiment. Whenever more than one factor is involved simultaneously, we call the study a FACTORIAL experiment. A study of the effects of both factor A (learning method) and factor B (Latin) on vocabulary scores would be a controlled factorial experiment with six treatment combinations. The teacher does not control the division of his students into two groups on factor B, but he does control their assignment to levels on factor A.

A. NOTATION FOR SINGLE-FACTOR EXPERIMENTS

In this volume we shall not go beyond single-factor experiments. Such experiments have only two variables; one of these is the "factor" (traditionally called the INDEPENDENT OR EXPERIMENTAL VARIABLE), and the other is the *observed* or DEPENDENT VARIABLE.

We shall always call our single factor "factor A." This term can then be retained when we extend these methods to factorial experiments. The number of levels on factor A will be called C, and these levels will appear as labels for the columns in our data table. Later, when you wish to add a factor B, its level may appear as labels for the rows; the number of levels on factor B may be called R. The data table for a two-factor experiment will bear some resemblance to a contingency table.

The number of observations in the entire set will be called N. In a single-factor experiment there are C subgroups, one for each level on factor A, and each subgroup contains a certain number of observations. We shall usually assume that this number is equal for all subgroups; experiments with subgroups of equal size have certain important advantages. In these cases, we can call the number in each subgroup n, and n will equal N/C. However, it is not always possible to have equal subgroups; when the subgroups differ in size, we shall designate the size of subgroup A_1 as n_1, that of subgroup A_2 as n_2, and so on. We can then refer to the size of a single subgroup (without specifying which one) as n_j, where the subscript j is understood as referring to any one of the ordinal numbers from 1 to C.

In Lesson 6 we began to speak of each individual score as X_i, where i represented any of the ordinal numbers from 1 to N. It will make our later discussion somewhat easier if we now choose the letter Y for our dependent variable, writing Y_{ij} for any one of the n observations in subgroup A_j. When you deal with two factors, there will usually be $R \times C$ subgroups, and the scores within each subgroup will be thought of as arrayed *within a cell* of the $R \times C$ data table. In single-factor experiments we can arrange the scores for a subgroup vertically *within a column* (see Table 13-1).

Table 13-1 Schema for a Single-Factor Experiment with C Levels on Factor A

Score	Level on Factor A						
	A_1	A_2	A_3	. . .	A_j	. . .	A_C
First	Y_{11}	Y_{12}	Y_{13}	. . .	Y_{1j}	. . .	Y_{1C}
Second	Y_{21}	Y_{22}	Y_{23}	. . .	Y_{2j}	. . .	Y_{2C}
Third	Y_{31}	Y_{32}	Y_{33}	. . .	Y_{3j}	. . .	Y_{3C}
. . .	. . .	. . .	. . .	. . .	. . .	. . .	. . .
ith	Y_{i1}	Y_{i2}	Y_{i3}	. . .	Y_{ij}	. . .	Y_{iC}
. . .	. . .	. . .	. . .	. . .	. . .	. . .	. . .
Last	Y_{n1}	Y_{n2}	Y_{n3}	. . .	Y_{nj}	. . .	Y_{nC}

13-1 Two-variable experiments have one independent and one dependent variable. Observations are made on the ________ variable; these observations may or may not be influenced by the ________ variable.

dependent
independent

13-2 The name "factor" is applied to the ________ variable, and in single-factor experiments it is called factor ________.

independent
A

13-3 When the effects of more than one factor are studied at the same time, we call the study a ________ experiment.

factorial

13-4 Each value or level on factor A occupies a (*column; row*) in the data table. The number of levels is designated by the letter ________.

column
C

13-5 If individuals are randomly *assigned* to levels on factor A, the experiment is a ________ experiment and the levels can be called ________.

controlled
treatments

13-6 X_i refers to a particular member of a set of X scores. When there are N such scores, i may be any whole number from ________ to ________.

1; N

13-7 A_j refers to a particular level on factor A. Since there are C levels in any study, j may be any whole number from ________ to ________.

1
C

13-8 Y_{ij} refers to a particular member of a set of Y scores in a single-factor experiment. The subscript ________ designates the subgroup to which the score belongs.

j

13-9 The subscript i designates a particular score within a subgroup. If there are n scores in a subgroup, i may be any whole number from 1 to ________.

n

B. THE DISTINCTION BETWEEN FIXED AND RANDOM FACTORS

Some experiments include a subgroup for every possible level on factor A. We may consider the vocabulary example as such an experiment; the teacher is interested in comparing only these three learning methods. In such a case, the number of subgroups C will be the same as the number of levels on factor A, and we look for specific differences among these particular levels.

Some factors, however, have a very large number of possible levels. For example, a single-factor experiment could be designed to test the effect of differences among teachers (factor A) on the achievement test scores of seventh-grade arithmetic pupils. If the study is to apply to an entire state school system, the number of possible levels is equal to the number of seventh-grade arithmetic teachers in the system. It will not be practical to include a subgroup for every level. A *random selection* of C levels, i.e., individual teachers, is made, and only these levels are represented in the study.

Factors like "teachers" which can only be sampled are called RANDOM FACTORS. If we should repeat (or *replicate*) the same experiment with a new set of observations, we would make a new random selection of levels, and the new subgroups would not represent the same set of levels on factor A. We are not interested in comparing the particular levels (individual teachers) which appear in our study. We want to be able to say that differences among these teachers show that pupil performance *all over the state* is influenced by the teacher factor. This conclusion concerns all possible levels; it is justified only when we select the levels to be studied in a random way.

Factors like "learning method" are called FIXED FACTORS. All the levels in which we are interested are actually placed under study. If we should replicate the study, the new subgroups would come from the same fixed set of C levels on factor A. We want to be able to state which of these levels (methods) gives the best results, without regard to any other methods.

Sometimes factors with a very large number of possible levels can still be studied in a systematic way. Take the factor "years of teaching experience" as an example. Among all the arithmetic teachers in a state, this factor might have 40 or more possible levels. If a random selection of levels is made, the factor becomes a random factor. However, these levels can be grouped into categories, such as 1–10, 11–20, 21–30, 31–40, and more than 40 years. If one subgroup is included for each of these systematically arranged categories, the factor becomes a fixed factor and we look for specific differences among these five levels.

The distinction between fixed and random factors is a very important one. As we set up our statistical tests for differences among subgroup means, we shall reason differently for fixed- and random-factor experiments. In a random-factor experiment, we must constantly bear in mind that we have randomly sampled not only our observations within subgroups but also the actual levels which distinguish the subgroups from one another.

Two different research organizations are asked to make parallel studies of the effect of certain factors on the amount of supplementary income earned by welfare recipients. Each study will examine a sample of N welfare recipients, and the Y scores will be the N amounts of money they have earned in a given period.

13-10 In one experiment factor A is "recipient's education." It has four levels: grade school, junior high, high school, and beyond high school. Both studies (*will; will not*) include the same levels on this factor. — *will*

13-11 The levels on this factor do not vary from one replication to another. It is said therefore to be a (*fixed; random*) factor. — *fixed*

13-12 Each research organization will seek a ________ sample of welfare recipients from each of these four levels and inquire about their earned incomes. This experiment (*is; is not*) a controlled experiment. — *random*; *is not*

13-13 If the four subgroups differ significantly in earned income, it will be concluded that "recipient's education" has an ________ on the amount of his earned income. — *effect*

13-14 In another experiment factor A is "welfare agency." There are 205 agencies in the region to be studied. Factor A thus has ______ possible levels. — *205*

13-15 Each organization is expected to study only 20 levels on this factor. Both studies probably (*will; will not*) include the same levels. — *will not*

13-16 Each organization will make a ________ selection of 20 agencies from the 205 which might be studied. — *random*

13-17 Since the levels studied vary in a random way from one replication to another, the factor is a ________ factor. — *random*

13-18 Once the selection of its 20 agencies has been made, each organization will then seek a ________ sample of welfare recipients from each of these 20 agencies. There will be ______ subgroups in its study. — *random*; *20*

13-19 If the 20 subgroups differ significantly in earned income, it will be concluded that the welfare agency has an ________ on the amount of earned income. This conclusion will apply to (*all agencies in the region; only the 20 agencies studied*). — *effect*; *all agencies in the region*

C. SUMMATION OPERATIONS IN A SINGLE-FACTOR EXPERIMENT

Let us now recall the procedures recommended in Lesson 7 for handling a set of scores. You were told in that lesson always to begin by obtaining ΣX_i and ΣX_i^2 simultaneously. Then you will be able to derive the sum of squares SS by taking $\Sigma X_i^2 - T^2/N$, where T (total) is ΣX_i.

These procedures also apply to the handling of data in a single-factor experiment. However, certain subtotals must also be obtained. Since we intend to compare the means of C subgroups, we shall need ΣX and ΣX^2 for each subgroup. Of course, we are now calling the scores Y_{ij}, where i represents the position of a score within its column and j represents the column.

Table 13-2 presents a numerical example. There are three levels on factor A, and each of the three subgroups has just five scores. Our analysis begins with finding ΣY_{ij}^2 and ΣY_{ij} $(= T_j)$ for each of the three columns. We can then find a column mean $\bar{Y}_j$ by taking T_j/n_j. The sum of the three T_j's gives the *grand total* $G = \Sigma Y_{ij}$, and G/N gives the grand mean $\bar{Y}$.

We can find a sum of squares for all N scores in the usual way, $\Sigma Y_{ij}^2 - G^2/N$. This sum of squares will now receive the label SS_Y, *total* sum of squares, to distinguish it from certain other sums of squares.

The grand total G may be written $\sum_1^C \sum_1^n Y_{ij}$. Such **DOUBLE SUMMATIONS** are to be read and performed in a particular order; the inner (or second) summation is performed first. $\sum_1^C \sum_1^n Y_{ij}$ is therefore to be read as an instruction to sum the Y_{ij}'s *within* a column, i.e., add the Y_{ij}'s as i goes from 1 to n, then, having done this inner summation for each column, to take the sum *across* the columns, i.e., add the column sums T_j as j goes from 1 to C. $\sum_1^C \sum_1^n Y_{ij}^2$ specifies the same sequence of operations on the squared values Y_{ij}^2.

Table 13-2 Numerical Example for a Single-Factor Experiment

	Levels on Factor A						
	A_1		A_2		A_3		
Score	Y	Y^2	Y	Y^2	Y	Y^2	
First	1	1	2	4	6	36	
Second	2	4	3	9	7	49	
Third	3	9	4	16	8	64	
Fourth	4	16	5	25	9	81	
Fifth	5	25	6	36	10	100	
T_j	15		20		40		$G = 75$
$\sum_1^n Y_{ij}^2$		55		90		330	$\sum_1^C \sum_1^n \bar{Y}_{ij}^2 = 475$
$\bar{Y}_j$	3		4		8		$\bar{Y} = 5$

13-20 $\sum_{1}^{n} Y_{i1}$ refers to the sum of all the Y scores in column ______. Table 13-2 shows its value as $T_1 =$ ______. *1; 15*

13-21 The summation term $\sum_{1}^{n} Y_{i1}^2$ refers to the sum of the ______ of all the Y scores in column 1. Before this sum can be taken, each Y score must be ______. *squares* *squared*

13-22 We also refer to $\sum_{1}^{n} Y_{i1}$ as T_1; the letter T stands for "column ______." Similarly, $\sum_{1}^{n} Y_{i2}$ can be called ______. *total; T_2*

13-23 There are n scores in a column; $\sum_{1}^{n} Y_{ij}$ is the sum of all scores in a particular ______. There are N scores in the whole table; $\sum_{1}^{N} Y_{ij}$ is the sum of all scores in the ______, and it is also G for "______ total." *column* *table* *grand*

13-24 When we add the scores one column at a time, we write $\sum_{1}^{N} Y_{ij}$ as $\sum_{1}^{C} \sum_{1}^{n} Y_{ij}$. This term is a double ______ term. *summation*

13-25 There are n scores to add *within* a column; $\sum_{1}^{n}$ refers to summation of ______ numbers, and it therefore concerns the summation of scores ______ a column. *n* *within*

13-26 There are C columns; $\sum_{1}^{C}$ refers to summation of ______ numbers, one for each of the ______. It signifies a summation of scores ______ the columns. *C* *columns* *across*

13-27 Before we can get just one number for each column, we have to perform the summation *within* the column. Therefore, to find $\sum_{1}^{C} \sum_{1}^{n} Y_{ij}$, we must start with the operation indicated by the (*first; second*) summation sign. *second*

13-28 $\sum_{1}^{n} Y_{ij}/n$ defines the mean of a ______, represented by the symbol ______. $\sum_{1}^{C} \sum_{1}^{n} Y_{ij}/N$ defines the ______ mean, represented by the symbol ______. *column* *$\bar{Y}_j$; grand* *$\bar{Y}$*

The algebraic manipulation of summation terms requires practice in the use of two simple rules. The flow chart on this page will help you to learn these rules. They are used here to show why $SS_Y = \Sigma Y_{ij}^2 - G^2/N$.

Summation rule I $\sum_1^N (A_i + B_i) = \sum_1^N A_i + \sum_1^N B_i$. When an algebraic term consists of the sum of two terms, its summation is performed in two steps. First, each of the two component terms is summed; then the two sums are added together.

Summation rule II If k is a constant, $\sum_1^N k = Nk$, and $\sum_1^N kA_i = k \sum_1^N A_i$. If k is a constant for all N scores, then the sum of k across all N is simply N times k. When k is a constant multiplying a variable such as A_i, then the summation is performed on A_i, and this sum is multiplied by k.

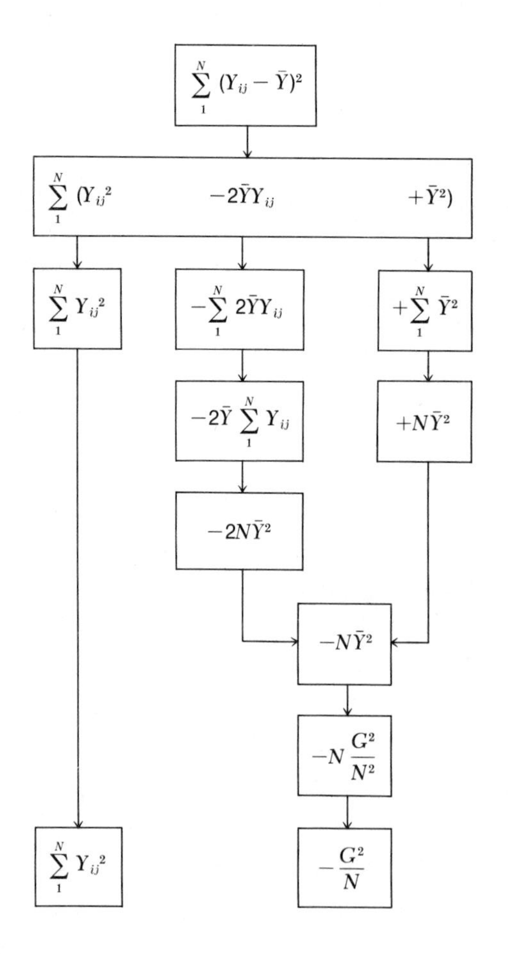

1 $SS_Y = \sum_1^N (Y_{ij} - \bar{Y})^2$

2 Expanded square

3 Summation rule I

4 Summation rule II; $\bar{Y}$ is a constant for all N elements.

5 $\bar{Y} = \frac{\sum_1^N Y_{ij}}{N}$, and so $\sum_1^N Y_{ij} = N\bar{Y}$

6 Collected terms

7 $\bar{Y} = \frac{G}{N}$

8

13-29 In summation rule I, A_i is a variable like X_i; there are N different values of A, and A_i represents all these in turn as i goes from ______ to ______.

1; N

13-30 $\sum_1^N (A_i + B_i)$ can be simplified into the sum of two separate summation terms, ______ and ______.

$\sum_1^N A_i$; $\sum_1^N B_i$

13-31 Step 2 of the flow chart on page 202 results in a similar term. $\sum_1^N (Y_{ij}^2 - 2\bar{Y}Y_{ij} + \bar{Y}^2)$ can be simplified into the sum of three separate summation terms, ______, and ______, and ______. (Watch their algebraic signs.)

$\sum_1^N Y_{ij}^2$; $-\sum_1^N 2\bar{Y}Y_{ij}$; $\sum_1^N \bar{Y}^2$

13-32 In summation rule II, k represents a number which is a ______ for all N scores. In steps 3 and 4 of the flow chart, ______ represents a number like k.

constant
$\bar{Y}$

13-33 According to summation rule II, $\sum_1^N k =$ ______. Since $\bar{Y}$ is a constant like k, $\sum_1^N \bar{Y}^2 =$ ______.

Nk
$N\bar{Y}^2$

13-34 If you have a column of 10 numbers and each of these numbers is 19, you will not add up the numbers. You will multiply 19 by ______. You could write this problem as $\sum_1^N k$, where $k =$ ______ and $N =$ ______.

10
19; 10

13-35 Summation rule II also indicates that $\sum_1^N kA_i =$ ______. Instead of multiplying each A_i by ______ before adding, you can add up the N values of A_i first, then multiply the sum by ______.

$k \sum_1^N A_i$
k
k

13-36 There is a term of this kind in step 3. It is ______; there are two constants in this term, ______ and ______.

$-\sum_1^N 2\bar{Y}Y_{ij}$
2; $\bar{Y}$

13-37 Using summation rule II, this term becomes ______ in step 4.

$-2\bar{Y}\sum_1^N Y_{ij}$

13-38 The flow chart shows that $SS_Y = \sum_1^N Y_{ij}^2 - G^2/N$ by substituting $N\bar{Y}$ for ______ (in step 5) and substituting G/N for ______ (in step 7).

$\sum_1^N Y_{ij}$
$\bar{Y}$

D. COMPONENT SUMS OF SQUARES

To obtain the sum of squares for the entire distribution, we take deviations from the grand mean $\bar{Y}$: $SS_Y = \sum_1^N (Y_{ij} - \bar{Y})^2$. If we treat each subgroup (column) as if it were a separate distribution, we can calculate a subgroup mean $\bar{Y}_j$ and a subgroup sum of squares SS_j. To obtain SS_j, the sum of squares *within* a column, we take deviations from the column mean; $SS_j = \sum_1^n (Y_{ij} - \bar{Y}_j)^2$.

The deviation of any score from the grand mean is $Y_{ij} - \bar{Y}$. It can be considered as the sum of two component deviations, $Y_{ij} - \bar{Y}_j$ and $\bar{Y}_j - \bar{Y}$.

$$Y_{ij} - \bar{Y} = (Y_{ij} - \bar{Y}_j) + (\bar{Y}_j - \bar{Y}) = Y_{ij} - \cancel{\bar{Y}_j} + \cancel{\bar{Y}_j} - \bar{Y}$$

Thus, any deviation score $Y_{ij} - \bar{Y}$ can be broken up into two components, a deviation of the score from its column mean $Y_{ij} - \bar{Y}_j$ and a deviation of the column mean from the grand mean $\bar{Y}_j - \bar{Y}$.

You may now be surprised to learn that the *squares* of these components, summed for all N elements, will add up to SS_Y. In mathematical symbols,

$$SS_Y = \sum_1^C \sum_1^n (Y_{ij} - \bar{Y})^2 = \sum_1^C \sum_1^n (Y_{ij} - \bar{Y}_j)^2 + \sum_1^C \sum_1^n (\bar{Y}_j - \bar{Y})^2$$

The flow chart on page 206 will show that this is a true equation.

We can give convenient names to the two component sums of squares. $\sum_1^C \sum_1^n (Y_{ij} - \bar{Y}_j)^2$ is derived from the deviations of scores around their column means; it thus represents variation *within* columns, and this sum (for all scores in all columns) is called the SUM OF SQUARES WITHIN COLUMNS SS_W. $\sum_1^C \sum_1^n (\bar{Y}_j - \bar{Y})^2$ is derived from the deviations of column means around the grand mean; it thus represents variation *between* columns, and this sum (again, for all N scores) is called the SUM OF SQUARES BETWEEN COLUMNS SS_B.

Table 13-3 shows how these components work out for our numerical example. The three sums of squares appear at the bottom: SS_B is 70, SS_W is 30, and SS_Y is 100. This table has been drawn up to show the exact values of each component for each of the 15 scores in the example; thus, for the first score in column 1, $(Y_{ij} - \bar{Y}_j)^2 = 4$, $(\bar{Y}_j - \bar{Y})^2 = 4$, and $(Y_{ij} - \bar{Y})^2 = 16$. You should notice two facts about this table:

1 Every one of the 15 scores contributes a squared deviation to each of the three sums of squares. There are 15 contributions to SS_W, and there are also 15 contributions to SS_B, even though each score in a particular column contributes exactly the same amount to SS_B.

2 For most of the individual scores, the two component squared deviations $(Y_{ij} - \bar{Y}_j)^2$ and $(\bar{Y}_j - \bar{Y})^2$ do not equal $(Y_{ij} - \bar{Y})^2$. Taking the first score, again, notice that $4 + 4 \neq 16$. When we say $SS_W + SS_B = SS_Y$, we are talking about the *sums* of the squared deviations, not about the contributions of individual scores to these sums.

Table 13-3 Sums of Squares for Numerical Example from Table 13-2

	A_1			A_2			A_3		
i	$(Y-\bar{Y}_1)^2$	$(\bar{Y}_1-\bar{Y})^2$	$(Y-\bar{Y})^2$	$(Y-\bar{Y}_2)^2$	$(\bar{Y}_2-\bar{Y})^2$	$(Y-\bar{Y})^2$	$(Y-\bar{Y}_3)^2$	$(\bar{Y}_3-\bar{Y})^2$	$(Y-\bar{Y})^2$
1	4	4	16	4	1	9	4	9	1
2	1	4	9	1	1	4	1	9	4
3	0	4	4	0	1	1	0	9	9
4	1	4	1	1	1	0	1	9	16
5	4	4	0	4	1	1	4	9	25
Sums	10	20	30	10	5	15	10	45	55

$$\sum_1^C \sum_1^n (Y-\bar{Y}_j)^2 = SS_W = 10 + 10 + 10 = 30$$

$$\sum_1^C \sum_1^n (\bar{Y}_j-\bar{Y})^2 = SS_B = 20 + 5 + 45 = 70$$

$$\sum_1^C \sum_1^n (Y-\bar{Y})^2 = SS_Y = 30 + 15 + 55 = 100$$

13-39 Remember the component sums of squares by the *deviations* which define them. $\sum_1^C \sum_1^n (Y_{ij}-\bar{Y}_j)^2$ is based on the deviation of a score from the ________ mean, and it thus represents variation (*between; within*) columns.

column
within

13-40 $\sum_1^C \sum_1^n (\bar{Y}_j-\bar{Y})^2$ is based on the deviation of a ________ mean from the ________ mean; it thus represents variation ________ columns.

column; grand
between

13-41 $\sum_1^C \sum_1^n (Y_{ij}-\bar{Y}_j)^2$ is called (SS_B; SS_W), the sum of squares ________ columns. $\sum_1^C \sum_1^n (\bar{Y}_j-\bar{Y})^2$ is called ________, the sum of squares ________ columns.

SS_W
within; SS_B
between

13-42 $SS_W + SS_B =$ ________. It is based on deviations of individual scores from the ________ mean.

SS_Y
grand

13-43 In Table 13-3, consider the first score in the *second* column, A_2. The contribution of this score to SS_W is ________; its contribution to SS_B is ________. The sum of these two contributions is ________, but its contribution to SS_Y is ________.

4
1
5; 9

13-44 There are 15 contributions to SS_W in this table. There are ________ contributions to SS_B. Every score in column A_2 contributes the same amount to ________.

15
SS_B

Because it is inconvenient, we do not usually calculate SS_W and SS_B in the manner shown in Table 13-3. We use the computing formulas in Table 13-4. The flow chart on this page shows how these computing formulas arise. It also demonstrates that $SS_W + SS_B = SS_Y$. In this demonstration we assume that all subgroups have the same n, that is, all $n_j = n$; the result is the same for unequal subgroups.

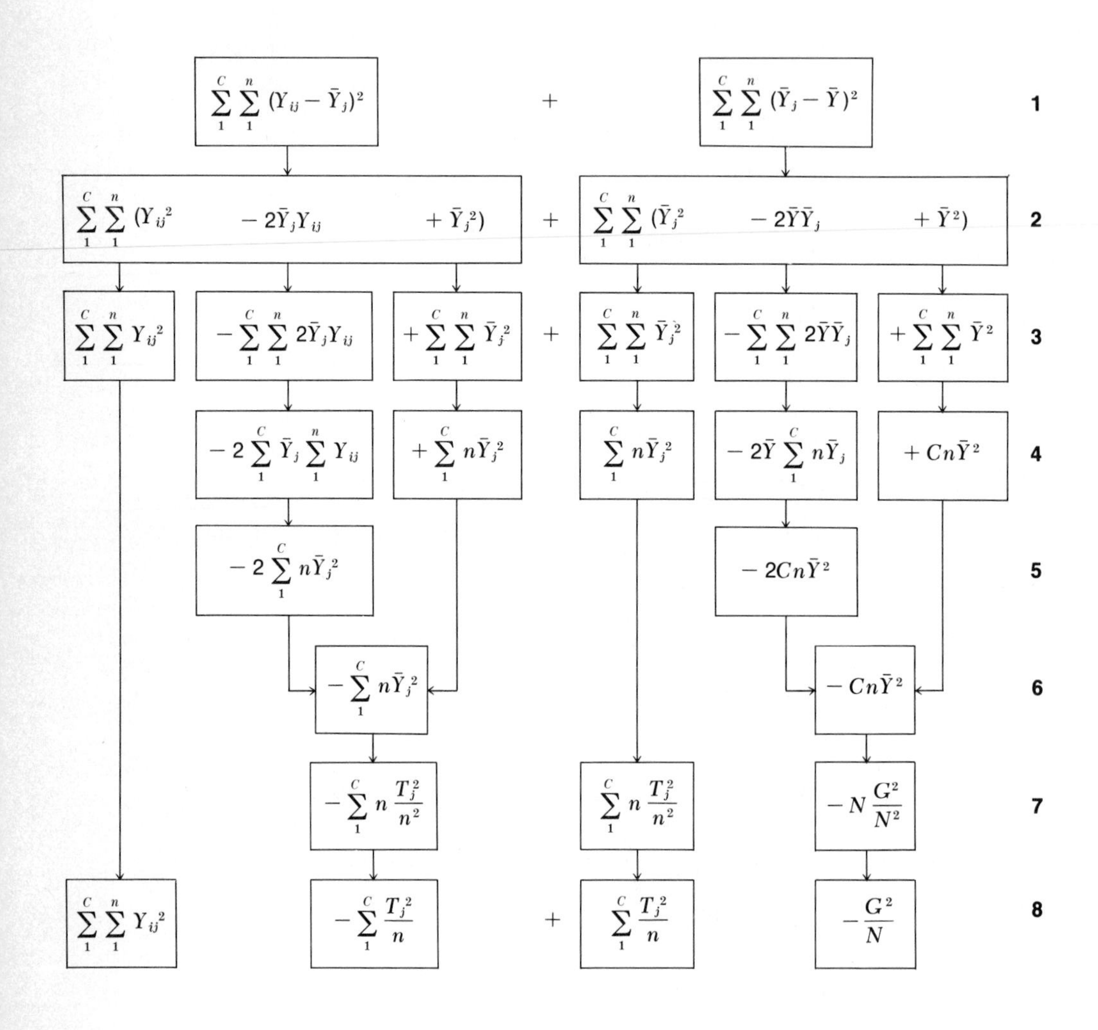

Table 13-4 Deviation Scores and Sums of Squares in a Single-Factor Experiment

Deviation	Sum of Squares	Name	Computing Formula
$Y_{ij} - \bar{Y}_j$	$\sum_1^C \sum_1^n (Y_{ij} - \bar{Y}_j)^2$	SS_W	$\sum_1^C \sum_1^n Y_{ij}^2 - \sum_1^C \frac{T_j^2}{n_j}$
$\bar{Y}_j - \bar{Y}$	$\sum_1^C \sum_1^n (\bar{Y}_j - \bar{Y})^2$	SS_B	$\sum_1^C \frac{T_j^2}{n_j} - \frac{G^2}{N}$
$Y_{ij} - \bar{Y}$	$\sum_1^C \sum_1^n (Y_{ij} - \bar{Y})^2$	SS_Y	$\sum_1^C \sum_1^n Y_{ij}^2 - \frac{G^2}{N}$

13-45 Step 1 (page 206) shows the two components of SS_Y. The one on the left is (SS_B; SS_W).

SS_W

13-46 When the squared deviation is expanded in step 2, each component becomes a sum of ______ terms.

3

13-47 Step 3 applies summation rule I. $\bar{Y}$ occurs in ______ of the 6 terms; $\bar{Y}_j$ occurs in ______ of them.

2
4

13-48 ($\bar{Y}_j$; $\bar{Y}$) is a constant for all N elements. ($\bar{Y}_j$; $\bar{Y}$) is a constant for all n elements within any column.

$\bar{Y}$; $\bar{Y}_j$

13-49 In step 4, $\bar{Y}_j$ is taken outside $\left(\sum_1^C; \sum_1^n\right)$; it remains inside ______. $\bar{Y}$ is taken outside *both* summations.

$\sum_1^n$
$\sum_1^C$

13-50 Two substitutions are made in step 5. Since $\sum_1^n Y_{ij}/n = \bar{Y}_j$, $\sum_1^n Y_{ij} =$ ______. Since $\sum_1^C n\bar{Y}_j/C = n\bar{Y}$, $\sum_1^C n\bar{Y}_j =$ ______.

$n\bar{Y}_j$; $Cn\bar{Y}$

13-51 After combining terms in step 6, a substitution is made for $\bar{Y}_j^2$ in step 7. The term substituted is ______. Also in step 7, ______ is substituted for ______.

$\frac{T_j^2}{n^2}$

$\frac{G^2}{N^2}$; $\bar{Y}^2$

13-52 After the fractions have been cleared in step 8, we have a total of ______ terms. The first two terms constitute the computing formula for ______; the last two terms are the computing formula for ______.

4
SS_W
SS_B

13-53 Since the middle two terms cancel, we are left with two terms, ______ and ______, whose sum is the computing formula for ______.

$\sum_1^C \sum_1^n Y_{ij}^2 \leftrightarrow -\frac{G^2}{N}$
SS_Y

REVIEW

13-54 This lesson is entitled Factorial Experiments because it prepares the way for the study of such experiments. However, we use the term "factorial experiment" only when the effects of more than one __________ are studied in the same experiment. *factor*

13-55 When the effects of only one factor are under study, we call the experiment a __________ experiment. *single-factor*

13-56 This single factor is called factor ______, and its levels are placed in separate __________ of the data table. *A* *columns*

13-57 The number of levels on factor A is the same as the number of __________ in the data table; this number is represented by the letter ______. *columns* *C*

13-58 The first level is called A_1, the second A_2, and so on. We represent any single level by the symbol ______, letting the subscript stand for any whole number from 1 to ______. A_j C

13-59 The number of observations in the entire experiment is N; the number in subgroup A_j is designated ______. When all the subgroups are of equal size, then the number in each subgroup is simply ______ $= N/C$. n_j n

13-60 The n observations in subgroup j are arranged vertically in column A_j. Each observation is designated Y_{ij}, where j represents the __________ to which the observation belongs and i stands for any whole number from 1 to ______. *subgroup (or column)* n *(or* n_j*)*

13-61 In a controlled experiment, the investigator is able to *assign* individuals at random to each of his __________. The French vocabulary example is a __________ experiment. *subgroups* *controlled*

13-62 Students assigned to A_1 use a different learning method from students assigned to A_2 or A_3. Since members of different subgroups get different treatments, the factor levels in this controlled experiment can be called __________. *treatments*

13-63 In the welfare study (page 199), the individuals (*are; are not*) assigned to different levels on the factor "recipient's education." Is this a controlled experiment? ______________ *are not* *No*

13-64 "Recipient's education" is a fixed factor in that study; every replication of the experiment will contain the same __________ on that factor. *levels*

13-65 "Welfare agency" is a random factor in that study. Every replication will contain a different __________ selection of factor levels. *random*

13-66 Whenever a factor has more possible levels than the intended number of subgroups, it may be treated as a __________ factor. *random*

13-67 "Teachers" is a __________ factor when the study concerns all the arithmetic teachers in the state. *random*

13-68 However, if conclusions are to be drawn only about the 5 arithmetic teachers in a city system, there are only ________ possible levels, and the factor will be a __________ factor. *5* *fixed*

13-69 Transform the expression $\sum_{1}^{N} (A_i - B_i)$ into a simpler form: ______. $\sum_{1}^{N} A_i - \sum_{1}^{N} B_i$

13-70 Simplify $\sum_{1}^{N} (3A_i - 6)$, taking $N = 10$. This expression becomes ______. $3 \sum_{1}^{N} A_i - 60$

13-71 $\sum_{1}^{C} \sum_{1}^{n} Y_{ij}$ is an example of a double summation. $\sum_{1}^{n}$ represents the summation (*across; within*) columns; $\sum_{1}^{C}$ represents the summation __________ columns. *within* *across*

13-72 The summation __________ columns must be performed first. The symbol for this summation is placed (*first; second*) in the double summation. *within* *second*

13-73 $\sum_{1}^{C} \sum_{1}^{n} (Y_{ij} - \bar{Y}_j)^2$ defines the sum of squares __________ columns. It describes the variation of scores around their __________, and it is designated as ______. *within* *column means; SS_W*

13-74 $\sum_{1}^{C} \sum_{1}^{n} (\bar{Y}_j - \bar{Y})^2$ defines the sum of squares __________ columns. It describes the variation of __________ means around the __________ mean, and it is designated as ______. *between* *column* *grand; SS_B*

13-75 We now call $\sum_{1}^{C} \sum_{1}^{n} (Y_{ij} - \bar{Y})^2$ the __________ sum of squares and designate it as ______. $SS_W + SS_B =$ ______. *total* *SS_Y; SS_Y*

PROBLEMS

1. For a study of the effect of a certain drug on mental arithmetic scores, factor A is "quantity of drug administered." This quantity may vary from 0 to 100 milligrams. Only six subgroups are feasible. Assuming that there are at least 100 levels of this factor, tell how it could be treated as a random factor. Would it be possible to treat it as a fixed factor?

2. Find the column totals T_j, the grand total G, and the sum $\sum_1^C \sum_1^n Y_{ij}^2$ for the data in the following table. Then derive SS_W, SS_B, and SS_Y, checking to see whether $SS_W + SS_B = SS_Y$.

A_1	A_2	A_3
95	91	88
90	87	82
86	84	79
85	80	75
83	78	74
81	76	72
79	75	70
78	74	69
77	73	69
76	72	67
73	70	64
71	68	63
67	65	60
66	62	57
61	58	54

LESSON 14 THE ADDITIVE MODEL IN SINGLE-FACTOR EXPERIMENTS

As we enter upon statistical tests for differences among two or more means, we should make sure that our thinking is still firmly based on the principles developed in Part One. It will pay us to spell them out as they apply to a single-factor experiment in which factor A is fixed.

In the fixed-factor vocabulary experiment we have three samples to compare. On H_1 we assume that all three come from the same population and that any differences we observe among them, including any differences among the three sample means, have arisen through sampling variability alone. This parent population has a mean μ and a variance σ^2, neither of which is precisely known.

If all the samples are of about the same size, we can consider their means $\bar{Y}_1$, $\bar{Y}_2$, and $\bar{Y}_3$ as belonging to the same RSD of means of samples of size n. Provided n is large enough and provided the population distribution is approximately normal, we know that the mean of this RSD is μ and that its variance is σ^2/n. We can estimate μ and σ from each sample; by combining or *pooling* these estimates, we can arrive at a satisfactory picture of the RSD which is needed for a statistical test of H_1.

Notice that several things have to be true of the experiment before we can apply this kind of reasoning:

1 The samples have to be similar to each other in certain respects. They must all be unimodal and not too skewed, or else we shall wonder whether it is reasonable to suppose they come from a single population which is approximately normally distributed. They must have similar n and similar s^2, or else we shall not be able to think of them as sharing a common RSD of means.

2 The samples must be randomly drawn, and they must be independent of each other. It will not do to have the same student contribute a score to more than one subgroup. We must be able to consider each subgroup as a random sample of size n from this particular parent population, and we must also be able to consider the entire set of N scores as a random sample of size N from the same population.

Normality of the parent population helps in arriving at an RSD. It is even more important, however, in suggesting the nature of the factors influencing the scores. Recall from Lesson 9D that evidence for a normal distribution is presumptive evidence of the operation of many independent, random, and additive influences. In Sec. A we begin considering the kinds of influences which may affect the individual values of Y_{ij}.

A. SCORE COMPONENTS IN THE ADDITIVE MODEL

When we assume that all N scores are drawn from a population with mean μ, we may suppose that all the scores would *equal* μ if there were no influences operating to produce variation. Each individual score can then be supposed to contain two components, the component μ plus a component representing the deviation from μ brought on by the unknown influences. When we are only interested in μ itself, we can consider all these unknown influences as factors which produce *error*, and we can think of the individual score as μ plus an ERROR COMPONENT. We consider each Y_{ij} as a sum of two components:

$$Y_{ij} = \mu + e_{ij}$$

where e_{ij} represents the value of the error component for a particular Y score. This error component e_{ij} is a *variance-producing* component. It is the sum of all the positive and negative effects of influences producing variation from μ.

The equation $Y_{ij} = \mu + e_{ij}$ says, "For every Y, there is some amount of error which has combined with μ to produce the value of Y." This statement can be made with some confidence whenever the observations appear to come from a normally distributed population. The population distribution makes it reasonable to suppose that each score is an additive combination of a large number of independent and random influences. When we are interested mainly in μ, all these influences are regarded as part of one term, e_{ij}.

However, we are actually interested in using the N observations to detect and demonstrate the effect of an independent or experimental variable, factor A. This factor may exert its influence in an additive way, just like the other influences contained in e_{ij}. If so, it is possible to remove factor A's effect from the error term and give it a separate term in the equation. If a_j represents the effect of factor A at level A_j, we can write Y_{ij} as the sum of three terms:

$$Y_{ij} = \mu + a_j + e_{ij}$$

Like any other additive component, a_j may have either a positive or a negative sign, or it may be zero. We write a_j rather than a_{ij} because this component (unlike e_{ij}) does not vary among scores in the same column. We shall call a_j the EFFECTS COMPONENT.

When we write the effect of factor A in this additive way, we rule out other ways in which the factor might be thought to influence the scores. For example, we rule out the possibility that it operates by multiplying μ or by changing any of the influences contained in e_{ij}. According to this equation, factor A divides the total population into a set of subpopulations. Each subpopulation has a mean $\mu + a_j$ which deviates from μ by the amount a_j. Each subpopulation has the same variance because e_{ij} produces the subpopulation variance and e_{ij} varies among scores in a random way unaffected by a_j.

Figure 14-1 shows how the equation directs us to think about the effects of factor A. If H_1 is true, there are no subpopulations with differing means; all $a_j = 0$, and there is only one parent population with mean μ and variance σ^2. If H_2 is true, there are at least two (Fig. 14-1*b* shows three) subpopulations; these subpopulations have the same variance, but they have different means. The mean of the subpopulation means is μ.

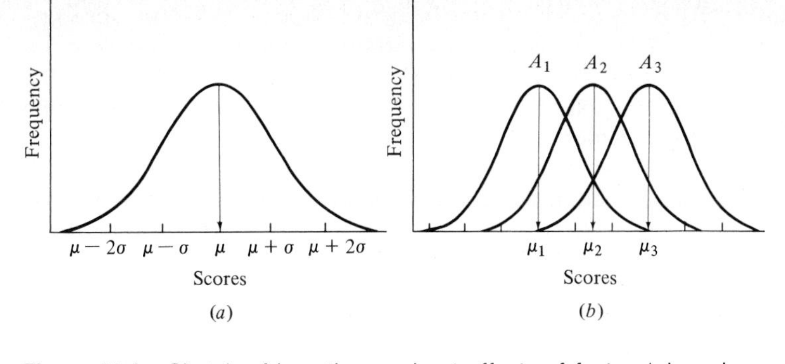

Figure 14-1: Sketch of hypotheses about effects of factor A based on the additive model. (*a*) The single population assumed in H_1; (*b*) a possible version of H_2, with three subpopulations having different means but equal variances.

14-1 We have drawn N scores from a population whose unknown mean is μ. If there are *no* influences operating to produce variation in this population, all the scores will be equal to ________.

μ

14-2 The scores are not all equal; they appear to be normally distributed. We can expect that the influences which produce this variation are ________, ________, and ________.

independent ↔ random ↔ additive

14-3 Since we are interested in factor A, we divide these influences into two kinds. One is the effect of ________; all the rest are regarded as ________ factors.

factor A
error

14-4 Each score is the sum of three components: the population mean μ, the effects component for its subgroup (called ________), and its own unique error component (called ________).

a_j
e_{ij}

14-5 If H_1 is true, a_j equals ________ for every subgroup and the true mean of every subgroup is equal to ________.

0
μ

14-6 If H_2 is true, there is at least one subgroup for which $a_j \neq 0$. The true mean of such a subgroup is ________.

$\mu + a_j$

14-7 The variance-producing component within a subgroup is (a_j; e_{ij}). Since a_j does not affect e_{ij}, the subgroups (*do; do not*) have different variances.

e_{ij}; do not

14-8 The effects component a_j produces variation among the ________ of the subgroups.

means

14-9 Variation *within* the subgroup is produced by the ________ component.

error

The equation $Y_{ij} = \mu + a_j + e_{ij}$ is the **ADDITIVE MODEL OF SCORE COMPONENTS**. It underlies the use of analysis of variance to compare two or more sample means. Therefore, analysis of variance cannot be used unless the experiment fits this model, just as the binomial test cannot be used unless the experiment fits a coin-toss model.

The additive model also applies to experiments in which factor A is a random factor, but there is an important difference in our picture of the situation. Suppose factor A is "teachers" and Y represents "arithmetic scores." With all arithmetic teachers in the state as possible levels of this factor, we treat factor A as random and select only a random sample of teachers whose pupils will form our subgroups. The same number n of pupils is randomly selected from each teacher's classes.

If H_1 is true, we can still picture the situation as shown in Fig. 14-1*a*. But we cannot draw a diagram to represent the situation when H_2 is true. Let us say we have $C = 8$ subgroups with sample means, but there are many more than 8 levels, each representing a subpopulation which may have a different mean. If H_2 is true, there is a whole *distribution* of such subpopulation means—something impossible to diagram in only two dimensions. We can still say that the mean of each subpopulation is $\mu + a_j$ and that μ is the mean of all the subpopulation means. If we let C^* represent the total number of possible levels of factor A, we can say that $\sum_1^{C^*} a_j = 0$: the sum of the deviations of *all* subpopulation means from μ is zero. But we have to remember that we shall not actually observe a sample from each subpopulation in any single experiment with a random factor A. When factor A is fixed, $\sum_1^{C} a_j = 0$ even when H_2 is true. But when factor A is random and H_2 is true, $\sum_1^{C} a_j$ may not be 0 for the subpopulations actually observed.

In random-factor experiments, we are not interested in comparing the specific subgroups that we happen to be studying. We look for differences among them only because these differences give a clue to the distribution of subpopulation means. We are really interested in the variance of that distribution. If H_1 is true, the means do not differ, the variance of $\mu + a_j$ is 0, and the variance of a_j is also 0. The variance of a_j is called **EFFECTS VARIANCE**, and it is written σ_A^2 to indicate its connection with factor A.

In fixed-factor experiments we are definitely interested in a fixed set of subpopulation means for their own sake. If H_1 is true, the means do not differ and each $a_j = 0$. If H_2 is true, at least one of the a_j's differs from 0. We could say, in this case, that σ_A^2 is not equal to 0, but we do not talk in terms of σ_A^2 in a fixed-factor experiment. It is more meaningful in such an experiment to talk about actual values of a_j.

In random-factor experiments, on the other hand, we are not interested in particular subpopulation means as such. We are interested in the distribution of subpopulation means, and it is more meaningful to talk about the variance of this distribution. If H_1 is true, $\sigma_A^2 = 0$; if H_2 is true, σ_A^2 is different from 0. We thus think directly in terms of effects variance.

14-10 When factor A is "learning method," there may be three subpopulations, whose means are $\mu + a_1$, ______, and ______.

$\mu + a_2$
$\mu + a_3$

14-11 If we can show that $a_1 > a_2 > a_3$, we can conclude that method (*1, 2, 3*) is the best method and method ______ is the worst.

1; 3

14-12 When factor A is "teachers," there may be as many subpopulations as there are ______ on this factor. We collect evidence on only ______ of these possible subpopulations.

levels
8

14-13 One subpopulation is made up of scores from Mrs. Smith's classes; we sample n of these scores. We also sample n scores from Mrs. Brown's classes. If we can show that the mean $\mu + a_j$ is higher for Brown than Smith, we can conclude that ______ is the more effective teacher.

Brown

14-14 However, we are not specifically interested in comparing Smith and Brown. We are interested in showing that differences among teachers in general have some ______ on pupil performance.

effect

14-15 Differences among our eight sampled subpopulations are not important in themselves; they are only important because they show that different levels on factor A can produce different subpopulation ______.

means

14-16 Thus, we are not interested in the particular means but in the *distribution* of such means and, in particular, in the size of the ______ of this distribution.

variance

14-17 We call this variance "______ variance" and symbolize it by ______.

effects
σ_A^2

14-18 σ_A^2 will be different from 0 whenever at least some values of ______ are different from 0.

a_j

14-19 We think in terms of effects variance when factor A is a ______ factor.

random

14-20 When we write H_1: all $a_j = 0$, we are dealing with a factor A which is ______; when we write H_1: $\sigma_A^2 = 0$, we are dealing with a factor A which is ______.

fixed
random

14-21 When factor A is fixed, we are interested in comparing the ______ of the subgroups we have studied, for their own sake. When factor A is random, we are interested only in the ______ of all possible subgroup means.

means
variance

B. ESTIMATING EFFECTS AND ERROR FROM DATA

The equation $Y_{ij} = \mu + a_j + e_{ij}$ is a theoretical equation describing the model we wish to apply to our data. We do not actually know the values of any of the terms on the right-hand side of this equation, but we can estimate them from the sample data.

Let us first transpose the term μ, writing our model as $Y_{ij} - \mu = a_j + e_{ij}$. Since $E(\bar{Y}) = \mu$, we can estimate μ from the grand mean of the entire distribution $\bar{Y}$. The deviation $Y_{ij} - \mu$ can therefore be estimated by $Y_{ij} - \bar{Y}$, the deviation of a Y score from the grand mean $\bar{Y}$.

The expectation $E(\bar{Y}_j)$ of a subgroup mean is $\mu + a_j$. Therefore, we can estimate the value of a_j for any particular level by taking the deviation of its subgroup mean from the grand mean $\bar{Y}_j - \bar{Y}$.

Now recall from Lesson 13 that any particular deviation score $Y_{ij} - \bar{Y}$ can be broken into two components, the deviation of Y_{ij} from its column mean $Y_{ij} - \bar{Y}_j$ and the deviation of its column mean from the grand mean $\bar{Y}_j - \bar{Y}$:

$$Y_{ij} - \bar{Y} = (Y_{ij} - \bar{Y}_j) + (\bar{Y}_j - \bar{Y})$$

The deviation $\bar{Y}_j - \bar{Y}$ contains the effect of the one influence, factor A, in which we are interested. The other component, $Y_{ij} - \bar{Y}_j$, combines all the other influences in which we are not interested for the moment. $Y_{ij} - \bar{Y}_j$ thus represents an estimate of e_{ij}, the error component.

At the end of Lesson 13 we saw that $Y_{ij} - \bar{Y}_j$ can be squared and summed for all Y_{ij}, yielding a sum of squares called SS_W, the within-columns sum of squares. Now we can add that SS_W reflects the influence of the error component alone. The other deviation $\bar{Y}_j - \bar{Y}$ can also be squared and summed for all Y_{ij}, yielding the between-columns sum of squares SS_B. We can now say that SS_B *contains* the influence of factor A upon the N scores in the distribution, but we cannot say that it reflects *only* the influence of factor A. SS_B would be a pure reflection of factor A's influence if each $\bar{Y}_j - \bar{Y}$ were exactly equal to a_j, but we know that sampling variability has operated to influence the sample means we actually observe. Therefore, SS_B contains some error along with the influence of factor A.

Let us now rewrite Table 13-4 (page 207) including the relation of each component to the additive model. Remember that $SS_W + SS_B = SS_Y$, so that SS_Y will also reflect the influence of both factor A and the other factors which are all "error" from our viewpoint.

Table 14-1 Components of the Additive Model and Their Estimates from Data

Component from Model	Estimate	Sum of Squares	Estimate Influenced by
e_{ij}	$Y_{ij} - \bar{Y}_j$	SS_W	Error only
a_j	$\bar{Y}_j - \bar{Y}$	SS_B	Effects and error
$Y_{ij} - \mu$	$Y_{ij} - \bar{Y}$	SS_Y	Effects and error

14-22 $\bar{Y}_1$ is actually known. It is the ________ of column A_1 in the data table. *mean*

14-23 Scores in column A_1 are a random sample from the subpopulation whose mean is ________. $\mu + a_1$

14-24 We do not know the true value of $\mu + a_1$. Our best estimate is ________. $\bar{Y}_1$

14-25 Our best estimate of μ is ________. Our best estimate of a_1 therefore is ________. $\bar{Y}$; $\bar{Y}_1 - \bar{Y}$

14-26 For any column, our best estimate of the effects component a_j is ________. $\bar{Y}_j - \bar{Y}$

14-27 Variation among scores within column 1 cannot be due to differences in factor A. All the scores in column 1 belong to level ________ on this factor. A_1

14-28 Variation within the column must then be due to the ________ components of the individual scores, not to the effects component. *error*

14-29 If we knew a_1 and μ, we could state exactly how big the error component for a score in column 1 must be. Since $Y = \mu + a_1 + e$, we could find e by subtracting ________ and ________ from Y itself. μ; a_1

14-30 We do not know μ, but we estimate it to be ________. We estimate a_1 to be ________. Therefore, we can estimate e, for the score Y in column 1, to be ________. $\bar{Y}$; $\bar{Y}_1 - \bar{Y}$; $Y - \bar{Y}_1$

14-31 For any score Y_{ij} in any column, we can estimate its error component e_{ij} to be ________. $Y_{ij} - \bar{Y}_j$

14-32 $Y_{ij} - \bar{Y}_j$ is a deviation whose square enters into the sum of squares (SS_B; SS_W). The square of $\bar{Y}_j - \bar{Y}$ enters into the other component sum of squares, ________. SS_W; SS_B

14-33 SS_W is therefore related closely to the ________ component in the additive model. *error*

14-34 The sum of squares which contains the effects component of the additive model is ________. SS_B

14-35 SS_B also contains some of the ________ component because $\bar{Y}$ and $\bar{Y}_j$ are only ________ of the true population and subpopulation means. *error*; *estimates*

C. THE MEAN-SQUARE ESTIMATES OF POPULATION VARIANCE

In this section we shall consider the situation assumed to be true under H_1. When H_1 is true, the distribution in Fig. 14-1*a* is the parent population for all samples and there are no subpopulations. This situation holds for all single-factor experiments, whether factor A is fixed or random.

Under H_1 the appropriate RSD is an RSD of the means of samples of size n. We can use the central limit theorem to find its parameters. The parameter μ gives us no trouble; it is best estimated as $\bar{Y}$, the grand mean of all N scores. To obtain its variance, which we must call $\sigma_{\bar{Y}}^2$ (read "small sigma, sub Y bar, squared") because it is a variance of the RSD of values of $\bar{Y}_j$, we first have to estimate population variance σ^2. We can make this estimate in more than one way.

Let us first take the method already known to you. In Lesson 7 you learned how to estimate the population variance σ^2 from sample data. Because $E(s^2) = [(N-1)/N]\sigma^2$, the estimated σ^2 is taken as $[N/(N-1)]s^2$. Usually the estimate is made directly from the sum of squares $\sum_1^N (Y_{ij} - \bar{Y})^2$. Since $s^2 = \text{SS}_Y/N$, the estimated $\sigma^2 = \text{SS}_Y/(N-1)$.

This estimated σ^2 is our first example of a **MEAN-SQUARE** estimate of variance. A mean square is a sum of squares divided by a number which represents the *degrees of freedom* contained in the sum of squares. SS_Y is a sum of N squared deviations, but only $N-1$ of these are independent; since $\sum_1^N (Y_{ij} - \bar{Y})$ must equal 0, the last deviation score is fully determined by the choice of the other $N-1$ deviation scores. Thus the set of deviation scores has $N-1$ rather than N degrees of freedom.

If it had not been necessary to *estimate* the population mean μ, the sum of squares would have had N degrees of freedom. We impose a restriction on SS_Y when we take $\bar{Y}$ as an estimate of the unknown parameter μ. The population sum of squares $\Sigma(Y_{ij} - \mu)^2$ does not have this restriction; all the deviation scores are independent in this case because the entire population is under study. No arbitrary selection of just N values has taken place, and there is no arbitrary restriction of the last deviation score to make $\sum_1^N (Y_{ij} - \bar{Y})$ equal 0.

We shall call this estimate MS_Y, the mean square based on SS_Y. This method of estimating σ^2 is not new. We have only added the concept of degrees of freedom and thus defined $\text{SS}_Y/(N-1)$ as a *mean-square* estimate of σ^2. But $\text{SS}_Y = \text{SS}_W + \text{SS}_B$; there are two other sums of squares, carefully separated in Lesson 13, which together make up SS_Y. Since each of these is a sum of squares, it is logically possible to convert each component into a mean-square estimate of variance. We have only to divide each one by the proper number of degrees of freedom. Our next step will be to show how df is determined for SS_W and SS_B.

14-36 Lesson 7 distinguished s^2 from σ^2. s^2 represents ______ variance; σ^2 represents ______ variance. *sample; population*

14-37 We defined s^2 as SS/N, where SS stood for the sum of squared deviations from the (*population; sample*) mean and N was the number of elements in the (*population; sample*). *sample* *sample*

14-38 The sample mean is now being represented by the letter ______, and we call it the ______ mean. *$\bar{Y}$; grand*

14-39 $\sum_{1}^{N} (Y_{ij} - \bar{Y})^2/N$ defines (s^2; σ^2). If N^* is the number of elements in the population, $\sum_{1}^{n*} (Y_{ij} - \mu)^2/N^*$ defines ______. *s^2* *σ^2*

14-40 The squared deviations which contribute to σ^2 are taken from the (*population; sample*) mean. *population*

14-41 When we do not know the exact value of σ^2, we can estimate it from sample statistics. The estimate, as you learned in Lesson 7, is SS divided by ______. *$N - 1$*

14-42 We are now giving this sum of squares a more specific name. Instead of calling it simply SS, we write the symbol ______ and call it the ______ sum of squares. *SS_Y; total*

14-43 SS_Y is the sum of squared deviations of all N scores from their ______ mean. Write the summation term which defines SS_Y: ______. *grand* *$\sum_{1}^{N} (Y_{ij} - \bar{Y})^2$*

14-44 $SS_Y/(N - 1)$ is an estimate of ______ variance. The quantity $N - 1$ represents the number of degrees of freedom possessed by ______. *population* *SS_Y*

14-45 SS_Y is a sum of ______ numbers. Only ______ of these are able to vary independently; ______ of them is restricted by the fact that $\sum_{1}^{N} (Y_{ij} - \bar{Y})$ must equal ______. *N; $N - 1$* *1* *0*

14-46 We impose this one restriction on SS_Y when we take ______ as an estimate of the unknown parameter ______. *$\bar{Y}$; μ*

14-47 A sum of squares divided by its df is called a mean square. $SS_Y/(N - 1)$ is a ______, since SS_Y has $N - 1$ ______ of ______. *mean square* *degrees; freedom*

14-48 Since $SS_Y/(N - 1)$ is a mean square, we give it the symbol ______. It is an ______ of population variance. *MS_Y; estimate*

Now we turn to the component sums of squares, SS_W and SS_B. SS_W is made up of squares of the deviations $Y_{ij} - \bar{Y}_j$. For each column, $\sum_1^n (Y_{ij} - \bar{Y}_j)$ must equal 0; if there are n_j scores in the column, the number of degrees of freedom is $n_j - 1$. One restriction is thus imposed for each column when $\bar{Y}_j$ is taken as the estimated value of the subpopulation mean $\mu + a_j$. There are N squared deviations in SS_W, but there are C restrictions; the total number of degrees of freedom is therefore $N - C$.

$SS_W/(N - C)$ is a mean-square estimate of population variance σ^2, and we shall call it MS_W (mean square within columns). It is a *pooled* estimate from C different samples, each of which is capable of yielding an independent estimate of σ^2. $\sum_1^n (Y_{ij} - \bar{Y}_j)^2/(n_j - 1)$ will give an estimate from a single sample. If these estimates were calculated separately and combined into one, the result would be the same estimate as MS_W.

SS_B is made up of N squared deviations, where the deviations are $\bar{Y}_j - \bar{Y}$. However, there are only C different values of $\bar{Y}_j$; the deviation $\bar{Y}_j - \bar{Y}$ is exactly the same for every score in column A_j. Once you have determined $\bar{Y}_j - \bar{Y}$ for one score in the column, you have fixed the value of this deviation for all scores in the column. SS_B therefore cannot have more than C degrees of freedom. Furthermore, $\sum_1^C (\bar{Y}_j - \bar{Y})$ must equal 0; the mean of the subgroup means equals the grand mean. Therefore, only $C - 1$ of the C different values of $\bar{Y}_j - \bar{Y}$ can be independent; the last one is fully determined, once the first $C - 1$ are known. SS_B thus has $C - 1$ degrees of freedom.

We can look at SS_B another way. It is an empirical sampling distribution of subgroup means. There are C means in the distribution, and each $\bar{Y}_j - \bar{Y}$ is a deviation of one of the means from the distribution mean $\bar{Y}$. But $\bar{Y}$ is an estimate of μ, the true mean of the RSD of subgroup means. Taking $\bar{Y}$ as an estimate of μ imposes one restriction on the sum of squares $\sum_1^C (\bar{Y}_j - \bar{Y})^2$, making its $df = C - 1$.

When we look at SS_B as a sample of subgroup means, we see that $SS_B/(C - 1)$ will be an estimate of σ^2 based on this sample of means. We call $SS_B/(C - 1)$ the mean square between columns, MS_B. We have taken the values of $\bar{Y}_j$ and treated them as a sample *from the RSD of sample means.* Instead of trying to estimate the variance of this RSD, as we did in Lesson 9, we have used the sum of squares from our set of sample means (SS_B) to estimate σ^2, the variance of the parent population. Thus MS_W and MS_B are independent estimates of σ^2. The variation of scores *within* columns cannot influence MS_B; the variation *between* columns cannot influence MS_W.

Since $SS_W + SS_B = SS_Y$, we have only two (not three) independent estimates of σ^2; MS_Y is dependent on each of the other two estimates. Table 14-2 summarizes the considerations leading to these mean-square estimates. Observe that the df for SS_W and SS_B add to make up the df for SS_Y. In Table 14-2, the columns SS and df are additive, while the column MS is not. MS_W plus MS_B does not equal MS_Y.

Table 14-2 Sums of Squares and Mean Squares in the Additive Model

SS	Definition	Number of Different Deviation Scores	Number of Parameters Estimated	df	Mean Square
SS_W	$\sum_1^C \sum_1^n (Y_{ij} - \bar{Y}_j)^2$	N	C	$N - C$	$\frac{SS_W}{N-C} = MS_W$
SS_B	$\sum_1^C \sum_1^n (\bar{Y}_j - \bar{Y})^2$	C	1	$C - 1$	$\frac{SS_B}{C-1} = MS_B$
SS_Y	$\sum_1^C \sum_1^n (Y_{ij} - \bar{Y})^2$	N	1	$N - 1$	$\frac{SS_Y}{N-1} = MS_Y$

14-49 $\sum_1^C \sum_1^n (Y_{ij} - \bar{Y}_j)^2$ defines ______. It is a sum of ______ numbers. The column means $\bar{Y}_j$ are taken as estimates of the unknown subpopulation ______ $\mu + a_j$.

SS_W; N

means

14-50 Each estimate of a parameter such as $\mu + a_j$ removes 1 degree of freedom. With C columns, we must estimate ______ subpopulation means, and we lose ______ degrees of freedom.

C

C

14-51 SS_W thus has ______ degrees of freedom. SS_W divided by ______ is a mean square.

$N - C$

$N - C$

14-52 $\sum_1^C \sum_1^n (\bar{Y}_j - \bar{Y})^2$ defines ______. It is also a sum of ______ numbers. However, the deviation $\bar{Y}_j - \bar{Y}$ is the same for all n_j scores in a particular ______.

SS_B

N

column

14-53 If we know $\bar{Y}_j - \bar{Y}$ for one score in a column, we know $\bar{Y}_j - \bar{Y}$ for a total of ______ scores. When we have C columns, there are only ______ different values of $\bar{Y}_j - \bar{Y}$, one for each ______.

n_j (or n)

C

column

14-54 Moreover, the mean of the subgroup means must equal the ______ mean. $\sum_1^C (\bar{Y}_j - \bar{Y})$ must equal ______.

grand; 0

14-55 There are C different deviations $\bar{Y}_j - \bar{Y}$, but only ______ of these are independently varying. SS_B thus has only ______ df, and MS_B = ______.

$C - 1$

$C - 1$

$\frac{SS_B}{C-1}$

14-56 SS_W has ______ df; SS_B has ______ df. The total df for *both* components is ______, which equals the df for ______.

$N - C$; $C - 1$

$N - 1$

SS_Y

D. EXPECTATIONS OF THE MEAN SQUARES

Section C developed the argument as if H_1 were actually true. But what if H_1 is not true? If factor A is a source of variation in Y, the effects of factor A will be found in SS_B. No such effects will be found in SS_W. By partitioning SS_Y into these two components, we have confined factor A's effects to the component SS_B.

We can now put these facts in terms of the mean-square estimates of variance, MS_W and MS_B. Since only the error component e_{ij} affects SS_W, we can state that MS_W is an estimate of ERROR VARIANCE σ_e^2. SS_B is affected by a_j as well as by e_{ij}. Hence MS_B is not a pure estimate of error variance unless H_1 is true. When H_1 is true, factor A has no effect, and all population variance is due to error. Then both MS_W and MS_B are simply estimates of population variance. When H_1 is false, population variance and error variance are no longer equal; some population variance is due to factor A. MS_W still estimates error variance alone. Since sampling error enters into the estimation of $\mu + a_j$, MS_B estimates a variance which is a mixture of error variance and the effects of factor A.

We can say a little more about the particular mixture of error variance and effects in MS_B when we consider the expectations of these mean squares. $E(MS_W)$ and $E(MS_B)$ are not difficult to derive algebraically. In this book, however, we shall discuss them only in general terms.

Recall, first, that the expectation of any statistic is the average of that statistic over all possible random samples from the same population (pages 104 to 105) If we imagine repeating the same experiment many times, we can imagine a statistic MS_W for each of the successive samples we observe. Each time we observe a new sample, we average its MS_W with those observed in previous samples. As the number of replications gets very large, the mean of these observed MS_W's will approach the mean of the MS_W's for all possible samples, namely, the expectation of the statistic MS_W.

Table 14-3 tells us that the expectation of MS_W is simply error variance σ_e^2 for both fixed and random factors. Since $E(MS_W) = \sigma_e^2$, we can consider MS_W to be an unbiased estimate of σ_e^2. Whenever we need an estimate of error variance in a single-factor experiment, we cannot do better than to take the observed value of MS_W.

The expectation of MS_B is more complex, and its value depends on whether factor A is fixed or random. You will not need to remember the mathematical details of these expectations. Remember the following facts:

1 When factor A is random, $E(MS_B)$ is a simple sum of error variance and $n\sigma_A^2$, that is, effects variance multiplied by n.

2 When factor A is fixed, $E(MS_B)$ is a sum of error variance and a term which varies directly with the sum of a_j^2. Since a_j is the deviation of a subpopulation mean from μ, $E(MS_B)$ increases as the size of these deviations increases.

The expectations $E(MS_W)$ and $E(MS_B)$ play a key role in statistical tests for the effects of factor A. Lesson 15 will show exactly how they are used.

Table 14-3 Expectations of the Mean Squares for Single-Factor Experiments

Mean Square	E(MS) when Factor A Is Fixed	E(MS) when Factor A Is Random
MS_W	$E(MS_W) = \sigma_e^2$	$E(MS_W) = \sigma_e^2$
MS_B	$E(MS_B) = \sigma_e^2 + \dfrac{\sum_1^C n_j a_j^2}{C-1}$	$E(MS_B) = \sigma_e^2 + n\sigma_A^2$
If H_1 is true	$\sum_1^C a_j^2 = 0$; $E(MS_B) = \sigma_e^2$	$\sigma_A^2 = 0$; $E(MS_B) = \sigma_e^2$

14-57 The additive model $Y_{ij} = \mu + a_j + e_{ij}$ shows two kinds of influences making Y_{ij} different from μ: a_j represents the influence of ________; e_{ij} represents the influence of ________.

factor A; error

14-58 Only one of the two influences enters into MS_W. MS_W is affected by (a_j; e_{ij}) but not by ______.

e_{ij}; a_j

14-59 MS_W is a mean-square estimate of variance. Since it is not affected by a_j, the variance which it estimates is the variance due to ________.

error

14-60 Error variance is represented by σ_e^2. Table 14-3 shows that the expectation of MS_W is always ______.

σ_e^2

14-61 Both a_j and e_{ij} affect MS_B. MS_B is also a ________ estimate of variance, but the variance which it estimates is a mixture. Some of it arises from (*effects*; *error*), and some of it arises from the ________ of factor A.

mean-square

error

effects

14-62 Table 14-3 shows that the expectation of MS_B is σ_e^2 only when ______ is true and factor A has ________.

H_1; *no effects*

14-63 When H_1 is true, $E(MS_W)$ and $E(MS_B)$ are both equal to ______.

σ_e^2

14-64 When H_2 is true, MS_B is a sum of two terms. One of these terms is always ______; the other term depends on the amount of effect exerted by ________.

σ_e^2

factor A

14-65 When factor A is fixed, we are always interested in the actual subpopulation means, and the amount of effect is indicated by the size of (a_j; σ_A^2).

a_j

14-66 When factor A is random, we are interested only in the ________ of the distribution of subpopulation means and the amount of effect is indicated by the size of ______.

variance

σ_A^2

REVIEW

14-67 This lesson explains the additive model of score __________. Whenever we are dealing with normally distributed scores, we can suppose them to be determined by a large number of independent and random influences operating in an __________ way.

components

additive

14-68 $Y_{ij} = \mu + a_j + e_{ij}$ is the __________ model. Each score Y_{ij} consists of μ, which is the unknown __________, and two components which cause Y_{ij} to differ from ________.

additive
population mean
μ

14-69 The error component is ________. The effects component is ________.

e_{ij}
a_j

14-70 Factor A determines the size of the component ________. Factor A does not influence the component ________ in any way.

a_j; e_{ij}

14-71 Variation *within* the columns or subpopulations is produced by the component ________. Variation *between* the columns is produced by ________.

e_{ij}
a_j

14-72 The variance of a subpopulation is therefore determined by ________. Since factor A (*can; cannot*) affect this component, all subpopulations will have (*different; the same*) variance.

e_{ij}; *cannot*
the same

14-73 Each subpopulation has a mean equal to ________. If factor A has no effect, all the a_j values equal ________ and every subpopulation mean is equal to ________.

$\mu + a_j$
0
μ

14-74 We are always interested in comparing *particular* subpopulation means when factor A is a (*fixed; random*) factor.

fixed

14-75 In such cases, we think in terms of the size of a_j. If H_1 is true, all the a_j values equal ________. If H_2 is true, (*all; some*) of the a_j values are different from 0.

0; some

14-76 Not all the levels of factor A are represented when the factor is (*fixed; random*).

random

14-77 In such experiments, we are not interested in comparing particular subpopulation means; instead, we try to learn something about the __________ of the distribution of subpopulation means.

variance

14-78 We call this variance $\sigma_A{}^2$ or __________ variance. When H_1 is true, $\sigma_A{}^2 =$ ________.

effects
0

14-79 We describe H_1 and H_2 in terms of σ_A^2 when factor A is ________. We describe them in terms of a_j when factor A is ________.

random
fixed

14-80 We do not know the precise values of the components in the additive model. We have to ________ them from our sample data.

estimate

14-81 $\bar{Y}$ serves as our estimate for ________. $\bar{Y}_j$ is our estimate of $\mu + a_j$, and $\bar{Y}_j - \bar{Y}$ serves as our estimate of ________.

μ
a_j

14-82 $Y_{ij} - \bar{Y}_j$ serves as an estimate of ________. When it is squared and summed for all N scores, it becomes a sum of squares called ________, the sum of squares ________ columns.

e_{ij}
SS_W; *within*

14-83 $\sum_1^C \sum_1^n (Y_{ij} - \bar{Y}_j)^2$ has ________ degrees of freedom. When SS_W is divided by its df, it becomes a ________.

$N - C$
mean square

14-84 MS_W, the mean square ________ columns, is an estimate of (*effects; error*) variance.

within
error

14-85 σ_e^2 is the symbol which represents ________ variance. The expectation of MS_W is ________ in all single-factor experiments.

error
σ_e^2

14-86 $\sum_1^C \sum_1^n (\bar{Y}_j - \bar{Y})^2$ is called ________, the sum of squares ________ columns. It has ________ degrees of freedom.

SS_B
between ; $C - 1$

14-87 $SS_B/(C - 1)$ defines the mean square ________ columns. Because $\bar{Y}_j - \bar{Y}$ is only an estimate of a_j, MS_B is affected by sampling variability. In addition to the effects of factor A, MS_B contains some ________ variance.

between
error

14-88 The expectation of MS_B is a sum of two terms. One of these is always ________.

σ_e^2

14-89 The other term is $n\sigma_A^2$ when factor A is ________; it is $\Sigma n_j a_j^2/(C - 1)$ when factor A is ________.

random
fixed

14-90 This second term will be 0 whenever (H_1; H_2) is true. In that case, $E(MS_B)$ will equal ________, just like $E(MS_W)$.

H_1
σ_e^2

14-91 SS_Y, the ________ sum of squares, has ________ df. This number is the sum of the df for ________ and ________.

total; $N - 1$
$SS_W \leftrightarrow SS_B$

PROBLEMS

1 For Prob. 2 on page 210, determine the degrees of freedom for SS_W, SS_B, and SS_Y. Check to make sure that the df for SS_W and SS_B add to give the df for SS_Y. Then calculate MS_W, MS_B, and MS_Y. Does $MS_W + MS_B = MS_Y$? Explain why the mean squares are not additive like the sums of squares and the degrees of freedom.

2. The following table gives subgroup n, T_j, and $\sum_1^n Y_{ij}^2$ for a one-way analysis of variance with five levels on factor A. Find the sums of squares and the mean squares for this table.

	A_1	A_2	A_3	A_4	A_5
n	30	34	32	33	35
T_j	789	697	752	1,297	784
$\sum_1^n Y_{ij}^2$	38,019	31,590	34,947	68,271	34,853

3. For the data in Prob. 2, above, give your best estimate for each of the following terms: μ, a_1, a_2, a_3, a_4, a_5. Is $\sum_1^C a_j = 0$?

LESSON 15
ONE-WAY ANALYSIS OF VARIANCE AND THE F TEST

Our single-factor experiments are now ready for a statistical test. We have formed certain statistics, called mean squares, to be used in the test. The general procedure we have already used in arriving at these mean squares is called ANALYSIS OF VARIANCE. In single-factor experiments, where the subgroup means to be compared are classified in only one dimension (one way), it is commonly called ONE-WAY analysis of variance.

To this point we have had in mind an H_1 stated in very broad and general terms. For the fixed-factor experiment, we assume that all three samples of vocabulary scores are random samples from the same population, in which there are no subpopulations with different means determined by factor A, "learning method." For the random-factor experiment, we assume that all eight samples of arithmetic scores are random samples from the same population, within which there is no distribution of subpopulations with different means determined by factor A, "teachers."

Since we have been assuming that the subgroups are all of approximately equal size, we can reason that their means will belong to the same RSD of sample means when H_1 is true. The subgroup means, in that case, will be expected to differ from each other only as much as sample means from the same RSD will differ. To complete our test, we need only find out how much difference should be expected among C sample means which belong to the same RSD. We need an RSD for the *differences* among C means of samples drawn from the same population.

As you will see in Lesson 16, we follow this line of reasoning quite explicitly when $C = 2$ and we have only two subgroup means to compare. In that case, we actually form a statistic called the estimated standard deviation of the RSD of differences between two means. We can do so because we have only one difference between means to evaluate, the difference $\bar{Y}_1 - \bar{Y}_2$. But in the vocabulary experiment, with $C = 3$, we have three such differences, $\bar{Y}_1 - \bar{Y}_2$, $\bar{Y}_1 - \bar{Y}_3$, and $\bar{Y}_2 - \bar{Y}_3$. In the arithmetic example with $C = 8$, we have $8!/2!6! = 28$ such differences. It will not do to test all these differences one at a time, even if we could bear the labor involved. We would then be acting as if all the differences were independent of each other when, in fact, no more than $C - 1$ of the differences can be independent. All the others are fully determined by the size of the first $C - 1$ differences.

Consequently, when C is greater than 2, we begin with an omnibus test, one which pools all these differences into a single statistic and asks whether the combined differences reach a significantly large value. This statistic is MS_B, the mean square based on differences between subgroup means. If MS_B

is small enough, the combined differences can all be regarded as due to sampling variability. If MS_B is large enough, we shall be able to reject H_1, concluding that at least some of the differences contain an effect of factor A.

A. HYPOTHESES FOR DECISION MAKING IN ONE-WAY ANALYSIS OF VARIANCE

The size of MS_B can be evaluated by comparing it with MS_W. The expectation of MS_W is equal to error variance σ_e^2. The expectation of MS_B is the sum of σ_e^2 and a term which will be zero when factor A has no effect. We could therefore state H_1, the hypothesis that factor A has no effect, as $MS_B = MS_W$. Or we can equally well make H_1 assert that the true value of the ratio MS_B/MS_W is 1.

When H_2 is true and factor A does have some effect, MS_B should be larger than MS_W. In a fixed-factor experiment, $E(MS_B) = \sigma_e^2 + \Sigma n_j a_j^2/(C - 1)$. With H_2 true, at least one of the values of a_j will be different from 0, and MS_B will contain something more than error variance. When at least one a_j is different from 0, we can expect at least one column mean $\bar{Y}_j$ to differ from the other column means by an amount too large to attribute to sampling variability (error) alone.

In a random-factor experiment, $E(MS_B) = \sigma_e^2 + n\sigma_A^2$. With H_2 true, σ_A^2 is not equal to 0, and MS_B is expected to contain effects variance along with error variance. When effects variance exists, a sampling from the possible levels of factor A is expected to result in an amount of variation among sample means which is too great to attribute to error alone.

In either case, then, H_2 leads us to expect that the larger of these two mean squares will be MS_B. We could state H_2, the hypothesis that factor A has some effect, as $MS_B > MS_W$. Or we can use the ratio MS_B/MS_W again, making H_2 assert that this ratio has a true value greater than 1.

We have already pointed out that the use of these mean squares leads to an omnibus test for effects of factor A. All possible variations among subpopulations are lumped into the statistic MS_B. The test cannot tell us which pairs of column means differ significantly from each other; it can only tell us that there is at least one significant difference present somewhere in the data.

In random-factor experiments such an omnibus test is all we need to have, for we are not actually interested in the particular subpopulations we have obtained by random sampling. Our interest is in the question of an overall effect of factor A, and that is just what the test in terms of MS_B and MS_W will give. However, we are likely to be interested in specific subpopulations whenever factor A is fixed. An omnibus test tells us whether all the treatments can reasonably be considered as equal, but it does not tell us *which ones* are significantly different.

Consequently, the omnibus test is ordinarily followed by additional tests on specific pairs of column means in fixed-factor experiments. These comparisons may be *planned* in advance of data collection or chosen *post hoc* after the data are seen. Methods for making planned and post hoc comparisons can be found in more advanced texts. (See references on page 369.)

15-1 In Lesson 2 you learned that H_1 is always a conservative hypothesis. An experimenter, wishing to avoid making an unwarranted claim, asserts as H_1 a hypothesis which (*does; does not*) make this claim.

does not

15-2 When one is looking for possible effects of factor A, it is risky to assume that factor A does have effects on Y. This claim will be ________ unless it is supported by evidence.

unwarranted

15-3 Therefore, the assumed hypothesis H_1 becomes the hypothesis that factor A (*does; does not*) have effects on Y. The alternate hypothesis H_2 is the hypothesis that factor A (*does; does not*) have such effects.

does not

does

15-4 The expectation of MS_W is always ________ in a single-factor experiment. Therefore, MS_W is always an estimate of ________ variance in such an experiment.

σ_e^2

error

15-5 The expectation of MS_B is also σ_e^2, provided that ________ is true.

H_1

15-6 Thus when we assume H_1, we assume that $E(MS_B) =$ ________ and that MS_B is also an estimate of ________ variance.

σ_e^2

error

15-7 If we assume H_1, we assume that $E(MS_W)$ and $E(MS_B)$ are both equal to σ_e^2 and that the true value of MS_B/MS_W is ________.

1

15-8 We do not expect MS_B/MS_W to be exactly equal to 1 in every experiment. Both MS_B and MS_W are subject to ________ variability; they will vary randomly around σ_e^2 even when ________ is true.

sampling

H_1

15-9 But if MS_B/MS_W differs very much from 1, we shall have reason to suspect that something besides ________ variability is at work.

sampling

15-10 $E(MS_B)$ equals error variance plus a term which reflects the ________ of factor A. When H_2 is true, this second term (*will; will not*) be equal to 0.

effects

will not

15-11 Thus when H_2 is true, we expect MS_B/MS_W to have a true value which is (*greater; less*) than 1.

greater

15-12 A test based on the statistic MS_B/MS_W (*will; will not*) indicate *which pairs* of subgroup means differ significantly from each other. For this reason, we call it an ________ test for effects of factor A.

will not

omnibus

B. THE *F* DISTRIBUTION

When we state H_1 and H_2 in terms of the ratio MS_B/MS_W, we can use a distribution of the statistic F as our RSD. This statistic, named in honor of the British statistician Sir Ronald Fisher, has a whole family of probability distributions. Each member of the family is a probability distribution of the ratio of two variance estimates when the estimates are based on two independent samples drawn from two normally distributed populations with the same variance. The members of the family differ according to the degrees of freedom contained in the two estimates. Since every such ratio has a df for the numerator and a df for the denominator, there is a different F distribution for every pair of values which these dfs can take.

Suppose we draw a sample of size n_1 from a normal population with mean μ_1 and variance σ^2; we then draw another sample from a normal population with mean μ_2 and the same variance σ^2. We can estimate the value of σ^2 from each sample independently; our two estimates will be $MS_1 = SS_1/(n_1 - 1)$ and $MS_2 = SS_2/(n_2 - 1)$. The F distribution for $n_1 - 1$ numerator df and $n_2 - 1$ denominator df is the RSD for the F ratio MS_1/MS_2. If MS_1 is larger than MS_2, F will be greater than 1; if MS_1 is smaller, F will be less than 1 but greater than 0. Since the F statistic is a ratio of two positive numbers, it will always be positive, i.e., greater than 0. The expectation of F differs a little from one distribution to another, but it is always close to 1; when the variance is truly the same for both populations sampled, the two estimates should not differ much from each other. Since estimates from smaller samples are less accurate (and thus more variable) than estimates from larger samples, F will also be more variable for smaller samples.

Turn now to Table A-3 (page 353), the table of critical values of F for $\alpha = .05$. Each cell in this table states the critical value of F at $\alpha = .05$ for a particular F distribution. If each distribution were given in full, there would be as many tables as there are cells in the table shown here. Compare this F table with Table A-2 (page 352) of critical values of χ^2. With χ^2 we could state the critical values at several α levels in the same table; with F we require an entire table to state the critical values for $\alpha = .05$. A χ^2 distribution has only one parameter distinguishing it from other χ^2 distributions, namely, its df; an F distribution is distinguished by two parameters, the numerator df and the denominator df.

Table A-3 shows critical values for only the *upper tail* of the F distribution, where the F statistic is greater than 1. It does not give values for the lower end of the distribution, where F is less than 1; when such values are needed, they can be derived from the ones given in the table. However, we shall deal in this book only with F statistics in which the numerator is expected to be larger than the denominator.

The critical value of F for df(40,120), that is, 40 numerator and 120 denominator df, is found in the column for 40 df and the row for 120 df. This value, 1.50, is the minimum value which F must attain in order to reach $\alpha = .05$. The probability is only .05 that F will equal or exceed 1.50 when the two populations sampled actually do have the same variance.

15-13 The statistic F is a ratio of two independent estimates of the same __________. The parent populations are assumed to be __________ distributed.

variance
normally

15-14 For the F ratios to be used in this book, the estimate which is expected to be larger will always be placed in the (*denominator; numerator*).

numerator

15-15 Since two estimates of the same variance should not differ much, the observed value of F is expected to be close to ________.

1

15-16 However, the two estimates are likely to differ by small amounts because of __________ variability.

sampling

15-17 If the larger of the two estimates is in the numerator, the observed value of F will be __________ than 1.

greater

15-18 Tables of the F distribution show the probability of obtaining an F as (*large; small*) as the stated value when the two populations sampled actually have the __________ variance.

large
same

15-19 Each column in the F table represents a different number of (*denominator; numerator*) df. Each row represents a different number of __________ df.

numerator
denominator

15-20 df(5,15) means that the numerator df = ________ and the denominator df = ________.

5
15

15-21 The critical value of F at $\alpha = .05$ for df(5,15) is ________. This is the (*maximum; minimum*) value which F can have if its probability is to be no more than .05.

2.90
minimum

15-22 With df(5,15) the probability is only .05 that F will equal or exceed ________ when the two populations have the same variance.

2.90

15-23 We can get a better estimate of σ^2 from a (*large; small*) sample than from a (*large; small*) one.

large
small

15-24 There is likely to be more difference between two estimates of the same variance when samples are (*large; small*).

small

15-25 F varies randomly around a value close to ________. It varies (*less; more*) when numerator and denominator df are small than when they are large. The critical values of F for $\alpha = .05$ are (*larger; smaller*) for small df, in the upper left region of the table.

1
more
larger

C. THE *F* TEST IN ONE-WAY ANALYSIS OF VARIANCE

When H_1 is true in a single-factor experiment, MS_W and MS_B are both estimates of the same population variance; under H_1, error variance and population variance are the same. Moreover, MS_W and MS_B are independent estimates of σ^2; MS_W is an estimate based on the variation of scores around their column means, while MS_B is an estimate based on the variation of column means around the grand mean. Therefore, the ratio MS_B/MS_W is an F statistic when H_1 is true. We can state H_1: $F = 1$, meaning that the *true* value of this F is 1. Under H_1 we expect the *observed* value of F to differ from 1 only as much as the ratio of two independent variance estimates would differ from 1.

H_2 asserts that the true value of F is greater than 1 (H_2: $F > 1$). We have no reason to expect MS_B to be less than MS_W; if anything, it will be greater. Since we have put MS_B in the numerator, we shall be interested in only the upper rejection region under the F distribution. Our F test will be one-tailed because we expect the F statistic MS_B/MS_W to be at least 1.

Although the F test in analysis of variance is one-tailed, this test has nothing to say about the location and direction of those mean differences which are significant. Remember that MS_B arises from a sum of squared differences. The ratio MS_B/MS_W is similar to the Pearson χ^2 statistic in this respect; both statistics lump together a set of squared differences without regard to their source or direction. Like the χ^2 test, the F test in analysis of variance is sensitive only to the *size* of the differences which it examines. Further analysis is necessary in fixed-factor experiments whenever we wish to know exactly which ones of the subgroups are significantly different.

Our numerical example from Lesson 13 may help to make this clear. From our three small samples we obtained $MS_W = 2.5$ and $MS_B = 35$. The F ratio is $35/2.5 = 14$. With $N = 15$ and $C = 3$, F has df(2,12); the critical value for $\alpha = .01$ is 6.93. We can thus reject H_1 at this level and conclude that factor A has some effect. However, we still do not know whether this effect creates differences among all three subgroups or whether it shows up only in the difference between A_3 ($\bar{Y}_3 = 8$) and the other two subgroups ($\bar{Y}_1 = 3$, $\bar{Y}_2 = 4$). Further analysis would be needed to determine whether A_1 and A_2 are significantly different.

Is it conceivable that one could ever get a ratio MS_B/MS_W which is *less* than 1? With relatively small samples, $F < 1$ might occur when H_1 is true. However, this outcome is rare, and you should not proceed to accept H_1 without further question. Check for possible mistakes in calculation. If there are none, ask yourself whether there may be some element in your experimental procedure which artificially increased the variation *within* your subgroups. Take any case of $MS_B/MS_W < 1$ as an indication that something may be amiss in the design or the analysis of your experiment.

15-26 An F statistic can be calculated in one-way analysis of variance. When H_1 is true, we have two independent estimates of population variance, ______ and ______.

$MS_W \leftrightarrow MS_B$

15-27 These estimates are independent because MS_W is based upon variation ______ the columns, while MS_B is based upon variation ______ the columns.

within
between

15-28 When H_1 is true, these are both estimates of σ_e^2, the variance due to ______. In that event, they are both estimates of population variance because there is no other source of variance when ______ is true.

error
H_1

15-29 When H_1 is true, these two estimates should be approximately ______. When H_2 is true, MS_B is likely to be the (*larger; smaller*) of the two.

equal
larger

15-30 The estimate which is expected to be larger should be the (*denominator; numerator*) of an F ratio. Therefore, the F statistic for one-way analysis of variance is ______.

numerator
$\frac{MS_B}{MS_W}$

15-31 H_1 asserts that the ______ value of this F statistic is 1.

true

15-32 H_2 asserts that the true value of F is ______ than 1.

greater

15-33 When the true value of F is greater than 1, we can conclude that MS_B is a mixture of error variance and ______ of factor A.

effects

15-34 If the F statistic is ______ than 1, you should recheck your calculations before concluding that H_1 is true.

less

15-35 If the French vocabulary example gives $MS_B = 175$ with 2 df and $MS_W = 50$ with 87 df, the F statistic will be ______.

3.5

15-36 This F ratio has df(______, ______). Does $F = 3.5$ lie in the rejection region for $\alpha = .05$ (see Table A-3)? ______. For $\alpha = .01$ (see Table A-4)? ______.

(2,87)
Yes
No

15-37 When H_1 is true, the probability of getting $F = 3.5$ with df (2,87) is between ______ and ______.

.05 ↔ .01

15-38 With three subgroup means, there are three possible differences between pairs of means. The significant F test tells us that (*all; at least one*) of these differences (are) (is) significant.

at least one

Table 15-1 Format for Calculating a One-Way Analysis of Variance

	A_1	A_2	A_3		
	1	2	6		
	2	3	7		
	3	4	8		
	4	5	9		
	5	6	10		
$\sum_{1}^{n} Y_{ij} = T_j$	15	20	40	$G = 75$	
$(T_j)^2$	225	400	1,600	$G^2 = 5{,}625$	$\frac{G^2}{N} = 375$
$\sum_{1}^{n} Y_{ij}^2$	55	90	330	$\sum_{1}^{C}\sum_{1}^{n} Y_{ij}^2 = 475$	

$$\text{SS}_W = \sum_{1}^{C}\left[\sum_{1}^{n} Y_{ij}^2 - \frac{(T_j)^2}{n}\right] = (55 - 45) + (90 - 80) + (330 - 320) = 30$$

$$\text{SS}_B = \sum_{1}^{C}\frac{(T_j)^2}{n} - \frac{G^2}{N} = \frac{225}{5} + \frac{400}{5} + \frac{1{,}600}{5} - 375 = 70$$

$$\text{SS}_Y = \sum_{1}^{C}\sum_{1}^{n} Y_{ij}^2 - \frac{G^2}{N} = 475 - 375 = 100$$

Table 15-1 illustrates a format convenient for calculating the sums of squares. Table 15-2 shows the mean squares and the F ratio; such a table is normally included in the report of a one-way analysis of variance.

Using a desk calculator, you can organize the calculation into six steps:

1 Arrange your raw scores in a table of C columns and n rows, as the Y scores are arranged in Table 15-1 (data from Table 13-2, page 200).

2 Write the symbols T_j, $(T_j)^2$ and ΣY_{ij}^2 at the lower left of the tables as they appear at the left of Table 15-1. Then you will be able to enter the value of each of these statistics *for each column*, in the rows opposite their symbols.

3 Write the symbols G, G^2, G^2/N, and $\Sigma\Sigma Y^2$ at the lower right of the table as they appear at the right of Table 15-1. Then you will be able to enter these values for the whole set of N scores in a convenient place.

4 Beginning with column A_1, take ΣY_{i1} and ΣY_{i1}^2 simultaneously for all the values in the column, and write down the sums at the bottom of the table opposite their symbols. Do the same for each column.

5 Fill in the statistics at the lower right, using the appropriate column sums to find their values.

6 Copy the computing formulas as they are shown at the bottom of Table 15-1. Enter the statistics from your table which these formulas require, and write the SS values as you calculate them. Notice that you obtain independent determinations of each of the three sums of squares; you can then use the fact that SS_Y must equal $\text{SS}_W + \text{SS}_B$ as a check on your accuracy.

Table 15-2 Mean Squares and *F* Ratio for Numerical Example from Lesson 13

Source	Sum of Squares	df	Mean Square	F
Between groups (treatments)	70	2	35	14 ($p < .01$)
Within groups (error)	30	12	2.5	
Total	100	14		

15-39 The symbol T_j stands for the ________ of all scores in the ________, A_j. In Table 15-1, $T_1 =$ ________.

sum (or total)
column; 15

15-40 The symbol G stands for the ________ of all scores in the distribution; it is the ________ total. In Table 15-1, $G =$ ________.

sum
grand
75

15-41 $\sum_{1}^{n} Y_{ij}^2$ is the sum of all the squared scores in a ________; $\sum_{1}^{C} \sum_{1}^{n} Y_{ij}^2$ is the sum of all the squared scores in the entire ________.

column
distribution (or sample)

15-42 You must know $\sum_{1}^{n} Y_{ij}^2$ for each column in order to calculate (SS_B; SS_W).

SS_W

15-43 For this example, $C =$ ________ and $N =$ ________. $N - C$ gives the df for (SS_B; SS_W); $C - 1$ gives the df for ________.

3; 15
SS_W; SS_B

15-44 A mean square is a sum of squares divided by its ________. MS_W for this example is ________; MS_B is ________.

df; 2.5
35

15-45 The F statistic in this case has ________ df for the numerator and ________ df for the denominator.

2
12

15-46 With these df, the critical value of F for $\alpha = .01$ is ________. Since $F = 14$, we ________ H_1 at $\alpha = .01$.

6.93; reject

15-47 Table 15-2 states that $p < .01$. This statement means that the ________ of observing an F as ________ as 14 when H_1 is true is ________ than .01.

probability; large
less

15-48 When we reject H_1, we accept the hypothesis that the ________ value of this F ratio is ________ than 1.

true; greater

D. ASSUMPTIONS IN ONE-WAY ANALYSIS OF VARIANCE

At the beginning of Lesson 14 we observed that the line of reasoning to be developed would apply only to experiments in which certain requirements are met. When these requirements are not met, the additive model of score components does not apply, and the use of one-way analysis of variance is not appropriate.

Now that you understand the line of reasoning on which the F test is based, we should look at these requirements again.

Requirement 1: That the raw scores within each column be normally distributed The additive model requires an approximately normal distribution. Moreover, the F test assumes that both mean-square estimates are based on samples from normally distributed populations. For MS_B, the "population" sampled is the population of sample means; its distribution is the RSD of means of samples of this size. Unless this RSD is normal, the requirements for an F test are not met.

However, because of the central limit theorem, we know that the RSD of sample means can be normal even when the population sampled is considerably skewed, especially when sample size is large. Therefore, this requirement of normally distributed subgroups is not critical *unless the subgroups are very small.*

Requirement 2: That the subgroups all have the same variance This requirement arises because we must be able to think of the subpopulations as all having the same error variance σ_e^2, which is estimated by MS_W. If the subpopulations have different variances, the additive model does not apply because MS_W itself contains some effects of factor A and is no longer a pure estimate of σ_e^2.

This requirement of equal *sample* variances is more serious for random-factor than for fixed-factor experiments. However, it is not critical as long as all the subgroups are of equal size. For this reason, equal subgroups are especially important in random-factor experiments.

Requirement 3: That all the N observations in all the subgroups be independent of each other This requirement arises because we must assume that the error component e_{ij} for each score Y_{ij} is not affected by the error component in any other score. If these error components are not independent, our whole line of reasoning about the estimates of σ_e^2 will be invalid. Therefore, this requirement is essential, and *it must not be violated.*

Requirement 3 is most likely to be violated in studies where repeated observations are made on the same experimental subjects. In such cases it is reasonable to wonder whether the error components for the two or more scores from the same subject do not contain some influences which are peculiar to the individual subject. The requirement of independence is most easily met when each subject contributes only one of the N scores. *When repeated observations on the same persons cannot be avoided, a different procedure must be followed for analysis of variance.*

15-49 One-way analysis of variance can be used to analyze a single-factor experiment only when that experiment fits the ________ model of score components.

additive

15-50 To fit the additive model, an experiment should meet three requirements. The most critical of these is the requirement that all N observations be ________ of each other.

independent

15-51 Independence of the observations is one of the hallmarks of a random sample. A sample is not random if the selection of one observation can affect the ________ of any other observation.

selection

15-52 When the same individual contributes a score to more than one subgroup, the scores from this individual (*are; are not*) independent of each other.

are not

15-53 The error component e_{ij} contains some influences arising from individual differences. If two scores are from the same person, the ________ components of these two scores are likely to be similar.

error

15-54 Unless we have N randomly varying error components, MS_W will not be an estimate of ________ variance based on $N - C$ degrees of freedom.

error

15-55 The additive model also requires that the subpopulations be ________ distributed. We can be most sure that this requirement is met when our actual subgroups are approximately ________.

normally

normal

15-56 However, some population skewness can be tolerated when the samples are large enough to give a normal ________ of sample means. From Lesson 9 you know that a sample n of ______ is considered large enough for this purpose.

RSD
30

15-57 Therefore this requirement is critical only when the subgroup n's are ________ than 30.

smaller

15-58 The additive model also requires that the subpopulations all have the same ________. We can be most sure that this requirement is met when our actual subgroups have about the same ________.

variance

variance

15-59 This requirement is somewhat more important when factor A is a (*fixed; random*) factor. However, it is not critical when the subgroups are all of equal ________.

random
size

E. THE *F* TEST FOR EQUALITY OF VARIANCES

The F test has many uses besides its application in analysis of variance. The following example may suggest the broader possibilities of the F statistic.

Some proponents of programmed instruction have claimed that this kind of instruction will raise all students to a more nearly equal level of performance than classroom instruction. Classroom instruction occupies the same time period for all students, and within that time period the students will reach varying levels of mastery of a subject. Programmed instruction, in which each individual can proceed at his own rate, might allow every individual to achieve the same high level of mastery. If so, the variability in achievement scores after programmed instruction should be less than the variability after classroom instruction.

Let us first notice that this claim cannot be tested by analysis of variance; it does not fit the additive model. Factor A, "type of instruction," is expected to have an effect on the *variance* of Y; it is supposed to create subpopulations with different variances. These subpopulations might or might not also have different means, but such a difference—if it exists—is irrelevant to the claim about variances which has been offered for testing.

Figure 15-1*a* shows the situation assumed under H_1. If H_1 is true, there are two subpopulations with means μ_1 and μ_2; these populations are both normally distributed, and they both have the same variance σ^2. Although the figure shows $\mu_2 > \mu_1$, the difference between the two means could take any value (including 0) without affecting H_1. If H_2 is true, the two subpopulations have different variances. As shown in Fig. 15-1*b*, H_2 asserts that the variance $\sigma_1{}^2$ is larger than the variance $\sigma_2{}^2$.

H_1 can be tested by taking a random sample from each of these two subpopulations. If we have 82 students, we should assign $n = 41$ at random to receive classroom instruction and the other 41 to receive programmed instruction. Then both groups take the same achievement test. When H_1 is true, scores on this test will provide two independent mean-square estimates of the same variance. If the two subpopulations are normally distributed, the ratio MS_1/MS_2, where MS_1 is the estimate from the classroom sample and MS_2 is the estimate from the programmed sample, is an F statistic. H_2 asserts that the true value of this statistic is greater than 1. We can use the F distribution for df(40,40) to evaluate the observed F ratio. This F test is one-tailed because H_2 is directional.

Such an experiment is an example of the F test for *equality of variances*. It can be used to test hypotheses about the difference between two variances whenever the population distributions can be assumed to be normal. It should be stressed, however, that the requirement of population normality is more critical in this use of the F test than in analysis of variance; the central limit theorem provides assistance in analysis of variance but not in this case. An F test for equality of variances should not be used with skewed population distributions unless sample size is quite large.

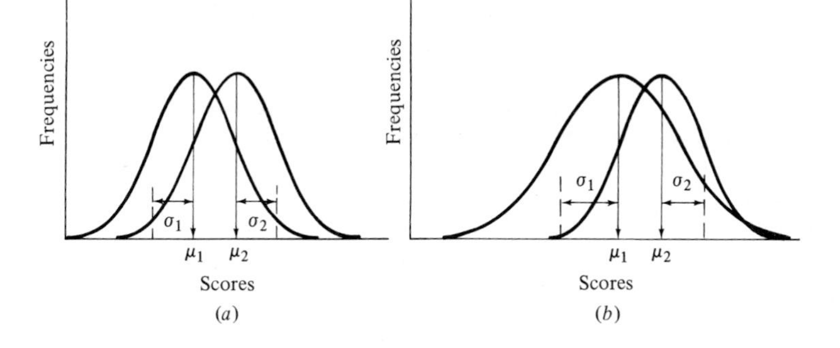

Figure 15-1: Sketch of hypotheses about effects of factor A in an F test for equality of variances. (*a*) A possible version of H_1 (that the subpopulations have equal variances); (*b*) a corresponding version of H_2 (that σ_1^2 is greater than σ_2^2).

15-60 Some experts claim that the variance of achievement scores after programmed instruction will be (*greater; less*) than the variance after classroom instruction. This claim should become (H_1; H_2) in a test comparing two actual samples.

less
H_2

15-61 The claim concerns only the difference between the __________ of the populations from which the samples are drawn. Any difference between the population means (*is; is not*) relevant to this claim.

variances
is not

15-62 H_1 in this case is the hypothesis that the two samples are drawn from populations with equal __________.

variances

15-63 If $\bar{Y}_2$ is the mean of scores after programmed instruction, $\Sigma(Y - \bar{Y}_2)^2$ is the ________ for those scores. $MS_2 = \Sigma(Y - \bar{Y}_2)^2/(n_1 - 1)$ is a mean-square estimate of the __________ of the population.

SS
variance

15-64 The F statistic will be the ratio of the two mean squares, MS_1 and MS_2. Since we are expecting ________ to be the larger estimate, we place it in the (*denominator; numerator*).

MS_1
numerator

15-65 The F statistic will be the ratio ________. H_2 asserts that its true value is ________. If each sample contains 41 scores, each of the mean squares will have ________ degrees of freedom.

$\frac{MS_1}{MS_2}$
>1
40

15-66 In order to reach the .05 level, F must be at least ________. If $MS_2 = 80$, MS_1 will have to be at least ________ in order to give an F significant at the .05 level.

1.69; 135.20

REVIEW

15-67 The F statistic is a ratio of two __________ estimates of the same __________.

independent
variance

15-68 The variance which is estimated must be the variance of a __________ population.

normal

15-69 Since its numerator and denominator are both positive numbers, F (*can; cannot*) be negative and (*can; cannot*) be 0; the F statistic will always be a __________ number.

cannot; cannot
positive

15-70 When the larger of the two estimates is placed in the numerator, F will be __________ than 1.

greater

15-71 Tables of the F distribution give critical values for only the __________ end of the distribution, i.e., for cases in which F will be __________ than 1. There is a different distribution for each combination of numerator and denominator __________.

upper
greater
df

15-72 To find a critical value of F, locate the column which represents the __________ df and the row which represents the __________ df.

numerator
denominator

15-73 The statement "$F(2,12) = 14$" means that the observed value of F is ________ with numerator df = ________ and denominator df = ________.

14; 2
12

15-74 The critical value of F (40, 120) is 1.50 for $\alpha = .05$. Therefore, F is expected to equal or exceed 1.50 in ________ percent of cases with df(40, 120).

5

15-75 Without looking at the F table, you can state that the critical value of F for df(10,20) is (*less; more*) than 1.50 because smaller samples are expected to give estimates which vary (*less; more*) than those from large samples.

more

more

15-76 One-way analysis of variance produces statistics which can form an F ratio. This ratio is ________.

$\frac{MS_B}{MS_W}$

15-77 MS_B and MS_W are estimates of the same variance when ________ is true. They are estimates of __________ variance.

H_1; error

15-78 To form an F ratio, the two estimates must be __________ of each other. MS_W arises from variation __________ subgroups; MS_B arises from variation __________ subgroups.

independent
within; between

15-79 Since MS_B is based on variation among subgroup means, we can also say that it arises from variation within the __________ of means of samples of size n.

RSD

15-80 The F test in analysis of variance is one-tailed because we expect _______ to be larger than _______ when H_1 is false.

MS_B; MS_W

15-81 When the F test is used for analysis of variance, H_1 asserts that the __________ value of F is _______; H_2 asserts that the __________ value is _______.

true; 1
true; >1

15-82 If we can accept H_2, we would usually like to know *which* subgroups are different in a (*fixed-; random-*) factor experiment. A test of $F = MS_B/MS_W$ is called an omnibus test because it (*does; does not*) give us this information.

fixed-

does not

15-83 One-way analysis of variance is based on three requirements. The most critical of these is the requirement that all __________ be __________.

observations; independent

15-84 Another requirement is that the C samples be __________ distributed. This requirement is not critical unless the samples are __________.

normally
small

15-85 Finally, the C samples should all have the same __________. This requirement is not critical when the samples are all of equal __________.

variance
size

15-86 When the hypothesis to be tested concerns a difference between two population *variances*, the additive model (*does; does not*) apply.

does not

15-87 If the two populations are __________ distributed, an F test can be used to test for equality of the two __________.

normally
variances

15-88 If the two variances are called σ_1^2 and σ_2^2, H_1 in this case asserts that _______ equals _______.

σ_1^2; σ_2^2

15-89 If σ_1^2 is expected to be the larger of the two variances, H_2 asserts that σ_1^2 _______.

$>\sigma_2^2$

15-90 If MS_1 is based on n_1 observations and MS_2 on n_2, the F ratio will be _______ with _______ numerator df and _______ denominator df.

$\frac{MS_1}{MS_2}$; $n_1 - 1$; $n_2 - 1$

15-91 The requirement that populations be normally distributed is (*less; more*) critical in the test for equality of variances than in analysis of variance.

more

PROBLEMS

1. Set up a summary table like Table 15-2 for the one-way analysis of variance in Prob. 1 on page 226.

2. Complete the analysis of variance for the data in Prob. 2 on page 226. What is the value of F? What are its degrees of freedom? Give its probability level on H_1.

LESSON 16
THE t TEST FOR TWO INDEPENDENT SAMPLES

We have been dealing since Lesson 13 with statistical tests of differences among sample means. At the beginning of Lesson 15 we spelled out a line of reasoning which we showed could be applied literally only when there are just two sample means to compare, i.e., when $C = 2$. We then went on to modify this reasoning for use with $C > 2$; the result was the F test for one-way analysis of variance. At the end of that lesson we emphasized again that this test requires independence of the N observations, in other words, that the C samples must be *independent* random samples.

When we have only two independent samples in a fixed-factor experiment, it is not necessary to use the F test. We have only one mean difference to evaluate, not several. If factor A has no effect, $\bar{Y}_1$ and $\bar{Y}_2$ will differ only as much as the means of two samples from the same population would be expected to differ.

How much difference should we expect when H_1 is true and the samples are from the same population? To answer this question, we need to derive an RSD for $\bar{Y}_1 - \bar{Y}_2$, the difference between two sample means when the samples come from the same population. This probability distribution is called the RSD of differences between sample means, or the RSD OF MEAN DIFFERENCES.

The mean of this RSD is zero. On the average we expect as many differences between the means of such pairs of samples to be positive as to be negative when the samples are randomly drawn from the same population. The variance of this RSD will be discussed in Sec. B.

One-way analysis of variance as discussed in Lessons 14 and 15 can be applied to any number of groups. It can also be applied when $C = 2$ provided that the requirements for analysis of variance are met. When only two sample means are to be compared, however, it is possible to make use of the RSD of mean differences. The resulting test is called a t TEST.

The t test will be discussed here in the light of its relation to one-way analysis of variance. It will make the meaning of analysis of variance clearer, and at the same time it will provide an alternative method for treating decisions about two sample means. Some students will prefer the t-test procedure when it is a proper alternative, and in Lesson 17 we shall take up a kind of experiment in which the t test *must* be used instead of analysis of variance.

Our line of study will begin with an example to which analysis of variance for $C = 2$ is applied. Then we shall develop the t test as it may be used for such cases. Finally, we shall apply the t test to this same example in order to see how the t statistic is related to the F statistic in one-way analysis of variance.

A. ONE-WAY ANALYSIS OF VARIANCE WHEN $C = 2$

An investigator studies the effect of age upon speed of reaction. His subjects must respond to a flash of light by pressing a key; the experimenter records the time interval between light and key response. Group 1 is made up of 12 persons between the ages of eighteen and twenty-one; group 2 contains 10 persons between sixty and sixty-five. Table 16-1 is a summary table for one-way analysis of variance on these samples.

Remembering the requirements for analysis of variance, we should observe that both groups are small enough to raise the question, "Is the distribution of scores in the population normal?" If it is not, we cannot assume that the RSDs of the sample means will be normal. In fact, since these data are time scores (reaction times), we ought to expect some degree of skewness in the population distribution; the skewness will be positive, since it is not physically possible for time scores to deviate as far in the direction of short reaction times as in the direction of longer reaction times. Let us suppose, however, that we find our two samples to be in fact reasonably symmetrical.

Since n_1 and n_2 differ, we have cause to worry about the requirement that the two subpopulations have equal variance. We find that SS for group 1 is 440; $SS_1/(n_1 - 1) = 440/11 = 40$, the estimated population variance for the population from which this sample is drawn. For group 2, $SS_2/(n_2 - 1) = 460/9 = 51.1$, the estimated population variance for the second sample. Although these variances are of the same order of magnitude, they certainly are not equal, and this fact should be kept in mind.

However, we do not have to worry about the third and most essential requirement for analysis of variance, the independence of the 22 scores with their independent and randomly varying error components. All the observations have been contributed by different persons. Provided that the investigator has been careful to draw his samples randomly from the age-group populations which he is interested in comparing, the requirement of independence is met by these data. It is possible to imagine ways in which the experimenter might unintentionally bias his samples; one of the commonest ways is by taking all the members of an already existing group as subjects. Thus, for statistical reasons alone, a careful reading of the procedure section of an experimental report can be essential to a proper evaluation of the reported results.

On H_1 both samples come from a population with mean μ; $\bar{Y}_1$ is an estimate of $\mu + a_1$, $\bar{Y}_2$ is an estimate of $\mu + a_2$. Since the factor is fixed, $\Sigma a_j = a_1 + a_2 = 0$ and $a_1 = -a_2$. If both a_1 and a_2 are 0, the true value of $\bar{Y}_1 - \bar{Y}_2$ is 0. The F test provides a decision between two hypotheses: H_1, that $F = 1$ because a_1 and a_2 are 0; and H_2, that $F > 1$ because a_1 and a_2 do not equal 0. Consulting Table 16-1, we find that the F ratio is so large that it would occur as a result of sampling variability (when H_1 is true and both mean squares contain only error variance) less often than 1 time in 100 such experiments. We can reject H_1 at the .01 level.

Table 16-1 One-Way Analysis of Variance for Reaction-Time Data, $C = 2$

Source	SS	df	MS	$F(1,20)$
Between groups	545.45	1	545.45	12.57 ($p < .01$)
Within groups	900	20	45	
Total	1,445.45	21		

16-1 One-way analysis of variance is a method of testing for differences between the ________ of samples. *means*

16-2 The F test can be used when there are more than 2 samples. The t test is a special method for making decisions about differences between the means of only ______ samples. *2*

16-3 Any two-sample experiment is likely to be a (*fixed-; random-*) factor experiment. If levels of factor A have to be sampled, study of only 2 levels (*is; is not*) likely to be sufficient. *fixed-* *is not*

16-4 The true mean of each sample is $\mu + a_j$. With only 2 levels, the true mean of subgroup A_1 is ______ and that of subgroup A_2 is ______. $\mu + a_1$ $\mu + a_2$

16-5 Since factor A is fixed, the number of possible levels of the factor is ______. The sum Σa_j is always ______ because μ is the mean of the subgroup means. *2; 0*

16-6 Since $\Sigma a_j = 0$, $a_1 + a_2 =$ ______ and $a_1 =$ ______. The true size of both effects components is the same, but they are of opposite ________. *0*; $-a_2$ *sign*

16-7 In one-way analysis of variance, we take $\bar{Y}$ as an estimate of ______. $\bar{Y}$ is always the mean of the column means. With only two column means, $(\bar{Y}_1 + \bar{Y}_2)/2 =$ ______. μ $\bar{Y}$

16-8 We take $\bar{Y}_j$ as an estimate of ______, and $\bar{Y}_j - \bar{Y}$ as an estimate of ______. $\mu + a_j$ a_j

16-9 If H_1 is true, all a_j equal ______. $\bar{Y}_1 - \bar{Y}$, which estimates a_j, is expected to have a true value of ______ when H_1 is true. *0* *0*

16-10 $\bar{Y}_2 - \bar{Y}$ is also expected to have a true value of 0 when H_1 is true, and the mean difference $\bar{Y}_1 - \bar{Y}_2$ is expected to have a true value of ______. *0*

16-11 SS_B has ______ df, and the F ratio has df(______,______) in Table 16-1. We can reject the hypothesis that a_1 and a_2 are equal to ______ at the .01 level. *1* *(1,20)* *0*

B. THE RSD OF THE DIFFERENCE BETWEEN TWO SAMPLE MEANS

The mean of the RSD of mean differences is 0. In order to find its variance, called σ_d^2, we consult the central limit theorem as it applies to this RSD:

When pairs of independent samples of sizes n_1 and n_2 are drawn from the same population, the RSD of the difference between means approaches a normal distribution as both n_1 and n_2 grow large, regardless of the form of the population distribution. The variance of this RSD is equal to the sum of two variances: the variance of the RSD of means of samples of size n_1 plus the variance of the RSD of means of samples of size n_2.

These facts can be derived from the central limit theorem because the mean difference $\bar{Y}_1 - \bar{Y}_2$ is an additive combination of two normally distributed variables. Thus, when population variance σ^2 is known, the RSD of mean differences will have a variance $\sigma_d^2 = \sigma^2/n_1 + \sigma^2/n_2$.

In most practical uses of the RSD of mean differences, the true value of σ^2 is unknown and must be estimated from sample statistics. The flow chart below distinguishes among the three variance parameters which are involved and shows how they can be estimated from sample statistics. Line 1 shows these sample statistics. Line 2 gives three estimates of population variance; σ^2 can be estimated from either sample alone or jointly from both samples. The POOLED estimate from both samples is equivalent to MS_W, and it is this estimate which is used in estimating the variance $\sigma_{\bar{Y}}^2$ of the RSD of sample means (line 3) and the variance σ_d^2 of the RSD of mean differences (line 4). Table 16-2 gives hypothetical statistics for the college-height example so that you may have practice in estimating σ_d^2.

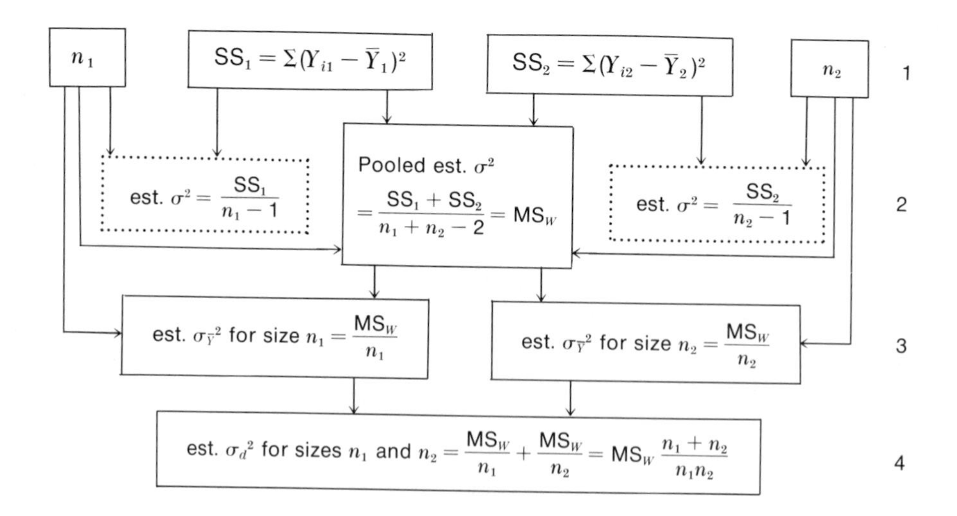

Table 16-2 Calculation of σ_d^2 for College-Height Example (A_1 = Alpha, A_2 = Omega)

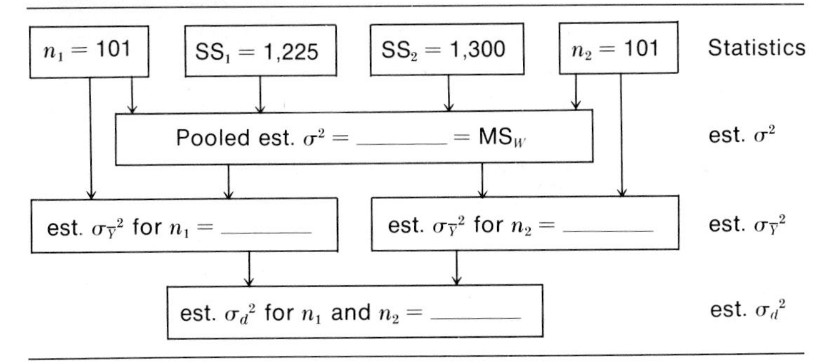

16-12 In Table 16-2 SS_1 has ________ df. The estimate of σ^2 obtained *from this sample alone* is ________.

100
12.25

16-13 We can pool the data from the two samples to obtain what is called a ________ estimate. When we add $SS_1 + SS_2$, we obtain what we call (SS_B; SS_W) in analysis of variance.

pooled
SS_W

16-14 SS_W always has $N -$ ________ df. For this SS_W, $N = n_1 + n_2$; SS_W has ________ df.

C
200

16-15 Thus ($SS_1 + SS_2/n_1 + n_2 - 2$) is the same as the mean square which we call ________ in analysis of variance.

MS_W

16-16 MS_W is the ________ estimate of population variance. Its value in Table 16-2 is ________. (Fill in the table.)

pooled
12.625

16-17 For samples of size n, the estimated variance σ_Y^2 of the RSD of sample means is MS_W divided by ________.

n

16-18 Since $n_1 = n_2$ in Table 16-2, the estimate of $\sigma_{\bar{Y}}^2$ for samples of size n_1 or n_2 is ________.

.125

16-19 σ^2 is the variance of the ________. $\sigma_{\bar{Y}}^2$ is the variance of the ________ of sample ________. For our statistical test, we need to know ________, the variance of the RSD of mean differences.

population
RSD; means
σ_d^2

16-20 The estimated σ_d^2 is the sum of two terms, MS_W/n_1 plus ________.

$\dfrac{MS_W}{n_2}$

16-21 For Table 16-2, estimated σ_d^2 is equal to ________.

.250

16-22 The last block on page 246 shows that $MS_W/n_1 + MS_W/n_2$ equals MS_W times the quantity ________.

$\dfrac{n_1 + n_2}{n_1 n_2}$

C. THE t STATISTIC

The statistic t, for any two sample means $\bar{Y}_1$ and $\bar{Y}_2$, is defined as follows:

$$t = \frac{\bar{Y}_1 - \bar{Y}_2}{\text{est. } \sigma_d}$$

where $\bar{Y}_1 - \bar{Y}_2$ is the observed difference between sample means and σ_d is the *standard deviation* of the RSD of mean differences.

In Sec. B we found that the estimate of σ_d^2 for samples of different size is

$$\frac{\text{MS}_W}{n_1} + \frac{\text{MS}_W}{n_2} = \text{MS}_W \frac{n_1 + n_2}{n_1 n_2} \qquad \text{where } \text{MS}_W = \frac{\text{SS}_1 + \text{SS}_2}{n_1 + n_2 - 2}$$

When both samples are of the same size n, this estimate becomes

$$\text{MS}_W \frac{2n}{n^2} = \frac{2\text{MS}_W}{n}$$

The denominator of the t ratio is the *square root* of this estimate.

The expected shape of the distribution of t depends upon the actual size of the samples. There is a whole family of t distributions, one for each number of df which the t ratio may have. A t ratio is based upon $n_1 + n_2$ independent observations, but two restrictions are introduced when the two parameters $\mu + a_1$ and $\mu + a_2$ are estimated from $\bar{Y}_1$ and $\bar{Y}_2$. Therefore, the t ratio has $n_1 + n_2 - 2$ df. When the samples are of equal size, t has $2n - 2$ df. The family of t distributions will be discussed in Sec. D.

For a t test the assumed hypothesis H_1 asserts that $t = 0$ because $\mu_1 - \mu_2 = 0$. μ_1 is the true mean of the subpopulation arising from level A_1; $\mu_1 = \mu + a_1$. μ_2 is the true mean for the subpopulation arising from level A_2. H_1 asserts that there is no real difference between these subpopulation means.

This hypothesis is similar to the H_1 in analysis of variance (that the true value of $F = 1$ because a_1 and a_2 both equal 0 and there is no difference among subpopulation means). However, the investigator planning a t test has a choice among alternate hypotheses. In analysis of variance, no matter what his conjecture about the direction of a mean difference, the investigator must always choose H_2: $F > 1$ because some a_j do not equal 0. The F ratio is insensitive to the directions of mean differences. The t ratio, on the other hand, deals with only a single mean difference. This difference may be positive or negative, and it retains its sign in the t statistic. Thus an investigator may choose any one of the following alternate hypotheses:

H_2: $t \neq 0$ because $\mu_1 - \mu_2 \neq 0$
H_2: $t > 0$ because $\mu_1 - \mu_2 > 0$
H_2: $t < 0$ because $\mu_1 - \mu_2 < 0$

The first of these alternate hypotheses is nondirectional; it requires a two-tailed test, and the rejection region will be divided between the two tails of the t distribution. The other two alternate hypotheses are directional; they require one-tailed tests, and the rejection region will be only at the upper or lower end of the distribution. The t test thus permits either one- or two-tailed tests of H_1.

16-23 The t statistic is a ratio. The numerator is the ________ between the two sample means. The denominator is the estimated standard deviation of the ________ of ________ for these samples.

difference
RSD
mean differences

16-24 A standard deviation is the ________ of a variance. If estimated $\sigma_d^2 = .25$, estimated $\sigma_d =$ ________. (Be careful of the decimal point.)

square root
.5

16-25 The mean difference observed in the college-height example is .25 inch. The t ratio is therefore ________.

.5

16-26 Imagine or sketch the RSD of mean differences for this example. Its shape is that of a ________ distribution, according to the central limit theorem.

normal

16-27 Its mean is ________. Its standard deviation is *approximately* ________ inch; remember that this is not an exact value but an ________ of σ_d.

0
.5
estimate

16-28 The tails of a normal distribution approach the horizontal axis at about +3 and −3 standard deviations. Your sketch will thus have its mean at ________ and its tails almost reaching the horizontal axis at ________ and ________ inches.

0
-1.5; $+1.5$

16-29 Now locate the *observed* mean difference $\bar{Y}_1 - \bar{Y}_2 = +.25$ inch, with respect to this RSD. Its position (*is; is not*) very far from the mean. Do you think it will turn out to be a significant difference?

is not

16-30 The t statistic is a ratio between a mean difference and the *estimated* ________ of its RSD.

standard deviation

16-31 The t ratio is based upon ________ independent observations. However, ________ subpopulation means have to be estimated in calculating the t statistic.

$n_1 + n_2$
2

16-32 Since there are ________ restrictions on its degrees of freedom, t has ________ df.

2
$n_1 + n_2 - 2$

16-33 The mean square MS_W enters into the calculation of t. The denominator of t is the ________ of $MS_W/n_1 + MS_W/n_2$.

square root

16-34 MS_W also has ________ df when $C = 2$. If both samples are of the same size, both MS_W and t have ________ df.

$n_1 + n_2 - 2$
$2n - 2$

16-35 The t ratio is the ratio of a mean difference to the ________ standard deviation of its RSD.

estimated

D. THE t DISTRIBUTIONS AND THEIR TABLE

Figure 16-1 shows four members of the family of t distributions, those for df of 1, 9, 25, and infinity. Each of the t distributions had to be calculated separately. The distributions were originally determined by an English mathematician who used the pseudonym Student, and the t statistic is accordingly sometimes called STUDENT'S t.

In Fig. 16-1 the dashed curve is for df = 1, while the solid curve is for df = infinity. Observe that the area under the curve for df = 1 and within ±1 standard deviation of the mean is *smaller* than the area under the corresponding part of the curve for df = infinity. Similarly, the distribution for df = 1 has a *larger* proportion of its area beyond $t = \pm 2$ than the distribution for df = infinity does. In general, when we compare these two curves, we find that the distribution for df = 1 has a relatively greater proportion of its area at the extremes, under the tails of the curve, and a relatively smaller proportion of its area at the center near the mean.

The t distributions for df = 9 and df = 25 lie between the distributions we have just been discussing. As the degrees of freedom increase with increasing sample size, the t distribution gradually comes to have a smaller proportion of its area under the tails and a larger proportion near the mean.

The t ratio for very large samples is actually a normally distributed z score. Recall that $z = (X_i - \bar{X})/s$; it is the ratio of a deviation score to the standard deviation of its distribution. For a distribution of mean differences, $\bar{Y}_1 - \bar{Y}_2$ is analogous to X_i, and the mean is 0; $(\bar{Y}_1 - \bar{Y}_2)/\sigma_d$ is the ratio of a deviation score to the standard deviation of its distribution, and it is therefore a z score. Since $t = (\bar{Y}_1 - \bar{Y}_2)/(\text{est. } \sigma_d)$, the difference between t and z disappears when sample size grows large enough to make estimated σ_d equal to σ_d. Because the RSD of mean differences is normal, the t distribution for very large samples is the normal probability distribution.

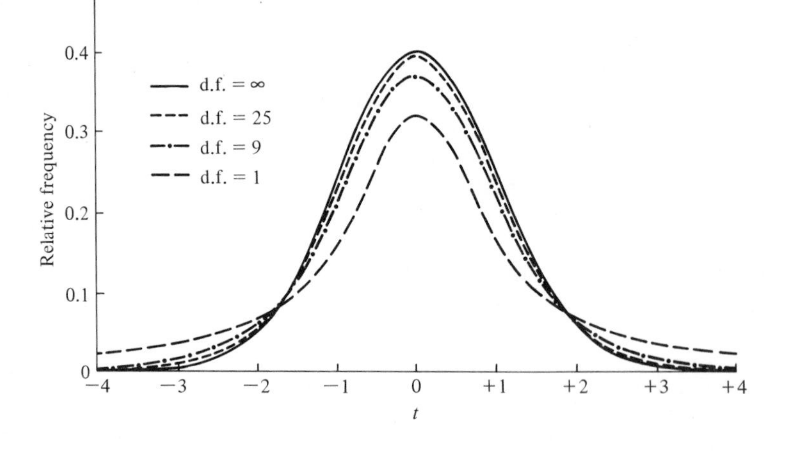

Figure 16-1: Distributions of t for various degrees of freedom. (*From D. Lewis,* Quantitative Methods in Psychology, *published by the author, Iowa City, 1948.*)

16-36 z is the ratio between a deviation score, such as $X_i - \bar{X}$, and the ________ of its distribution. *standard deviation*

16-37 We have been discussing the RSD of mean differences. This distribution has a mean of ________. Therefore, $\bar{Y}_1 - \bar{Y}_2$ is a ________ like $X_i - \bar{X}$. *0* *deviation score*

16-38 The standard deviation of the RSD of mean differences is ________. This parameter has to be ________ from sample statistics. *σ_d; estimated*

16-39 t is the ratio between a mean difference and the ________ standard deviation of its distribution. *estimated*

16-40 The ratio $(\bar{Y}_1 - \bar{Y}_2)/\sigma_d$ fits the definition of a ________. The ratio $(\bar{Y}_1 - \bar{Y}_2)/(\text{est. } \sigma_d)$ fits the definition of a ________ statistic. *z score* *t*

16-41 If estimated $\sigma_d = \sigma_d$, the t statistic becomes the same as a ________. *z score*

16-42 Not all z scores are normally distributed. But if the distribution of X_i is normal, the probability associated with $z = (X_i - \bar{X})/s$ can be obtained from the ________ table. *normal-curve*

16-43 According to the central limit theorem, the RSD of mean differences is always ________ distributed when the population sampled is approximately normal. *normally*

16-44 Thus, when the t statistic can be regarded as a z score, the probability distribution of t will be identical with the ________ distribution, provided the population of Y scores is normal. *normal*

16-45 For this reason, the probability distribution of t is identical with the normal distribution when the samples are very ________ — technically, when df = infinity. *large*

16-46 However, the t distribution for df = 25 (see Fig. 16-1) is almost identical with the normal probability distribution. For samples large enough to give $n_1 + n_2 - 2$ at least equal to 30, the ________ distribution is a good approximation to the t distribution. *normal*

16-47 For df < 30, the t distribution (compared with the normal distribution) has a larger proportion of its area under the ________ of the curve. *tails*

The usual table of t distributions is a critical-values table, similar to the χ^2 table in that a line of critical values is devoted to each df. You should compare the t table on page 355 (Table A-5) with the table of critical values of z which you derived (pages 156 to 158) from your study of the normal-curve table. Like your z table, Table A-5 lists critical values for both one- and two-tailed tests.

Observe how the critical value of t varies with increasing df within the column for $\alpha = .05$ (two-tailed). As df increase, the critical value of t declines. The smallest value which t reaches in this column is 1.960; but notice that when df = 30, the critical value of t has already reached 2.042. The critical t's follow a similar pattern in every column; as df increase, the value of t at any given probability level decreases. This change is rapid when the values of df are small, and it becomes less rapid as the values of df become larger. Notice that the critical values of t in the last row (for df = infinity) are exactly the same as the critical values of z in Table 10-4 (page 158).

Distributions of t are thus close relatives of the normal distribution. The reason t distributions differ can be made clearer by considering the expectations of estimated σ_d^2 and of estimated σ_d. The expectation of the estimate of σ_d^2, that is, $E\ (\text{MS}_W/n_1 + \text{MS}_W/n_2)$, is always equal to the parameter σ_d^2; the average of the estimates of σ_d^2 from all possible samples is always equal to the *variance* of the RSD of mean differences. But whenever the samples are small, the expectation of the estimate of σ_d, that is, $E\ (\sqrt{\text{MS}_W/n_1 + \text{MS}_W/n_2})$, is less than the parameter σ_d. This difference arises from the fact that the square-root transformation is nonlinear; the square root of $\text{MS}_W/n_1 + \text{MS}_W/n_2$ does not have a distribution shaped like the distribution of $\text{MS}_W/n_1 + \text{MS}_W/n_2$ itself. Taking the square root of our estimate of σ_d^2 thus introduces a systematic error of underestimation.

The amount of underestimation is greater when samples are very small than when $n_1 + n_2 - 2$ is above 30. By calculating the amount of underestimation likely for particular sample sizes, it is possible to determine how much the distribution of the t ratio should deviate from a normal distribution. The result of such calculations is the family of t distributions.

It is important for you to notice that this line of reasoning rests on the assumption that the quantity itself is normally distributed. The quantity we are talking about in this lesson is $\bar{Y}_1 - \bar{Y}_2$; we are safest in assuming a normal distribution for $\bar{Y}_1 - \bar{Y}_2$ when the parent population is truly normal or when sample size is large. Therefore, we ought to hesitate about using the t ratio whenever the two samples are small and the parent population is extremely different from normal. For a population distribution which is J-shaped, or U-shaped, or multimodal, or rectangular, one ought not to expect even relatively large samples to have means whose RSD is normal. In such cases, the RSD of mean differences will not be normal, and the table of critical values of t will not be valid for a t statistic calculated from sample data. Lesson 18 will present an alternative to the t test for these cases.

16-48 The RSD of $\bar{Y}_1 - \bar{Y}_2$ can be expected to be normal only when the two quantities $\bar{Y}_1$ and ______ are normally distributed. — *$\bar{Y}_2$*

16-49 The distribution of $\bar{Y}_1$ is an RSD of ______ of samples of size n_1. According to the central limit theorem, this RSD will be normal for small samples only when the ______ distribution is not very different from normal. — *means*; *population*

16-50 The ratio between a normally distributed quantity and its *known* standard deviation is a ratio which itself has a ______ distribution. — *normal*

16-51 The ratio between a normally distributed quantity and its *estimated* standard deviation is called a ______ statistic. Its probability distribution is found in the table of critical values of ______. — *t*; *t*

16-52 The ratio between a quantity which is *not* normally distributed and its known standard deviation has a probability distribution found in the (*normal-curve table; t table; neither of these tables*). — *neither of these tables*

16-53 We ought to hesitate about using the t statistic whenever the two samples are very ______ and the population distribution is more than moderately different from ______. — *small*; *normal*

16-54 In Table A-5 (page 355), the column headings give the probability that t will reach or exceed the ______ values listed in the columns. Headings are given for ______- and ______-tailed tests. — *critical*; *one*; *two*

16-55 In the first row, for a t distribution with df = ______, the critical value at $\alpha = .05$ (two-tailed) is ______. — *1*; *12.706*

16-56 The sum of the area beyond $t = -12.706$ at the lower end and the area beyond $t = +12.706$ at the upper end must be ______. The area beyond $t = +12.706$ is only ______. — *.05; .025*

16-57 If df = 1 and $t = +12.706$, H_1 can be rejected at $\alpha =$ ______ by a one-tailed test. — *.025*

16-58 In the college-height example, we have $t = .5$, df = 200. We consult the t table for df = ______. — *infinity*

16-59 The lowest critical value of t in the table, at df = infinity, is ______. Therefore, $t = .5$ is not significant even at the highest α levels in the table ($\alpha =$ ______, two-tailed, and $\alpha =$ ______, one-tailed). — *1.282*; *.20*; *.10*

E. THE t DISTRIBUTIONS COMPARED WITH THE F DISTRIBUTIONS

To show the relationship between t and F, let us return to the problem of reaction times which we analyzed by one-way analysis of variance in Sec. A. We can apply a t test to the same example, and you will do so on page 255. The t ratio you obtain will be taken to the t table (Table A-5); to make a two-tailed test with $\alpha = .05$, you will consult the row for df = 20 and find that the critical value of t is 2.086.

Now recall that the F ratio for these same data must be taken to the F distribution with df (1,20). In Table A-3 (page 353) you will find the critical value of F for df (1,20) at $\alpha = .05$; it is 4.35. Take a moment to find the square of the critical value of t for this same α level, 2.086^2.

What you have just illustrated will hold true in all cases when you compare the square of the critical t value for k df with the critical F value for df $(1,k)$. $F(1,k)$ equals t^2 for k df at the same α level. The rejection boundary for t^2, treated as an F statistic and taken to the F table for df$(1,k)$, will be the square of the rejection boundary for the t statistic itself at k degrees of freedom.

Why does this relationship between t^2 and F exist? We can show that these two statistics are mathematically identical when $n_1 = n_2$ and that they are approximately the same even when n_1 and n_2 are not equal. You already know that the denominator of the t ratio is the square root of $\text{MS}_W\ [(n_1 + n_2)/n_1n_2]$. Therefore, we can already write the following equation:

$$t^2 = \frac{(\bar{Y}_1 - \bar{Y}_2)^2}{\text{MS}_W[(n_1 + n_2)/n_1n_2]} = \frac{(\bar{Y}_1 - \bar{Y}_2)^2 n_1n_2/(n_1 + n_2)}{\text{MS}_W}$$

Since $F = \text{MS}_B/\text{MS}_W$, we need to see what relation exists between MS_B and $(\bar{Y}_1 - \bar{Y}_2)^2\ n_1n_2/(n_1 + n_2)$.

Consider now how you would go about calculating SS_B for just two independent samples. When the samples have different sizes n_1 and n_2, $\text{SS}_B = n_1(\bar{Y}_1 - \bar{Y})^2 + n_2(\bar{Y}_2 - \bar{Y})^2$. But in this case $\bar{Y} = (\bar{Y}_1 + \bar{Y}_2)/2$, and $2Y = \bar{Y}_1 + \bar{Y}_2$; therefore, $\bar{Y}_1 - \bar{Y} = \bar{Y} - \bar{Y}_2$, and $\text{SS}_B = (n_1 + n_2)\ (\bar{Y}_1 - \bar{Y})^2$.

Furthermore, $\bar{Y}_1 - \bar{Y} = 1/2\,(\bar{Y}_1 - \bar{Y}_2)$, so that $\text{SS}_B = [(n_1 + n_2)/4]\ \ (\bar{Y}_1 + \bar{Y}_2)^2$. Since SS_B has $C - 1 = 1$ df, MS_B is also equal to $[(n_1 + n_2)/4]\ (\bar{Y}_1 - \bar{Y}_2)^2$. When $n_1 = n_2$, the fraction $n_1n_2/(n_1 + n_2)$ (from the numerator of t^2) is exactly equal to $n/2$ and to $(n_1 + n_2)/4$. Thus, when $n_1 = n_2$, t^2 is exactly equal to MS_B/MS_W. Even when $n_1 \neq n_2$, the fraction $(n_1 + n_2)/4$ is approximately the same as $n_1n_2/(n_1 + n_2)$. The approximation becomes poorer as the sizes n_1 and n_2 become more widely different. t^2 and F will not have precisely the same numerical value in such a case, but they will still be similar enough to give the same statistical decision in the vast majority of problems.

The t test is thus a special case of one-way analysis of variance. If your test is two-tailed, it makes no practical difference whether you calculate t and use a t table or calculate MS_B/MS_W and use an F table. Even for a one-tailed test, you can calculate either t or $\sqrt{F}$ and take the statistic to a t table with $n_1 + n_2 - 2$ df.

Table 16-3 Data for a *t* Test on the Reaction-Time Example

	Group 1	Group 2
n	12	10
$\bar{Y}$ (milliseconds)	230	240
SS (milliseconds^2)	440	460

16-60 For Table 16-3, the estimated $\sigma^2 =$ ______. — *45*

16-61 For samples of size n_1, estimated $\sigma_{\bar{Y}}^2 =$ ______. For samples of size n_2, estimated $\sigma_{\bar{Y}}^2 =$ ______. Estimated $\sigma_d^2 =$ ______, the sum of the two estimates of $\sigma_{\bar{Y}}^2$. — *3.75*; *4.5*; *8.25*

16-62 The square root of 8.25 is 2.87. The t statistic (with sign) is ______. — *−3.483*

16-63 For $\mathrm{H}_2: \mu_1 - \mu_2 \neq 0$ and $\alpha = .01$, find the t distribution for df = ______. The test must be a ______-tailed test. — *20; two*

16-64 The critical value of t is ______. H_1 can be ______ by this test. — *2.845*; *rejected*

16-65 Now treat your t^2 as an F statistic. The square of your t statistic is ______. The F statistic on the same data is 12.57 (Table 16-1, page 245). t^2 is not exactly equal to F in this case because ______ is not exactly equal to ______. — *12.131*; *n_1; n_2*

16-66 Take $t^2 = 12.131$ to the F table on page 354. The critical value of $F(1,20)$ for $\alpha = .01$ is ______; H_1 can be ______. — *8.10; rejected*

16-67 With $\mathrm{H}_2: \mu_1 - \mu_2 < 0$ and $\alpha = .01$, the test would be ______-tailed. The critical value of t would then be ______. — *one*; *2.528*

16-68 Could you test H_1 against $\mathrm{H}_2: \mu_1 - \mu_2 < 0$ by using the F table? ______, because the F statistic (*does; does not*) contain information about the direction of the observed difference between the two means. — *No; does not*

16-69 If you have calculated F in a two-sample case, can you make a one-tailed test without dividing $\bar{Y}_1 - \bar{Y}_2$ by estimated σ_d to obtain t? ______, you can take the ______ of F as a t statistic. — *Yes; square root*

16-70 This t statistic will have the same df as the (*denominator; numerator*) of the original F. If $n_1 = n_2$, its value will be (*approximately; exactly*) the same as $(\bar{Y}_1 - \bar{Y}_2)/(\text{est. } \sigma_d)$. — *denominator*; *exactly*

REVIEW

16-71 The t test for independent samples is a special case of analysis of variance in which $C =$ ______ and factor A is a (*fixed; random*) factor.

2
fixed

16-72 In such a case, there are only ______ sample means to be compared, and there is only ______ mean difference to be considered.

2
1

16-73 To make a decision about the significance of this mean difference, we may calculate an F statistic. As in any other one-way analysis of variance, H_1 asserts that the true value of F is ______ because all $a_j =$ ______.

1; 0

16-74 The F table permits a test of H_1 against only one H_2, namely, the hypothesis that the true value of F is ______ because there is at least 1 ______ different from 0.

> 1
a_j

16-75 This H_2 (*does; does not*) indicate the direction of the mean difference $\bar{Y}_1 - \bar{Y}_2$.

does not

16-76 Instead of calculating an F statistic, we may calculate a ______ statistic for the two independent samples.

t

16-77 This t statistic is the mean difference $\bar{Y}_1 - \bar{Y}_2$ divided by the ______ of the RSD of mean differences.

estimated standard deviation

16-78 This t statistic provides a test between H_1: $\mu_1 - \mu_2 =$ ______ and an H_2 which may be either nondirectional or ______.

0
directional

16-79 As in any other statistical test, a directional H_2 is admissible only when the expected direction of the difference is known (*after; before*) the data are seen.

before

16-80 As in one-way analysis of variance, the most critical requirement for a t test is that all observations be ______. We stress this requirement when we call this test the "t test for two ______ samples."

independent
independent

16-81 It is also important that the shape of the parent population should not deviate too far from the shape of a ______ distribution. This requirement is critical only when the samples are ______.

normal
small

16-82 The third requirement is that the samples should have equal ________. This requirement is not critical when the samples are of approximately ________ size.

variances
equal

16-83 Consider an experiment with 2 samples of 15 observations each. If F is calculated, the F statistic is the ratio ________ with df = ________.

MS_B/MS_W
(1,28)

16-84 If t is calculated, the t statistic will be the ratio ________ with df = ________.

$\frac{\bar{Y}_1 - \bar{Y}_2}{est.\sigma_d}$
28

16-85 The square (t^2) of this t statistic will exactly equal the ________ statistic because n_1 and n_2 are ________.

F; equal

16-86 The t statistic is also related to the statistic z in the normal-curve table. While z is a ratio between a normally distributed quantity and its ________, t is a ratio between a ________ distributed quantity and its ________.

standard deviation
normally; estimated standard deviation

16-87 If the quantity in the numerator of the t ratio is ________ distributed, the probability distribution of t is found in the t table for df = $n_1 + n_2 -$ ________.

normally
2

16-88 The t distribution for df = infinity is the same as the ________ distribution.

normal

16-89 The t distribution for df = 30 is also very similar to the normal distribution. The difference becomes greater as the number of df becomes ________.

smaller

16-90 Consult the t table on page 355. For df = 10, the critical value at $\alpha = .05$ (one-tailed) is ________.

1.812

16-91 If an investigator used the normal-curve table instead (treating his t as if it were a value of z), he would believe the critical value to be ________.

1.645

16-92 The area under the normal curve beyond $z = +1.645$ is equal to ________. The area beyond $t = +1.645$ under the t distribution for df = 10 is (*larger; smaller*) than .05.

.05
larger

16-93 A t distribution with df less than 30 has a (*larger; smaller*) proportion of its area under its tails, compared with the normal distribution.

larger

16-94 The critical values of t for df less than 30 will thus always be (*less; more*) extreme than the critical values of z at the same α level.

more

PROBLEMS

1. Determine whether the number of hours of study per day is significantly different for the two samples described below:

	Students with Grade Averages of B+ or Better	Students with Grade Averages of B− or Less
n	27	25
$\bar{Y}$ (hours)	4.15	4.38
SS (hours^2)	24.37	42.25

2. The t test may also be used to test hypotheses about the value of the population mean when only a single sample is available. For example, suppose we have a single sample with $n = 10$ and $\bar{Y} = 50$. We wish to test H_1: $\mu = 47$ against H_2: $\mu \neq 47$. We would like to know the probability, given H_1, of obtaining a random sample from this population with the statistics observed in our sample. We are prepared to reject H_1 if the probability is less than $\alpha = .05$.

A t ratio with $n - 1 = 9$ df can be formed by taking $t = (\bar{Y} - 47)/(\text{est. } \sigma_{\bar{Y}})$. With SS = 99.2250, calculate the estimated population σ^2, the estimated $\sigma_{\bar{Y}}^2$ for samples of size n, and the estimated standard error of the mean $\sigma_{\bar{Y}}$ for samples of size n drawn randomly from this population. Then calculate the t statistic and determine whether H_1 can be rejected.

Within what limits (below 47 and above 47) could $\bar{Y}$ vary without our having to reject H_1 at $\alpha = .01$?

LESSON 17
TESTS FOR TWO RELATED SAMPLES

Unless you have good reason to expect the difference between $\bar{Y}_1$ and $\bar{Y}_2$ to be in a particular direction, the choice between t and F tests for two *independent* samples is purely a matter of personal preference. The F test is not sensitive to direction of difference; the t test is. With a directional alternate hypothesis, the t test is to be preferred because it allows a one-tailed test of H_1.

There is one kind of experiment in which a t test may be appropriate while an F test is absolutely forbidden. This is the experiment with two *related* samples, such as the reading experiment used in Lesson 10 to provide data for a sign test. When the same group of n persons is studied under two conditions, the resulting two samples of scores are not independent of each other; each pair of scores from the same individual may be related, particularly with respect to the error components e_{ij}. In such a case the related samples arise from REPEATED OBSERVATIONS on the same individuals.

Samples will also be related when different groups are studied under each condition if each individual in subgroup A_1 is matched with a particular individual in subgroup A_2. The related samples are then said to consist of MATCHED PAIRS of individuals. Pairs matched for IQ score are sometimes used in education experiments; the idea is to rule out (or *control*) the large amount of variability in performance which arises from differences in intelligence, so that the influence of factor A will not be obscured. For the same reason, matched pairs might be used to control other prominent sources of variation, such as age, previous training, and socioeconomic status.

If it is applied in the way indicated in this lesson, the t test may be used to study differences between such related (or matched) samples. No form of the F test exists for such a case. Of course, the usual requirements for a t test must be met. The selection of the first sample, i.e., the selection of the persons to be studied under condition A_1, must be random. The error components *within* each sample of scores must be independent of each other. Furthermore, the parent population must be reasonably normal. As we have already seen, the t test is relatively insensitive to departures from population normality when sample size is large, but the existence of t tables for very small samples has perhaps done much to encourage the use of t when samples are quite small. In such cases it is important to pay close attention to whether the population distribution is reasonably normal.

When the two related samples are small, and when normality of the population is in doubt, another kind of test should be used. The alternative recommended here is the Wilcoxon test, presented in Sec. E.

A. THE *t* TEST FOR TWO RELATED SAMPLES

With only two related samples, a matched pair of scores can be converted into a DIFFERENCE SCORE, $Y_{i1} - Y_{i2}$. Let us call this score V_i. The new score for person (or pair) i represents the difference in that person's performance under the two levels of factor A. Since the n scores within each group are independent of each other, the new distribution V_i is a set of n independent observations. We have already used the signs of such difference scores in Lesson 10B (page 150), where we designated the difference score as $A - B$. At that time we did not use the numerical value of the difference score at all.

For this set of difference scores, we can determine a mean $\bar{V}$ and a sum of squares SS_V. If factor A has no effect, the difference scores will differ from zero only because of sampling variability. There will be about as many positive difference scores as negative ones, and the expectation of $\bar{V}$ is 0.

We can imagine a large number of such means $\bar{V}$, each of them obtained in just the same way: by measuring the same (or equivalent) objects twice, obtaining the difference scores, and taking the arithmetic mean of these scores. When H_1 is true, the distribution of these means will constitute an RSD OF MEANS OF DIFFERENCE SCORES with a mean $\mu_{\bar{V}} = 0$. It will be an approximately normal distribution if the population distribution is normal or if the sample n is sufficiently large with a moderately nonnormal population. It will have a variance $\sigma_{\bar{V}}^2$ which will be estimated from SS_V in the usual way, by taking the estimated population variance $SS_V/(n - 1)$ and dividing it by n. The estimated $\sigma_{\bar{V}}^2$ thus equals $SS_V/n(n - 1)$. Be very careful not to confuse $\sigma_{\bar{V}}^2$, the variance of the RSD of means of difference scores, with σ_d^2, the variance of the RSD of differences between means of two independent samples.

With this RSD in mind we can proceed much as we did in Lessons 8 and 10. We have a particular $\bar{V}$, and we wish to know the probability that it comes from the population specified by H_1, a population with no effect of factor A. Means of samples of this size drawn from this population have an RSD with a mean of 0 and a variance $\sigma_{\bar{V}}^2$. We form the ratio $\bar{V}/(\text{est. } \sigma_{\bar{V}})$, just as we formed z scores in making binomial tests; but in this case we are using an estimated standard deviation instead of an exact standard deviation, and we consequently recognize that our ratio is a t statistic. This statistic

$$t = \frac{\bar{V}}{\text{est. } \sigma_{\bar{V}}}$$

has $n - 1$ df and must be evaluated by consulting the t distribution for $n - 1$ df.

The t test for two related samples enables us to get around the problem of related error components. Since only one distribution, V_i, is analyzed, this t test is actually easier to calculate than the t test for two independent samples. Table 17-2 (page 263) will help you to avoid confusing these two forms of the t test.

17-1 If we have two *independent* samples, each with n observations, we actually have a total of ______ independent observations. — $2n$

17-2 But when we have two matched samples, each with n independent observations, the set of $2n$ scores contains only ______ independent observations. Selection of the second member of a __________ pair is completely determined by selection of its partner. — n; *matched*

17-3 Each matched pair of scores can be converted into a __________ score by taking $Y_{i1} - Y_{i2}$, where Y_{i1} is the score of person i under level ______ of factor A and Y_{i2} is that same person's score under level ______. — *difference*; *1*; *2*

17-4 The difference scores V_i have a mean called ______. There are ______ independent scores in the V_i distribution. — $\bar{V}$; n

17-5 To obtain a probability distribution for $\bar{V}$, we imagine an RSD of the __________ of samples of n difference scores, all drawn from the same difference-score population. — *means*

17-6 Since the expectation of $\bar{V} =$ ______ when H_1 is true, the mean $\mu_{\bar{V}}$ of the RSD of $\bar{V}$ is equal to ______. If the parent population is fairly normal, the shape of this RSD will be __________. — *0*; *0*; *normal*

17-7 This RSD of $\bar{V}$ is analogous to the RSD of $\bar{Y}$; the variance of an RSD of means, $\sigma_{\bar{Y}}^2$, is the population variance σ^2 divided by the __________ of the sample. — *size*

17-8 For a sample of size n, $\sigma_{\bar{Y}}^2$ is estimated from SS_Y by the equation: estimated $\sigma_{\bar{Y}}^2 =$ ______. — $\frac{SS_Y}{n(n-1)}$

17-9 For a distribution of V_i, the sum of squares will not be $SS_Y = \Sigma(Y_i - \bar{Y})^2$. It will be $SS_V =$ ______. — $\Sigma(V_i - \bar{V})^2$

17-10 The estimated variance of the RSD of *means of difference scores*, for samples of size n, will be SS_V divided by ______. We call this statistic ______. — $n(n-1)$; $\sigma_{\bar{V}}^2$

17-11 From two matched samples, we obtain an observed mean $\bar{V}$ for the set of n difference scores. The ratio $\bar{V}/(\text{est. } \sigma_{\bar{V}})$ is the ratio of a __________ distributed quantity to its __________ standard deviation. — *normally*; *estimated*

17-12 Such a ratio is a ______ statistic. With n independent observations, its df = ______. — t; $n - 1$

B. INDEPENDENT VS. RELATED SAMPLES: A DESIGN QUESTION

Table 17-2 compares the t test for related samples with the t test for independent samples. Notice particularly the difference in degrees of freedom. If $n_1 = n_2 = n$ for the two independent samples, the df for a t test will be $2n - 2$. The df for two related samples of size n will be $(n - 1)$; $(n - 1)$ degrees of freedom are lost by taking paired scores.

Compare an independent-samples experiment in which $n_1 = n_2 = 15$ with a related-samples experiment in which $n = 15$. Both experiments will require the same amount of work in collecting the data; 30 tests must be given and scored in either case. The t ratios obtained will be compared with two different t distributions; for independent samples, df = 28, while for related samples, df = 14. Table 17-1 below shows that the critical value of t must always be larger for the related samples to attain the same level of significance as the independent samples.

Table 17-1 Comparison of Critical Values of *t* for Independent and Related Samples ($n = 15$)

Samples	df	$\alpha = .05$		$\alpha = .01$	
		One-Tailed	Two-Tailed	One-Tailed	Two-Tailed
Independent	28	1.701	2.048	2.467	2.763
Related	14	1.761	2.145	2.624	2.977

When an investigator is designing his experiment, he may plan to use either independent or related samples for comparing two treatments. If he uses related samples, he will have to obtain a higher critical value of t for samples of a given size. Yet investigators often prefer the related-samples design. What is the reason for this preference?

Usually it is the hope that matched pairs will reduce the size of σ_e^2. Any difference due to factor A will tend to be obscured by a large error variance; the smaller the actual effect of factor A, the more easily it may go undetected when error variance is large.

When persons are the objects being studied, a considerable proportion of the influences in e_{ij} are likely to fall into the category of *individual differences*, i.e., in background, learning ability, test-taking ability, and other personal characteristics. Some of these individual-difference factors are the same for both scores in a matched pair; the difference score will then reflect only the factors (including factor A) which are different from one measurement to the other within a pair. The error components for V scores (differences between matched pairs) should therefore be somewhat smaller than the error components for Y scores from independent samples, and σ_e^2 should also be smaller. Thus, when individual differences make up an important part of the error component, the related-samples design may have a real advantage over independent samples.

Table 17-2 Comparison of t Tests for Independent and Related (Matched) Samples

Samples	Statistics Needed	t Statistic	df
Independent	$SS_1, n_1, \bar{Y}_1; SS_2, n_2, \bar{Y}_2$	$\dfrac{\bar{Y}_1 - \bar{Y}_2}{\text{est. } \sigma_d} = \dfrac{\bar{Y}_1 - \bar{Y}_2}{\sqrt{\left(\dfrac{1}{n_1} + \dfrac{1}{n_2}\right) \dfrac{SS_1 + SS_2}{n_1 + n_2 - 2}}}$	$n_1 + n_2 - 2$
Related	$SS_V, n, \bar{V}$	$\dfrac{\bar{V}}{\text{est. } \sigma_{\bar{V}}} = \dfrac{\bar{V}}{\sqrt{\dfrac{SS_V}{n(n-1)}}}$	$n - 1$

17-13 If only 30 observations are to be collected, 30 individuals may be selected and assigned at random to A_1 and A_2. There will then be ______ df for the t statistic. *28*

17-14 On the other hand, 15 individuals may be selected at random, and each individual may be measured under both A_1 and A_2. There will then be ______ df for the t statistic. *14*

17-15 With the same total number of observations, a related-samples design will always have (*half; twice*) as many df as an independent-samples design. *half*

17-16 When $\alpha = .01$ and the test is one-tailed, the critical value of t for independent samples is ______. For related samples, the critical t is ______. *2.467* *2.624*

17-17 For a t test at the same α level, the critical value of t will always be (*larger; smaller*) for related samples than for independent samples with the same total number of observations. *larger*

17-18 However, σ_e^2 may be (*larger; smaller*) for a set of difference scores than for a set of unrelated raw scores. *smaller*

17-19 Some factors contributing to the error component e_{ij} are peculiar to a particular individual. Such factors are likely to (*change; remain constant*) on the two occasions when the same person is measured. *remain constant*

17-20 Factors peculiar to the individual (*are; are not*) likely to be part of e_{ij} in a difference score V_i. They (*are; are not*) likely to be part of e_{ij} in a raw score Y_i. *are not* *are*

17-21 A related-samples design may therefore (*decrease; increase*) the amount of error variance, especially when ______ differences play an important part in determining the amount of error variance. *decrease* *individual*

C. FACTORS AFFECTING THE POWER OF A *t* TEST

The power of a t test is described exactly as we have described the power of a binomial test in Lessons 5 and 10. Having located a critical value of t in the distribution conditional upon H_1, we assume an exact H_2 and find the probability $1-\beta$ which lies beyond the critical value under the H_2 distribution. To obtain a power function, $1-\beta$ must be calculated for a series of possible exact H_2's.

The exact calculation of $1-\beta$ for a t test is more complex than for the binomial test, and it requires tables of special (noncentral) t distributions. However, for sufficiently large samples the normal-curve table provides a satisfactory approximation. We shall discuss the power of t tests by using relatively large samples to provide illustrations.

Figure 17-1 shows the way to visualize calculation of $1-\beta$ for two related samples of 100 observations. With df = 99, the t distribution will be almost the same as a normal distribution. A one-tailed test is assumed, and the left-hand distribution is conditional upon H_1; its mean is 0, and $\sigma_{\bar{V}}$ is taken as 2. At $\alpha = .05$, the critical value of t is 1.65; the critical value of $\bar{V}$ is $1.65\sigma_{\bar{V}} = 3.30$. The right-hand distribution is conditional upon H_2: $\mu_{\bar{V}} = 4$, where it is assumed that factor A has enough effect to give a true population mean of difference scores equal to 4. The shaded area is $1-\beta$, which equals .637 for this H_2.

The power of a t test will increase with increases in sample size and with increasing difference between H_1 and the true H_2, just as the power of a binomial test does. As in the binomial case, a one-tailed t test is relatively more powerful than a two-tailed t test, provided the true H_2 is in the expected direction. In the t test, however, power is affected also by an additional factor, the size of σ_e^2. Figure 17-2 shows the same pair of hypotheses as Fig. 17-1: that $\mu_{\bar{V}} = 0$ (H_1) and that $\mu_{\bar{V}} = 4$ (H_2). However, with a smaller error variance, σ^2 and $\sigma_{\bar{V}}^2$ will be smaller; $\sigma_{\bar{V}}$ is assumed to be 1 in Fig. 17-2. The power of the t test is correspondingly much larger; in Fig. 17-2, $1-\beta = .991$.

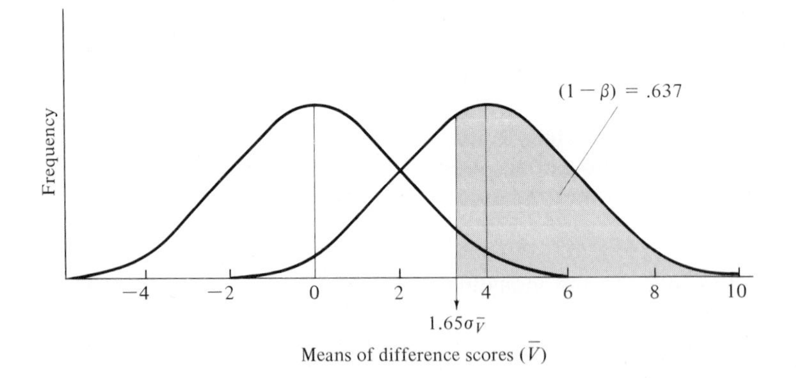

Figure 17-1: Random sampling distributions of the means of difference scores with $\mu_{\bar{V}} = 0$ (H_1, *left curve*) and with $\mu_{\bar{V}} = 4$ (H_2, *right curve*). The shaded area shows the power of this one-tailed t test (df = 99) at $\alpha = .05$ when $\sigma_{\bar{V}} = 2$.

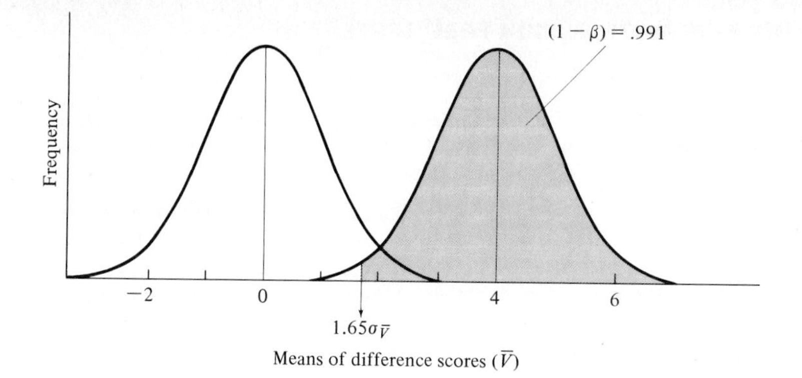

Figure 17-2: Random sampling distributions as in Fig. 17-1, but with $\sigma_{\bar{V}} = 1$. Shaded area shows the power of this one-tailed t test to distinguish H_1: $\mu_{\bar{V}} = 0$ (*left curve*) from H_2: $\mu_{\bar{V}} = 4$ (*right curve*) at $\alpha = .05$.

17-22 Figure 17-1 pictures two different RSDs of means of ________ scores. $\mu_{\bar{V}}$ for the distribution on the left is ________. This RSD is conditional upon ________, and its standard deviation $\sigma_{\bar{V}}$ is taken as ________. *difference* *0; H_1* *2*

17-23 The power of a t test is $1 - \beta$; this is the probability that we shall accept a particular (H_1; H_2) when it is actually true. *H_2*

17-24 Fig. 17-1 illustrates the power of a t test to lead to a correct acceptance of H_2: $\mu_{\bar{V}} =$ ________. *4*

17-25 Since $\alpha = .05$ and the test is one-tailed, we shall reject H_1 and accept H_2 if the observed $\bar{V}$ lies in the ________ region at the upper end of the H_1 distribution. The boundary of this region is the ________ value of $\bar{V}$ for $\alpha = .05$. *rejection* *critical*

17-26 With df = 99 the critical value of t is 1.65. The critical value of $\bar{V}$ is ________. *+3.30*

17-27 The probability $1 - \beta$ is the area under the (H_1; H_2) distribution and to the (*left; right*) of $\bar{V} = +3.30$. *H_2* *right*

17-28 In Fig. 17-2, the two hypotheses are still the same, but the assumed value of ________ has changed from 2 to ________. The critical value of t is still +1.65, but the critical value of V is now ________, and the area $1 - \beta$ is (*larger; smaller*). *$\sigma_{\bar{V}}$; 1* *+1.65; larger*

17-29 The power of a t test *increases* with the following changes: with (*decreases; increases*) in sample size; with (*decreases; increases*) in the difference between H_1 and H_2; and with (*decreases; increases*) in error variance. *increases* *increases* *decreases*

D. THE ROLE OF DISTRIBUTION-FREE TESTS

Both the t test and analysis of variance depend upon assuming a normal distribution of the parent population. This requirement is not critical unless the samples from each subpopulation are small; even a moderate size for each sample will ensure that the RSD of means is normal. However, use of these tests, particularly the use of a one-tailed t test, may lead to incorrect decisions when the RSD of means is actually skewed. This situation can arise even with large samples when the parent population is extremely different from normal.

You must wonder exactly how extreme the departure from normality must be in order to make a t test or an F test unsafe. Unfortunately, there is no simple rule of thumb to guide the beginner, and even experts are likely to disagree about borderline cases. The choice, however, is not between assuming population normality and having no test at all. For all the experiments we have been discussing, there exist other statistical tests which make no assumptions at all about the distribution of the parent population. Such tests are said to be DISTRIBUTION-FREE. Whenever you are in any doubt about the normality of the RSD, you should simply use the appropriate distribution-free test.

We shall discuss one such test in this lesson, the Wilcoxon test for two related samples. In Lesson 18 we take up a distribution-free test for two independent samples (the Mann-Whitney U test) and a distribution-free version of one-way analysis of variance. All these tests escape assumptions about population normality by converting score distributions (interval data) into ranks (ordinal data). Since this conversion involves some loss of information, it is important to point out that the distribution-free tests have almost the same power as the t and F tests. Where the slight loss of power might be critical, it can be compensated by relatively small increases in sample size.

The sign test discussed in Lesson 10 is also a distribution-free test for related samples. Its power is likely to be considerably less than either the t test or the Wilcoxon. Therefore, the sign test is ordinarily used only (1) when no other test can be applied or (2) when a quick estimate of significance is desired before performing further analysis.

Distribution-free tests may also be called NONPARAMETRIC tests. These tests make no assumptions about the shape of the population distribution because they do not rest upon estimates of population parameters; hence the term "nonparametric." The t test and other cases of analysis of variance have to assume approximate normality of the population distribution; otherwise, estimation of population variance would not be possible from sample data. When this parameter cannot be estimated satisfactorily, use of these PARAMETRIC tests is not justified. The distribution-free tests of course avoid the estimation of such parameters. As you will see, they deal with the parameters of a population of rank numbers, the integers from 1 to N. These tests are "nonparametric" only in the sense that they do not depend upon estimated parameters of the score population.

17-30 Analysis of variance and the t test both require estimation of population ___________. *variance*

17-31 For a t test, estimation of σ^2 is necessary before the standard deviation of the RSD can be estimated. This standard deviation is the ___________ of the t ratio. *denominator*

17-32 Unless the RSD is ___________ distributed, its variance and standard deviation cannot be estimated. *normally*

17-33 If we assume that the RSD is normal when it is actually skewed, the estimate of its standard deviation will be *too low*. Then the t statistic will have a (*higher; lower*) value than the data can justify. *higher*

17-34 Wrongly assuming a normal RSD will give a spuriously high t statistic. Consequently, the probability of rejecting H_1 incorrectly is (*decreased; increased*). *increased*

17-35 When a statistical test is applied properly, the probability of a type I error is determined by the decision rule; it is the probability called _______. α

17-36 If a t test is applied when the RSD is not normal, the probability of a type I error can no longer be calculated; however, this probability is likely to be (*greater; less*) than α. *greater*

17-37 Remember also that the t statistic is a ratio between a ___________ distributed quantity and its estimated standard deviation. If the quantity $\bar{Y}_1 - \bar{Y}_2$, or the quantity $\bar{V}$, is distributed in some other way, the ratio will not really be a _______ statistic and its probability distribution will be unknown. *normally* t

17-38 Tests which do not require the assumption of a normal distribution are called distribution-___________ tests. *free*

17-39 Whenever normality of the RSD is in doubt, one should use a ___________ test instead of a t test in making decisions about two samples. *distribution-free*

17-40 Distribution-free tests are also called nonparametric tests because they do not require the estimation of certain ___________ of the score population. *parameters*

17-41 Tests which do require the estimation of such parameters are called parametric tests. The t and F tests are ___________ tests. *parametric*

E. THE WILCOXON MATCHED-PAIR SIGNED-RANKS TEST

The data in Table 17-3 will be used to illustrate a Wilcoxon test with small samples. During the 1968 presidential campaign a professor gave a questionnaire on attitudes concerning gun-control legislation to nine students in his seminar. The assassination of Robert Kennedy occurred the following week, and the professor immediately had the same students complete the same questionnaire again. Table 17-3 shows the scores on this questionnaire before and after the assassination; high scores indicate a favorable attitude toward gun-control legislation.

The signed difference scores are calculated by subtracting the second score for each individual from his first score. Column 5 shows the *ranks* of these difference scores; the ranking is done on the *absolute values* of the difference scores. The smallest difference, regardless of sign, receives the rank of 1 (person 4), the next smallest receives the rank of 2 (person 7), and the largest difference receives the rank of 9 (person 2). After the ranking has been completed, the sign of the original difference score is assigned to its rank, hence the title *signed-ranks test.* The data to be analyzed are actually the signed ranks of difference scores. We calculate a statistic T which is the sum of the negative ranks or the sum of the positive ranks, whichever is smaller.

Table 17-3 Favorableness toward Gun-Control Legislation before and after Assassination of Robert Kennedy

Person	Before	After	Difference	Rank of Difference	Rank with Less Frequent Sign
1	20	25	−5	−4	
2	46	75	−29	−9	
3	18	22	−4	−3	
4	71	70	+1	+1	+1
5	35	45	−10	−6	
6	30	51	−21	−8	
7	41	43	−2	−2	
8	29	23	+6	+5	+5
9	49	66	−17	−7	
					$T =$ 6

The professor's assumed hypothesis asserts that the assassination made no difference in attitudes on the gun-control issue within his seminar. He expects to be able to reject this hypothesis and to accept H_2, that the assassination has increased the favorableness of his students toward gun-control legislation. In terms of the Wilcoxon test itself, H_1 asserts simply that the sum of the positive ranks is equal to the sum of the negative ranks. Within the limits of sampling variability, this hypothesis will be true when the assassination has not affected the attitudes sampled by this questionnaire.

17-42 The two sets of 9 scores each ("before" and "after" in Table 17-3) are first converted to one set of 9 ________ scores. *difference*

17-43 These difference scores are then ranked without regard to ________. The smallest difference is ______, shown by person number 4; this difference receives the rank number ______. *sign; +1* *1*

17-44 The largest difference is ______, shown by person 2; this difference is given the rank number ______. *−29* *9*

17-45 Of the 9 differences, ______ have positive signs and ______ have negative signs. These signs are carried over to the ________ of the differences in column 5. *2* *7* *ranks*

17-46 The sum of *all 9* rank numbers is the sum of the first 9 integers. It is given by $N(N+1)/2$ for $N = 9$; this sum is ______. *45*

17-47 The sum of the ranks with the less frequent sign is ______. The sum of the remaining ranks is 45 − ______ = ______. The statistic T is therefore equal to ______ because T is the (*larger; smaller*) of these two sums. *6; 6* *39; 6* *smaller*

17-48 With a set of eight signed ranks as follows: 1, −2, −3, 4, −5, 6, 7, −8, the sum of positive ranks is ______ and the sum of negative ranks is ______. *18* *18*

17-49 The positive and negative ranks are quite evenly distributed through this set; the two samples (*do; do not*) appear to differ in any way reflected by the signed ranks. *do not*

17-50 H_1 is the hypothesis that the ________ of the positive ranks equals the ________ of the negative ranks. If H_1 is true, any deviation from this expectation is due to random ________ variability. *sum* *sum* *sampling*

17-51 When there are just eight ranks in the set, the sum of all eight ranks is ______. If H_1 is true, the expected value of the statistic T is just half this sum, or ______. *36* *18*

17-52 With nine ranks in Table 17-3, the expected value of T under H_1 is ______. This expectation, $E(T)$, is exactly $N(N+1)$ divided by ______. *22.5* *4*

17-53 Since the observed value of T is 6, we need an RSD of the statistic ______ in order to decide whether to reject H_1 in this example. *T*

When H_1 is true, $E(T)$ is always $N(N+1)/4$, or half the sum of the first N consecutive integers $N(N+1)/2$. Small variations from this expected value should occur frequently as a result of sampling variability; large variations should occur more rarely. An exact calculation of the probability distributions of T has been made for values of N between 6 and 25; Table A-6 (page 355) gives certain critical values from these distributions.

If N is larger than 25, the statistic T has an approximately normal distribution with a mean $\mu_T = N(N+1)/4$ and a variance $\sigma_T^2 = [N(N+1)(2N+1)]/24$. The ratio between $\mu_T - T$ and its standard deviation is therefore a normally distributed z score. For any observed value of T, such a z score can be calculated, and this z score can be taken to the normal-curve table to determine its probability under either a one-tailed or a two-tailed test. (For a review of procedures in using the normal-curve table, see Lesson 10, pages 148 to 153.)

In applying the Wilcoxon test, you should remember the following additional rules:

1 A difference score which equals 0 is simply ignored. If there are nine pairs and one pair gives zero difference, then N is taken as 8.

2 When two or more difference scores are of the same size, follow the conventional rule for tied ranks: give the tied scores the average of the two or more ranks which they occupy. Thus, three difference scores tied for ranks 4, 5, and 6 will all be assigned rank 5; four difference scores tied for ranks 7, 8, 9, and 10 will all be assigned rank 8.5. The presence of tied ranks does not affect the RSD of T in any important way.

The power of distribution-free tests is best described in comparative terms. The power of test B (here, the Wilcoxon) is compared to the power of a parametric test A which might be used for the same data, provided its requirements are met. In this case, test A is the t test for related samples.

The ratio N_A/N_B is called the POWER-EFFICIENCY RATIO of test B: N_B is the sample size required in order for test B to reach the same power as test A with sample size N_A. The power-efficiency ratio of the Wilcoxon test in relation to the t test is $3/\pi = .955$. Translated into practical terms, this power-efficiency ratio means that a Wilcoxon test with 22 pairs has the same power as a t test with 21 pairs, since 21/22 is approximately equal to 3/3.14. We can say that the POWER EFFICIENCY of the Wilcoxon test is about 95 percent, compared with the t test.

We have discussed the Wilcoxon test in connection with two related samples because it has its greatest usefulness in making decisions about such samples. There is a form of the Wilcoxon test for use with two independent samples. It is not discussed here; instead, the more widely used Mann-Whitney U test, suited for two independent samples, will be taught in Lesson 18.

17-54 In Table 17-3, the observed value of T is 6. Consult Table A-6 (page 355) of critical values of T. Each row gives critical values for a different ________ of T; the relevant row is the row for $N =$ ______.

RSD
9

17-55 The observed value $T = 6$ is significant at the ______ level for a two-tailed test and at the ______ level for a one-tailed test.

.05
.025

17-56 H_2 for Table 17-3 asserts that attitudes will be more favorable on second testing. This H_2 requires a ________-tailed test.

one

17-57 In order to reach significance at the .01 level, T would have to be (*larger; smaller*). The critical values are therefore the (*maximum; minimum*) values which T may have if it is to fall inside a rejection region.

smaller
maximum

17-58 When N is larger than 25, the RSD of T is approximately ________ distributed.

normally

17-59 With $N > 25$ we use a normalized T distribution with a mean $\mu_T = N(N+1)/4$ and a variance $\sigma_T^2 =$ ______.

$\frac{N(N+1)(2N+1)}{24}$

17-60 $\mu_T - T$ is a deviation score whose standard deviation is the square root of ______.

σ_T^2

17-61 Therefore, the ratio ______ can be treated as a ________ distributed z score. Its probability can be obtained from the ________ table.

$\frac{\mu_T - T}{\sigma_T}$
normally
normal-curve

17-62 The power of the Wilcoxon test is described by comparing it with the power of the ______ test for two related samples.

t

17-63 This comparison is stated as the power-efficiency ratio N_A/N_B. N_A is the sample size for test A, which is always the (*parametric; nonparametric*) test.

parametric

17-64 N_B is the sample size which is required for test B, the ________ test, to have the same power as test A with sample size ______.

nonparametric
N_A

17-65 N_B is always somewhat (*larger; smaller*) than N_A, and the power-efficiency ratio is always somewhat (*greater; less*) than 1.

larger
less

17-66 The power efficiency of the Wilcoxon test is $3/\pi = .955$ in relation to the t test. The Wilcoxon test with $\pi = 3.14$ related pairs has the same power as a t test with ______ pairs.

3

REVIEW

17-67 Two methods of teaching algebra are to be compared: 30 students will be taught by each method, and their final examination scores will be compared. If 60 students are assigned at random to the two methods, the two samples will be (*independent; related*).

independent

17-68 Aptitude for math varies widely among the students from whom our samples will be drawn. If we therefore select 30 students at random, then find another 30 whose previous arithmetic grades exactly match those for the first group, these two samples will be (*independent; related*) samples.

related

17-69 We can use a t test for both plan M (matched pairs) and plan I (independent samples). If method 1 is expected to be superior to method 2, H_2 will be a __________ hypothesis and these t tests will be __________-tailed.

directional
one

17-70 In plan I we shall obtain (*1; 2*) sample means, and the statistic which may or may not reach significance is ______.

2
$\bar{Y}_1 - \bar{Y}_2$

17-71 In plan M we shall convert each matched pair of scores into a __________ score. The statistic which may or may not reach significance is the __________ of this set of scores, which we are calling ______.

difference
mean
$\bar{V}$

17-72 To decide about $\bar{Y}_1 - \bar{Y}_2$, we derive an RSD of __________ between two sample __________.

differences; means

17-73 To decide about $\bar{V}$, we derive an RSD of __________ of __________ scores.

means
difference

17-74 The RSD of mean differences will be normal if the RSDs of $\bar{Y}_1$ and $\bar{Y}_2$ are normal because $\bar{Y}_1 - \bar{Y}_2$ is then an additive combination of two __________ distributed variables.

normally

17-75 Since sample size is large, the RSD of $\bar{Y}$ will be normal for each sample, *provided that* the __________ from which these scores are drawn does not deviate extremely from a normal distribution.

population

17-76 Likewise the RSD of $\bar{V}$ will be normal provided that the parent population of __________ scores is not too different from normal.

difference

17-77 If $\bar{Y}_1 - \bar{Y}_2$ is a normally distributed quantity, the ratio $(\bar{Y}_1 - \bar{Y}_2)/(\text{est. } \sigma_d)$ is a ______ statistic with df = ______.

t; 58

17-78 If $\bar{V}$ is normally distributed, the ratio $\bar{V}$/(est. $\sigma_{\bar{V}}$) is a t statistic with df = ______. *29*

17-79 If plan M is chosen and the RSD of $\bar{V}$ is not expected to be normal, the proper test to use is the ______ test for two related samples. *Wilcoxon*

17-80 The Wilcoxon test will not use the actual set of 30 ______ scores. The form of their population distribution will therefore be irrelevant to this test. *difference*

17-81 The Wilcoxon test uses only the ______ of the difference scores, together with their algebraic ______. The rank of 1 is given to the (*largest; smallest*) difference score. *ranks* *signs* *smallest*

17-82 The statistic T is the sum of the ______ ranks or the sum of the ______ ranks, whichever is ______. *positive ↔ negative; smaller*

17-83 The critical value of T for a particular α level is a (*maximum; minimum*) value of T. *maximum*

17-84 For $N > 25$, the RSD of T is approximately ______, and critical values are obtained from the ______ table. *normal* *normal-curve*

17-85 The power of a t test increases when error variance is made (*larger; smaller*). *smaller*

17-86 Plan M might be chosen over plan I because M is expected to have a ______ error variance. If it does, plan M will have (*greater; less*) power than plan I. *smaller* *greater*

17-87 The power of any test (*decreases; increases*) with increases in sample size. The power of the Wilcoxon test for related samples may be made equal to the power of a t test for related samples by taking slightly ______ samples. *increases* *larger*

17-88 The ratio $N_A/N_B = .955$ tells how much larger the sample must be. N_A is the sample size for the parametric test, which in this case is the ______ test. N_B is the necessary sample size for the ______ test. *t* *Wilcoxon (or nonparametric)*

17-89 N_A/N_B is called the ______-efficiency ratio. Test B is slightly ______ efficient than test A because B takes a ______ sample to achieve the same power. *power* *less* *larger*

17-90 To achieve the same power with a Wilcoxon test as with a t test on 30 difference scores, we shall need about ______ difference scores. *31 or 32*

PROBLEMS

1. The following table gives scores on equivalent forms of the same test taken before (A_1) and after (A_2) a special audio-visual instructional program. Determine whether these data support the conclusion that this special program improved test performance. Use (*a*) a sign test, (*b*) the Wilcoxon test, and (*c*) a t test.

Person	A_1	A_2	Person	A_1	A_2
1	87	98	12	72	79
2	89	93	13	80	78
3	83	90	14	73	77
4	93	89	15	77	76
5	85	87	16	70	75
6	81	86	17	66	73
7	82	85	18	69	71
8	78	84	19	71	70
9	79	83	20	58	67
10	75	81	21	64	61
11	74	80			

2. Compute μ_T and σ_T (to one decimal place), and determine the probability of the observed T under H_1 in the following cases:

a. $N = 30$, $T = 107$, one-tailed test.

b. $N = 26$, $T = 60$, $T = 70$, two-tailed tests; $T = 100$, one-tailed and two-tailed tests.

3. Use the normal approximation with $N = 9$, and determine the critical values of T for a one-tailed test at $\alpha = .025$, $\alpha = .01$, and $\alpha = .005$. Compare with the critical values in Table A-6 (page 355). How good is the normal approximation when N is as small as 9?

4. In order to find out whether two French vocabulary tests are equally difficult, a teacher administers both tests to the same 12 persons. Since the scores (in percent correct) are obviously not normally distributed, use a nondirectional Wilcoxon test on the following data:

Person	Test 1	Test 2	Person	Test 1	Test 2
1	96	94	7	81	80
2	95	96	8	79	76
3	93	94	9	75	79
4	89	91	10	70	65
5	87	86	11	67	61
6	86	89	12	51	58

LESSON 18
DISTRIBUTION-FREE TESTS FOR TWO OR MORE INDEPENDENT SAMPLES

To introduce a distribution-free test for two independent samples, let us return to an example which is already familiar. In Lesson 11 we used χ^2 to compare a sample of young women psychologists with a sample of young men psychologists. The sociologist studying this problem was said to have given his 54 subjects a questionnaire on the relative importance of hereditary and environmental factors in shaping personality. We imagined that he then dichotomized each of his samples into two categories (environmentalist and nativist) on the basis of a previously chosen cutoff score of 50 out of 100 points. The scores themselves were not analyzed; they were simply used to form a 2×2 contingency table.

Now that you have studied the t test, you will recognize that the sociologist could have calculated a t statistic from his original questionnaire scores. However, a t test will be valid only if the population of scores on this questionnaire is normally distributed. Unless a questionnaire has undergone years of technical development, this assumption of normality is rarely justified. The assumption can be avoided by using the MANN-WHITNEY U TEST, a distribution-free test which has a power-efficiency ratio of .95 compared with the t test.

The idea of the U test is very simple. Let the whole set of 54 observations be put into a single rank order; the person, man or woman, with the highest score in the environmentalist direction receives the rank number 1, and all other persons receive rank numbers which progress toward 54. If the samples of 30 women and 24 men have been randomly drawn from the same population, the men and women ought to be reasonably well interspersed in this series. The U statistic summarizes the degree to which the two samples actually are interspersed at random in the single ordinal series.

You will see the logic behind the test more clearly if we consider smaller samples. With 5 men and 4 women, the total N is 9; there are $9!/5!4! = 126$ (recall page 58) different orders in which 5 men and 4 women can be arranged (without regard to the identity of particular men or women). When chance alone is operating, each of these orders is equally likely to occur. The probability of getting the order MMMMMWWWW is 1/126; there is no other order which would be more extreme (less interspersed) than this one. If this order occurs, you can immediately reject at $\alpha = .01$ the hypothesis that these are random samples from the same population. With less extreme outcomes such as MMWMMMWWW, one is faced with the additional problem of finding and enumerating all equally extreme outcomes. The U statistic enables us to find the probability of equally extreme orders more quickly and conveniently.

A. THE U STATISTIC

Table 18-1 shows how the U statistic arises from two small samples. In this table $N = 15$; a rank of 1 signifies the most environmentalistic.

Table 18-1 The *U* Statistic for a Hypothetical Experiment on Environmentalism-Nativism

	Rank															
	1	2	3	4	5	6	7	8	9	10	11	12	13	14	15	
Sex of person with this rank	M	M	M	W	M		M W W		M	M	W	M	W	W	W	
Number of W's following this M	7	7	7		6		5		4	4		3				$U_M = 43$
Number of M's following this W				5			$3\frac{1}{2}$ $3\frac{1}{2}$				1		0	0	0	$U_W = 13$

After we have written the rank numbers from 1 through 15 together with the sex of the person holding each rank, we begin to calculate the statistic U_M, the U statistic for the M sample. We consider each M in turn; to obtain its contribution to U_M, we add 1 for each W coming later in the series and $\frac{1}{2}$ for each W tied at the same position as this M. In this case, there is a tie at rank 7 between one M and two W's, and $U_M = 43$. We then consider each W in turn, adding 1 for each M that is later in the series and $\frac{1}{2}$ for each M tied at its position. The sum of the resulting 7 numbers is $U_W = 13$.

When the two samples arise from the same population, the two statistics U_M and U_W should equal each other, except for such differences as might arise because of sampling variability. H_1 is therefore the hypothesis $U_M = U_W$. But U_M and U_W are not independent statistics. The sum $U_M + U_W$ must equal $n_1 n_2 = 56$, where n_1 is the size of the smaller sample. Once U_M is known to be 43, U_W is also known; it is $56 - 43 = 13$. We therefore need to study the RSD of only one of these statistics.

For convenience we always work with the smaller of the two statistics (in this case U_W) and call it U. The other statistic is called U'. We expect U and U' to be about equal in most pairs of samples drawn randomly from the same population. As U grows smaller, U' grows larger and the difference between them increases. The greater this difference, the less likely it is that we can continue to accept H_1. Therefore, we look for quite small values of U to indicate that the difference between U and U' is great enough for rejection of H_1.

It is easier to count the U statistic for the smaller sample. Always work from the smaller sample; if the statistic obtained is greater than $n_1 n_2/2$, you have obtained U' and you can then get U by taking $n_1 n_2 - U'$.

18-1 In Table 18-1 we examine each M in turn, counting the number of W's which ________ it. For the M with rank 3, there are ________ such W's; for the M with rank 12, there are only ________ such W's.

follow
7
3

18-2 One of the M's is tied for rank 7. There are ________ W's which literally follow this M; there are ________ W's which are tied with it for the same rank. Allowing 1/2 for each of the tied W's, we write for this M the number ________.

4
2
5

18-3 When we have made this count for all eight M's, we take the total and call it U_M. In this case, $U_M =$ ________.

43

18-4 There is an analogous statistic for the W's, called U_W. If we consider each W in turn, counting the number of ________ which ________ it, the total of these numbers will give U_W.

M's
follow

18-5 Since the size of the smaller sample is n_1, n_1 is ________ and n_2 is ________ for this example. $U_W + U_M$ must equal n_1n_2, or ________.

7
8
56

18-6 Once we know that $U_M = 43$, we know that $U_W =$ ________ $- 43 =$ ________.

56
13

18-7 The U statistic is always the (*larger; smaller*) of the two statistics. The ________ one is called U'.

smaller
larger

18-8 In this example U has the numerical value ________ and U' has the value ________.

13
43

18-9 Consider another example. Sample A has 4 cases; sample B has 5. Since n_1 is the size of the ________ sample, $n_1 =$ ________ and $n_2 =$ ________. The sum $U + U' =$ ________.

smaller
4; 5; 20

18-10 The observed order is BBABBBAAA. Determine the U statistic for the smaller sample. The statistic is ________, because the A at rank 3 is followed by ________ B's. All the other A's are followed by ________ B's.

3
3
0

18-11 Since $U + U' = 20$, the statistic you obtained from the smaller sample is (U; U'). The statistic for the larger sample is ________, and it has a numerical value of ________.

U
U' ; 17

18-12 Now suppose the observed order to be AAABBBBAB. The U statistic for the smaller sample has a numerical value of ________. In this case, the U statistic for the smaller sample is (U; U').

16
U'

B. THE RSD OF *U*

Let us develop an RSD for the U statistic from a pair of very small samples. Suppose we have only 2 women and 3 men in the environmentalism study. Since $n_1 n_2 = 6$, the maximum value U can have is 3; U can equal 0, 1, 2, or 3, while U' can take the corresponding values 6, 5, 4, or 3, respectively.

Table 18-2 shows how these values of U and U' can arise. Since there are 2 W's and 3 M's, we know that there are $5!/2!3! = 10$ different orders in which these 5 cases can occur. According to H_1, each of the 10 orders is equally likely, and the probability of any one of the 10 orders is .1.

Table 18-2 Probabilities of Different Outcomes when $n_1 = 2$ and $n_2 = 3$

Order	Value of U_W	Probability of Outcome
MMMWW	0 (U)	.1
MMWMW	1 (U)	.1
MMWWM	2 (U)	.1
MWMMW	2 (U)	.1
WMMMW	3 (U)	.1
MWMWM	3 (U)	.1
WMMWM	4 (U')	.1
MWWMM	4 (U')	.1
WMWMM	5 (U')	.1
WWMMM	6 (U')	.1

Remember that the investigator expected the women to rank farther to the right (higher on nativism). His H_1 is that $U_M = U_W$, that is, that the true value of $U_W = 3$ because the groups do not differ. He has a directional H_2, that U_W is less than 3. His test will therefore be one-tailed; he is interested only in the probabilities of low values of U_W, not high values.

Table 18-2 shows that the probability of getting $U_W = 0$ is .1. In order to get a probability even this small, the observed order will have to be MMMWW. Since there is no order as extreme as this one in the same direction, the probability of this observed result is .1. Of course, this probability will not enable us to reject H_1 at $\alpha = .05$, even if the result MMMWW should occur.

Suppose that the H_2 is nondirectional. If he suspected only that men and women would differ on this issue, without having a clear prediction about the direction of difference, he would have to assert H_2: $U_W \neq 3$. His test would then be two-tailed, and he would have to consider probabilities of high as well as low values of U_W. Even if MMMWW occurs, there is one other equally extreme result to be considered: WWMMM. In a two-tailed test, he must find the probability that U_W is *either* at least as small as 0 *or* at least as large as 6. The probability of a result at least as extreme as MMMWW is not .1 but .2 when the test is two-tailed.

18-13 If the sample of 2 women and 3 men is analyzed according to the recommended procedure, the investigator will first determine the U statistic for the (*larger; smaller*) sample; this statistic is (U_M; U_W).

smaller
U_W

18-14 Any one of the first six outcomes in Table 18-2 will give a U statistic of 3 or less. If he has obtained any one of these outcomes, he will call his U statistic (U; U'). If he has obtained any of the last four outcomes, he will call the statistic ______.

U
U'

18-15 Table 18-2 gives a probability distribution for all 10 possible outcomes. The probability of getting $U = 1$ is ______. But since there are 2 outcomes which will give $U = 2$, the probability of getting $U = 2$ is ______.

.1

.2

18-16 Write down the probability distribution for the seven possible values of U_W:

U_W	Probability
0	______
1	______
2	______
3	______
4	______
5	______
6	______

.1
.1
.2
.2
.2
.1
.1

18-17 With H_2: $U_W < 3$, suppose we observe the outcome MMWMW. $U_W =$ ______; there is 1 at least equally extreme result in the same direction, namely, $U_W =$ ______. The probability of observing a U_W at least as extreme as 1, when H_1 is true, is ______.

1
0

.2

18-18 With H_2: $U_W \neq 3$, we observe this same outcome MMWMW. But we must now consider also the probability of $U_W =$ ______ and $U_W =$ ______; these results are at least equally extreme but in the opposite direction.

5 ↔ 6

18-19 With this nondirectional H_2, the probability of observing a U_W at least as extreme as 1 when H_1 is true is ______.

.4

18-20 The probability of the observed result under a two-tailed test is exactly ______ as great as its probability under a one-tailed test.

twice

18-21 A one-tailed test requires a prediction, before seeing the data, about the ______ of the difference.

direction

In answering questions about the probability of results as extreme as MMWMW, you had to add probabilities taken from your probability distribution for U_W. It would have been more convenient to write the probability distribution in a cumulative form, i.e., to write the probability of obtaining U_W *at least as small* as a given value.

Table 18-3 shows the standard form in which the probability distributions of U are presented for small samples. This table is for all cases in which the larger sample contains three observations. Column 1 gives the RSD of U when $n_1 = 1$, column 2 gives the RSD when $n_1 = 2$, and column 3 gives the RSD when $n_1 = 3$. Column 2 is of course the same distribution you derived in Frame 18-16, but it is presented here in cumulative form.

Included in column 2 are three values enclosed in parentheses. These probabilities do not normally appear in a U table because they apply to values of U'. They are included here in order to help you compare this cumulative distribution with your probability distribution in Frame 18-16. Columns 1 and 3 of Table 18-3 are in the standard form with probabilities of U' omitted.

At first it may seem to you that the probabilities associated with U' will be needed for certain one-tailed tests. Suppose in our example that the investigator had expected women to be more *environmentalist* than men; he would then have predicted that the W's would fall farther to the left than the M's in his ordinal series, and the outcome most favorable to his hypothesis would have been WWMMM (with $U_W = 6$). Since H_2 would then be $U_W > 3$, it might appear that he will need the probabilities of values of U_W greater than 3, values which will technically be called values of U'.

A little reflection will show, however, that the hypotheses H_1 and H_2 will not be affected by this shift provided they are stated in terms of U, the smaller statistic. Let H_1 assert $U = 3$, and H_2 assert simply $U < 3$. When he expects $U_W < 3$, U is expected to be the statistic U_W; when he expects $U_W > 3$, U is expected to be the statistic U_M. In either case, since U is the smaller of the two U statistics, U is expected to be less than 3 when H_2 is true. No probabilities associated with U' will be needed.

But aren't the probabilities associated with U' needed for a two-tailed test? Technically, they are; yet there is still no need to present them in the U table. The RSD of U is always symmetrical, and the probability of a result in one direction is equal to the probability of an equally extreme result in the opposite direction. Therefore, we recommend the following easy rule: if you have a predicted direction of difference, your test is one-tailed; the probability of an observed result will be the probability found in the table. If you have no predicted direction of difference, your test is two-tailed; the probability of an observed result is *twice* the probability found in the table.

Table 18-3 Probabilities Associated with Values as Small as Observed Values of *U* when $n_2 = 3$†

U	$n_1 = 1$	$n_1 = 2$	$n_1 = 3$
0	.250	.100	.050
1	.500	.200	.100
2	.750	.400	.200
3		.600	.350
4		(.800)	.500
5		(.900)	
6		(1.000)	

† U' in parentheses.

18-22 Consider two samples: A has 3 cases and B has 3. $n_1n_2 =$ ________, and the maximum value U can take is the whole number ________. Any larger value would be a value of U'.

9
4

18-23 Look at Table 18-3 for $n_1 = 3$. Probabilities are given for values of U from ________ through ________. Probabilities associated with values of U' (*are; are not*) given here.

0; 4
are not

18-24 Assume first that you have a prediction about the direction of difference. Your test will be ____________ -tailed.

one

18-25 If AAABBB occurs, U will equal ________.

0

18-26 The probability of a result this extreme, by a one-tailed test, is ________. You (*can; cannot*) reject H_1 at $\alpha = .05$.

.05; can

18-27 Now assume that you have no prediction about the direction of difference. Your test will be ____________-tailed, and the probability of a result as extreme as $U = 0$ is ____________ the tabulated probability of $U = 0$.

two
twice

18-28 The probability of a result this extreme, by a two-tailed test, is ________. You (*can; cannot*) reject H_1 at $\alpha = .05$.

.10; cannot

18-29 If the observed order is AABABB, $U =$ ________.

1

18-30 The probability of a result as extreme as $U = 1$ is ________ by a one-tailed test and ________ by a two-tailed test.

.100; .200

18-31 Turn to Table A-7 (pages 356–357). It gives U tables for six values of n_2, from $n_2 =$ ________ through $n_2 =$ ________.

3; 8

18-32 You have an experiment with $n_1 = 4$ and $n_2 = 5$, and your observed U is 2. What is the probability under H_1 of getting a result this extreme when the test is one-tailed? ________.

.032

C. USE OF *U* TABLES AND THE NORMALIZED *U* DISTRIBUTION

Tables like the one reproduced as Table 18-3 are available for cases in which n_2 is not greater than 8. A glance at Table A-7 (pages 356–357) will show that such tables will become prohibitively large as n_2 increases; for n_2 between 9 and 20, we make use of critical-values tables, such as Tables A-8 to A-11 (pages 358–359). When n_2 is larger than 20, the RSD of U is approximately normal and a normalized U distribution can be used.

Let us illustrate use of the critical-values tables by assuming larger samples in the environmentalism study. With 12 women and 15 men, $n_1 = 12$ and $n_2 = 15$. The expected value of U is $n_1n_2/2 = 90$. In terms of U, H_1 asserts $U = 90$, while H_2 asserts $U < 90$. However, if we wish to make it quite clear in our statement of H_2 that we have a directional hypothesis, we can write H_2: $U_W < U_M$, knowing that we shall test this hypothesis by a one-tailed test on the value of U. A nondirectional hypothesis would be written H_2: $U_M \neq U_M$, to be tested by a two-tailed test on the value of U.

Turn to page 358, and look at Table A-8 (critical values of U for a one-tailed test at $\alpha = .05$). In the row for $n_1 = 12$ and the column for $n_2 = 15$, the critical value given is 55. Unless the observed U is 55 or less, it will not meet the .05 level on a one-tailed test. These critical values are maximum values for U; if U is larger than the critical value, it will not fall into a rejection region.

A different table is required at each α level. If we wish to make a one-tailed test at $\alpha = .01$, we must use Table A-10 (page 359). The critical value will now be somewhat smaller; it is 42 for our example. Thus, for samples of $n_1 = 12$ and $n_2 = 15$, U must be no larger than 42 if it is to meet $\alpha = .01$ and no larger than 55 if it is to meet $\alpha = .05$.

Now observe that the table for a one-tailed test at $\alpha = .05$ will also serve when the test is two-tailed and $\alpha = .10$. Since the probability by a two-tailed test is always twice the probability of the same result by a one-tailed test, the one-tailed critical values at $\alpha = .05$ are also the two-tailed critical values at $\alpha = .10$. If we wish to make a two-tailed test at $\alpha = .05$, we must use Table A-9 (page 358). The critical values in this table are also valid for a one-tailed test at $\alpha = .05/2 = .025$.

When samples are large, it is convenient to have a short method for calculating the U statistic. Instead of writing out ranks and counting up U for the smaller sample, as we did in Table 18-1, we can determine the ranks of all $n_1 + n_2$ elements in a single ordinal series and then use the equation

$$U = n_1n_2 + \frac{n_1(n_1 + 1)}{2} - \Sigma R_{i1}$$

where ΣR_{i1} is the sum of the ranks of the n_1 members of the smaller group. This equation will sometimes give U'; if so, U is then found by taking $n_1n_2 - U'$.

18-33 In an experiment with $n_1 = 3$ and $n_2 = 6$, the α level is .05. From Table A-7 (page 356), write the values of U which will lie in the rejection region when the test is one-tailed: ______, ______, ______.

0
1; 2

18-34 For the same experiment, write the values of U (ignoring U') which will lie in the rejection region when the test is two-tailed: ______, ______.

0; 1

18-35 For samples of these sizes, what is the *largest* value U can have in a one-tailed test if it is to meet $\alpha = .05$? ______. What is the largest value U can have in a two-tailed test and still meet $\alpha = .05$? ______.

2
1

18-36 Since 2 is the largest value U can have if it is to meet $\alpha = .05$ in a one-tailed test, 2 is the ______ value of U for a one-tailed test at $\alpha = .05$ and for a two-tailed test at $\alpha =$ ______.

critical
.10

18-37 In Sec. A, with samples of 7 women and 8 men, we found $U = 13$. Use the table on page 357 to make a one-tailed test on this observed value of U. The probability of a result as extreme as $U = 13$ is ______.

.047

18-38 If $\alpha = .05$, we can ______ H_1 by this one-tailed test. The probability of a type I error is ______.

reject
.047

18-39 The critical value of U for samples of sizes $n_1 = 7$ and $n_2 = 8$ at $\alpha = .01$ is ______ for a one-tailed test. The critical value for a two-tailed test is ______.

8
6

18-40 Consider the rank-order AAABABB. By the method used in Sec. A, $U =$ ______.

1

18-41 Apply the equation $U = n_1 n_2 + n_1(n_1 + 1)/2 - \Sigma R_{i1}$ to the same rank order. $n_1 =$ ______, $n_2 =$ ______, and the ranks of the smaller sample are ______, ______, ______.

3; 4
4; 6; 7

18-42 The equation gives $U = 12 + 6 -$ ______ $=$ ______.

17; 1

18-43 In Table 18-1, the ranks of the smaller (women's) sample were 4, 7, 7, 11, 13, 14, and 15. With $n_1 = 7$ and $n_2 = 8$, the equation gives $U =$ ______.

13

18-44 If $n_1 = 10$ and $n_2 = 20$, the sum of U and U' is ______, and the expected value of U is ______. The critical value of U for a one-tailed test at $\alpha = .001$ is ______.

200
100
32

Just as binomial probability distributions become more and more like the normal distribution as N grows very large, so also do the RSDs of the U statistic. When n_2 is larger than 20, the U statistic is converted to a z score and the critical values of U are obtained from the normal-curve table.

In order to make the z-score transformation, we must find the mean and standard deviation of the normalized RSD of U for a particular pair of sample sizes. The mean is of course the expectation of U, and we know that $E(U) = n_1n_2/2$. The variance of the normalized RSD is

$$\sigma_U{}^2 = \frac{n_1n_2(n_1 + n_2 + 1)}{12}$$

For $n_1 = n_2 = 30$ and $U = 280$, the mean of the normalized distribution is $900/2 = 450$; its variance is 4,575. The deviation of U from the mean is $U - 450 = -170$, and $z = -170/\sqrt{4{,}575} = -170/67.6 = -2.5$. You may remember that a z score as large as 2.5 has a probability less than .01 in a two-tailed test.

When tied ranks occur, this normal approximation provides a test which is very slightly conservative. The variance obtained from $n_1n_2(n_1 + n_2 + 1)/12$ assumes no tied ranks, and it will be slightly larger than the variance of a normalized U distribution in which many tied ranks occur. The variance can be calculated in such cases by an equation which takes account of ties. For this equation, we define a quantity t, the length of a run of ties; that is, t represents the number of identical scores in a set. When only two scores are tied, $t = 2$; when three scores are tied, $t = 3$. Then the variance of the normalized distribution is found by removing a correction factor from the term $n_1 + n_2 + 1$:

$$\sigma_U{}^2 = \frac{n_1n_2}{12}\left\{n_1 + n_2 + 1 - \frac{\Sigma[(t^3 - t)/12]}{(n_1 + n_2)(n_1 + n_2 - 1)}\right\}$$

The factor $(t^3 - t)/12$ must be calculated for each set of ties. Table 18-4 shows how $(t^3 - t)/12$ increases with the number of ties in a set. The sum $\Sigma[(t^3 - t)/12]$ will thus depend both upon the number of sets of ties and upon the length of the sets. Unless the samples are small and the number of ties is large, the effect of the correction factor will be negligible.

Table 18-5 presents some artificial data for practice in the use of the normalized U distribution. The samples ($n_1 = 10$ and $n_2 = 12$) are quite small; you will be able to compare the results of a normal-curve test with the critical values in Tables A-8 to A-11. Since the original scores are limited to a 7-point scale, the number of ties is quite large, and you will also be able to use the correction factor for ties.

Table 18-4 Effect of Length of Tie on the Correction Factor in Calculating $\sigma_U{}^2$

t	$\frac{t^3 - t}{12}$	t	$\frac{t^3 - t}{12}$
2	0.5	5	10
3	2	6	17.5
4	5	7	28

Table 18-5 Frequency Distributions of Two Independent Samples to Illustrate Mann-Whitney *U* Test

Score	Sample 1	Sample 2	Rank	Sample 1 Frequency × R_{i1}	t
7	1	0	1	1	
6	4	3	5	20	7
5	2	2	10.5	21	4
4	1	0	13	13	
3	1	2	15	15	3
2	1	4	19	19	5
1	0	1	22	0	
				$\sum R_{i1} = 89$	

18-45 In Table 18-5, the score of 6 occurs for ________ persons in sample 1 and ________ persons in sample 2. The length of this tie is $t =$ ________.

4
3
7

18-46 Since these 7 persons occupy rank numbers 2 through 8, each of them is assigned a rank number ________.

5

18-47 R_{i1} stands for the ____________ of person i in sample ________. ΣR_{i1} for this table is ________. The value of U for the smaller sample is $U =$ ________.

rank
1; 89
86

18-48 The expectation of U for samples of these sizes is ________. Therefore, the statistic whose value is 86 is (U; U').

60; U′

18-49 The statistic U for Table 18-5 is ________. According to the U table, the critical value for $\alpha = .05$ is ________ (one-tailed).

34
34

18-50 Now calculate $\sigma_U{}^2$ for the normalized U distribution with $n_1 = 10$ and $n_2 = 12$. The factor $n_1 n_2/12 =$ ________.

10

18-51 Calculate $\Sigma[(t^3 - t)/12]$ for Table 18-5, using Table 18-4. Its value is ________.

45

18-52 $45/(22)(21) = .0974$, or approximately .1. With the correction for ties, $\sigma_U{}^2$ will equal 10 times ________, or ________. Without the correction for ties, $\sigma_U{}^2 =$ ________.

22.9; 229
230

18-53 The square root of 229 is 15.13. Since $E(U) =$ ________, the value of z for $U = 34$ is $24/15.13 =$ ________.

60
1.74

18-54 The probability of getting such a small U is ________, by the normal-curve table. H_1 (*can*; *cannot*) be rejected at $\alpha = .05$ by a one-tailed test.

.041
can

D. ONE-WAY ANALYSIS OF VARIANCE BY RANKS

When more than two samples are to be compared, one may use a distribution-free test which is an extension of the U test. As in the U test, there must be a set of N observations falling into C independent subgroups, and all N observations are put into a single ordinal series. The rank numbers of subgroup 1 are then arrayed in column 1 of the data table, with the rank numbers of subgroup 2 in column 2, and so on. Analysis-of-variance procedures are applied to these rank numbers just as if they were scores; however, these procedures are simplified by the fact that the N observations are actually the first N consecutive integers.

Table 18-6 shows the raw data for such a test, with the assigned rank numbers in parentheses. These rank numbers are analogous to Y_{ij}, and we shall designate them R_{ij}. The hypothesis to be tested is that the ranks within each subgroup constitute a random sample from the set of integers 1 through N. If so, the mean ranks for the columns, $\bar{R}_1$, $\bar{R}_2$, and $\bar{R}_3$, should all be equal. This hypothesis is analogous to $H_1: U = U'$ in the U test.

The test of this hypothesis is called the KRUSKAL-WALLIS H TEST. We define SS_B, the sum of squares between columns, exactly as we would in ordinary analysis of variance:

$$SS_B = \sum_{1}^{C} \frac{T_j^2}{n_j} - \frac{G^2}{N}$$

T_j is, as usual, the column total; in this case, it is $\sum_{1}^{n} R_{ij}$ for the column. We define SS_T as we would define SS_Y when raw scores are used:

$$SS_T = \sum_{1}^{N} R_{ij}^2 - \frac{G^2}{N}$$

The statistic H is a ratio which must be carefully distinguished from the F statistic:

$$H = \frac{SS_B}{MS_T}$$

where $MS_T = SS_T/(N-1)$, the mean square obtained when SS_T is divided by its $N-1$ degrees of freedom. Notice that H is *not* a ratio of two mean squares; its numerator is a sum of squares, SS_B. Notice also that the denominator is analogous to MS_Y in ordinary analysis of variance—not to MS_W, the denominator of the F statistic. The H statistic is not in any sense an F ratio.

The RSD of H is approximated by a chi-square distribution with $C-1$ df. When there are only three samples with not more than five cases each, the chi-square distribution is not used; a special table of the H distribution is available for such small samples.†

†See Sidney Siegel, *Nonparametric Statistics For the Behavioral Sciences,* pp. 282–283, McGraw-Hill Book Company, New York, 1956.

Table 18-6 Data for a One-Way Analysis of Variance by Ranks

	Factor A		
	A_1	A_2	A_3
	95 (1)	90 (2)	75 (10)
	87 (3)	83 (4)	69 (18)
	81 (5)	80 (6.5)	68 (19.5)
	80 (6.5)	77 (9)	67 (22.5)
	79 (8)	74 (11)	66 (26)
	72 (12.5)	72 (12.5)	66 (26)
	70 (16)	71 (14)	66 (26)
	70 (16)	70 (16)	65 (28.5)
	67 (22.5)	68 (19.5)	65 (28.5)
	67 (22.5)	67 (22.5)	63 (30)
T	(113.0)	(117.0)	(235.0)

$\sum R_{ij} = G = \frac{N(N+1)}{2} = 465$ $\qquad$ $\sum R_{ij}^2 = \frac{N(N+1)(2N+1)}{6}$

18-55 In Table 18-6, the score of 95 has a rank number of _______, and the score of 75 has a rank number of _______. The numbers in parentheses are the (*ranks; scores*). The procedure of analysis of variance is to be applied to the _________.

1; 10
ranks
ranks

18-56 R_{ij} is the _________ of person ______ in column ______. T_1 is the sum of the _________ in column _______.

Its value in Table 18-6 is _______.

rank; i
j (or A_j); ranks;
1 (or A_1)
113

18-57 Without doing the arithmetic write out the *terms* you need to evaluate SS_B: $SS_B = ($_______ + _______ + _______$) -$ _______.

$\frac{113^2}{10}; \frac{117^2}{10}; \frac{235^2}{10}; \frac{465^2}{30}$

18-58 $SS_B = 960.8$. This term will become the (*denominator; numerator*) of the H statistic.

numerator

18-59 The value of ΣR_{ij}^2, by the formula given at the bottom of Table 18-6 is _______; $G^2/N = 7{,}207.5$. $SS_T =$ _______, with df = _______, and $MS_T =$ _______.

9,455; 2,247.5
29; 77.5

18-60 The H statistic for Table 18-6 has a value of _______. Since there are _______ levels of factor A, this statistic is distributed approximately like chi square for df = _______.

12.4
3
2

18-61 According to the chi-square table on page 352, the probability of H at least as large as 12.4 is (*greater; less*) than .01.

less

18-62 We can reject, at $\alpha = .01$, the hypothesis that the average _________ for the three columns in Table 18-6 are all equal.

ranks

REVIEW

18-63 If two independent samples are made up of scores which are not ________ distributed, the t test should not be used. *normally*

18-64 Such samples can sometimes be treated as categorical (qualitative) data. The significance test to be used in this case is the ________ test. χ^2

18-65 Such samples can usually be treated as ordinal data. The two sets of scores are converted into a single set of ________, and the Mann-Whitney ________ test is used. *ranks; U*

18-66 The power-efficiency ratio of the U test is .95 in comparison with the t test. This ratio indicates that the U test with $n_1 + n_2 = 40$ will have the same power as a t test with $n_1 + n_2 =$ ________. *38*

18-67 The meaning of the U test can be illustrated with two samples of two and four cases, respectively. We call the size of the ________ sample n_1 and that of the ________ sample n_2. *smaller; larger*

18-68 If the observed order is AAABAB, there is only one order which could be more extreme in the same direction, the order ________. *AAAABB*

18-69 The number of combinations of 6 things where 2 are of one kind and 4 are of the other is $6!/2!4! =$ ________. There are therefore ________ orders in which these 6 observations might have occurred. *15* *15*

18-70 The probability of the order AAABAB is ________, and the probability of the order AAAABB is ________. *1/15* *1/15*

18-71 The probability of an order at least as extreme as the one observed, and in the same direction, is ________. *2/15*

18-72 Instead of making this kind of direct calculation, we can accomplish the same purpose by calculating the U statistic. When the order is AAABAB, the U statistic for the B sample is ________, because the first B in the series is followed by ________ A and the remaining B is followed by ________ A's. *1; 1* *0*

18-73 There is a U statistic also for the A sample; it will be (*larger; smaller*) than 1, and it will be called ________. *larger; U′*

18-74 Since $U + U'$ always equals the product ______, U' for our example equals ______. | *n_1n_2* *7*

18-75 When the two samples are randomly drawn from the same population, we expect U and U' to be ______. The expectation of U is half of the product ______. | *equal* *n_1n_2*

18-76 When n_2 is between 3 and 8, tables of U give the probability, for each value of U, that U will be at least as (*large; small*) as that value. | *small*

18-77 If we have a prediction about the direction of difference, the test will be ______-tailed. With no such prediction, the test is ______-tailed. | *one* *two*

18-78 The probability of the observed result is *twice* the probability found in the table when the test is ______-tailed. | *two*

18-79 We can also find U by taking

$$U = n_1n_2 + \frac{n_1(n_1 + 1)}{2} - \Sigma R_{i1}$$

where ΣR_{i1} is the sum of the ______ for the smaller sample. | *ranks*

18-80 When n_2 is between 9 and 20, tables of U give only the ______ values of U for certain α levels. | *critical*

18-81 When n_2 is greater than 20, we use a ______ approximation to the RSD of U. The (*mean; variance*) of this approximation is slightly affected by the number of ties in the rank series. | *normal* *variance*

18-82 When there are more than two independent samples and their distribution is not normal, the Kruskal-Wallis ______ test can be used. | *H*

18-83 This test applies the procedure of analysis of variance to the ______ of the scores in a single ordinal series. | *ranks*

18-84 The H statistic is the ratio SS_B/MS_T, where SS_B is analogous to (SS_B; SS_W; SS_Y) in ordinary analysis of variance, and MS_T is analogous to (MS_B; MS_W; MS_Y). | *SS_B* *MS_Y*

18-85 For samples of more than five cases each, H has a distribution approximately like that of a ______ variable with df = ______. | *chi-square* *$C - 1$*

PROBLEMS

1. Two cross-country teams of 12 men each compete in an athletic contest. The order, from first to last, in which the individual team members arrived at the finish line was ABAABABAABABAAABBABBABBB. Use the U test to determine whether you can reject, at $\alpha = .05$, the hypothesis that there is no difference in performance between teams A and B.

2. Two classes in geometry take the same test before and after being taught by different methods. The improvement in score for the 22 students in class I is shown in the first row; that for the 24 in class II is shown in the second row. Determine by the normalized U test whether a difference significant at the .05 level can be supported by these data; the expected direction of difference is in favor of class I. Does the use of the variance-correction factor for ties make any difference in your statistical decision?

I: 21, 21, 20, 19, 17, 17, 17, 15, 14, 12, 12, 11, 10, 9, 9, 7, 7, 7, 7, 7, 5, 5
II: 19, 18, 18, 16, 16, 16, 13, 13, 10, 8, 8, 8, 6, 6, 6, 6, 4, 4, 4, 3, 3, 2, 2, 2

3. The table below gives the 54 raw scores for the nativism-environmentalism example as it was introduced in Lesson 11. What is the general shape of the raw-score distribution for the men? For the women? Assign ranks and calculate U for these data. Give the probability of the observed U by a one-tailed test.

Men ($n_2 = 30$)		Women ($n_1 = 24$)	
95	69	91	35
93	66	89	34
92	59	87	33
91	54	83	31
90	43	75	31
89	42	68	29
89	38	61	29
89	35	56	29
88	35	49	28
87	34	48	
85	32	45	
81	30	41	
78	29	40	
77	29	37	
73	28	36	

4. Use one-way analysis of variance by ranks on the score distributions in Prob. 2 on page 210. How does the result of this distribution-free test compare with the result of the F test (page 242)?

PART FOUR
Linear Regression and Correlation

LESSON 19
LINEAR FUNCTIONS: AN OPTIONAL REVIEW

We have already assumed in previous lessons that you are familiar with the rectangular coordinate system, consisting of a horizontal, or X, axis meeting a vertical, or Y, axis at a point designated as the ORIGIN of the system. This origin is given the value zero on both axes. X values greater than 0 are plotted consecutively to the right of the origin, and negative values of X are plotted to the left. Y values greater than 0 are plotted above the origin, and those less than 0 are plotted below the origin. Since experimental data are usually positive numbers, we are often interested in only one quarter, or QUADRANT, of the space defined by these intersecting axes, namely, the quadrant to the right and above the origin, where positive values of both X and Y are found.

You will recall that the pair of values which define a particular point in the XY plane are called the COORDINATES of that point. The X coordinate is always stated first, followed by the Y coordinate. The point designated (2,3) has an X coordinate of 2 and a Y coordinate of 3.

A linear equation relating two variables, X and Y, has the general form $Y = mX + b$. In this equation, m and b are CONSTANTS. Suppose that $m = 2$ and $b = 0$; then we can write the equation as $Y = 2X$. Figure 19-1 is a graph of the function $Y = 2X$; in the table at the right of the graph you will find the coordinates of three points, shown as dots in the graph, which satisfy this equation and which therefore lie on the straight line representing the function.

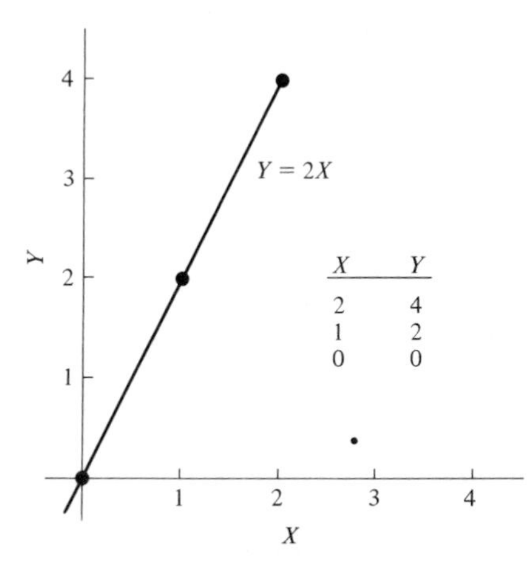

Figure 19-1: Graph of the linear function $Y = 2X$.

We can write the function $Y = 2X$ in the form $Y/X = 2$. The ratio Y/X is itself a constant for such an equation; its value does not change, no matter what values Y and X assume. Figure 19-2 shows two other linear functions, $Y = X$ and $Y = 3X$. In Fig. 19-2*a* the constant Y/X is equal to 1; in Fig. 19-2*b*, $Y/X = 3$. Observe that the straight line rises more steeply when Y/X has a higher value.

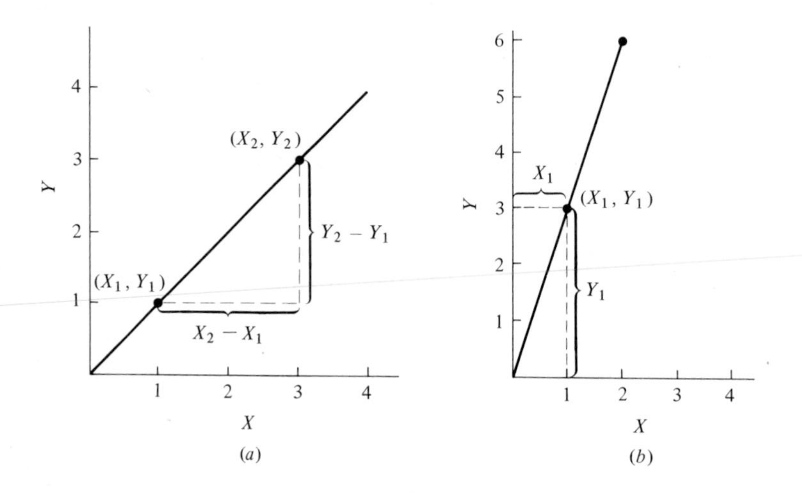

Figure 19-2: (*a*) Graph of the function $Y = X$. (*b*) Graph of the function $Y = 3X$.

This constant Y/X is the same as the constant m in our basic linear equation, as you can see if we write $Y = mX$ and $Y/X = m$. It is called the SLOPE CONSTANT. The value of m indicates the slope of the straight line representing the equation.

The value of m can be easily determined from the graph of any straight-line function, even when the equation for that function has not yet been stated. Two points whose coordinates can be easily read from the graph are chosen, each lying on the straight line. The difference between the Y coordinates is $Y_2 - Y_1$; the difference between the X coordinates is $X_2 - X_1$. The ratio $(Y_2 - Y_1)/(X_2 - X_1)$ is the same as the slope constant m, as you can verify in the figures above. In Fig. 19-2*a* the ratio is 2/2, and the slope constant $m = 1$. In Fig. 19-2*b* the ratio is 3/1, and $m = 3$.

The slope of a linear function is the ratio of the change in Y between two points to the change in X between those same two points. Whenever the graph of the equation passes through the origin (0,0), the slope constant can be found simply by the ratio Y/X for any point on the line. The point (0,0) can always be considered as the point (X_1,Y_1) in such cases, and any other point on the line can serve as (X_2,Y_2); the ratio $(Y_2 - Y_1)/(X_2 - X_1)$ thus becomes $(Y_2 - 0)/(X_2 - 0) = Y_2/X_2$.

But this simplification is possible only when the function passes through (0,0). In order for the function to pass through the origin, it is necessary that the other constant, b, equal 0. All the equations we have discussed so far have had $b = 0$. They are equations of the form $Y = mX$, and the point (0,0) always satisfies such an equation.

19-1 The basic equation for a linear function is ______. This equation has two constants, ______ and ______. — $Y = mX + b$; $m \leftrightarrow b$

19-2 The equation for a linear function becomes $Y = mX$ when (and only when) the value of the constant b is ______. In this case, the constant m is equal to the ratio ______. — *0*; $\frac{Y}{X}$

19-3 If $Y = 2X$, the ratio Y/X equals ______. For every change of 1 unit in X, Y changes by ______ units. — *2*; *2*

19-4 If $Y = 3X$, $Y/X =$ ______. When $X = 1$, Y must equal ______. If X increases from 1 to 3, Y must increase not by only 2 units but by ______ units. Thus if $X = 3$, $Y =$ ______. — *3*; *3*; *6; 9*

19-5 In Fig. 19-2*a*, the point (X_1,Y_1) has the coordinates (______, ______), and (X_2,Y_2) is (______, ______). — *(1,1); (3,3)*

19-6 The change $Y_2 - Y_1$ in Y is equal to ______. The change in X is equal to ______. The *ratio* of the change in Y to the change in X is ______. — *2*; *2*; *1*

19-7 In order to determine the slope of a straight line, we take two points on that line and determine the change in the ______ coordinate between those two points and the change in the ______ coordinate between the same points. — $Y \leftrightarrow$; X

19-8 Then we find the ratio between these two changes, always putting the change in (X; Y) in the numerator. — Y

19-9 In Fig. 19-2*b*, the point (X_1,Y_1) has the coordinates (______, ______). The line also passes through the points (0, ______) and (2, ______). — *(1,3)*; *0; 6*

19-10 Since the line passes through (0,0), which is called the ______, we can find its slope from the ratio $Y_1/$______. — *origin*; X_1

19-11 Whenever the change in X equals the change in Y, the slope of the line is equal to ______. — *1*

19-12 If X changes more slowly than Y, the slope of the line will be (*greater; less*) than 1. — *greater*

19-13 If the slope is less than 1, we know that Y is changing more (*rapidly; slowly*) than X. — *slowly*

19-14 In $Y = mX + b$, m is called the ______ constant. — *slope*

Figure 19-3*a* illustrates two linear functions, one with $b = 0$ and one with $b = 2$. Both have the same slope. The functions are parallel to each other, but the line $Y = 0.5X + 2$ lies exactly 2 units higher than $Y = 0.5X + 0$. Since the line $Y = 0.5X + 2$ intercepts (crosses) the Y axis at the point (0,2), we say that the Y INTERCEPT of this function is equal to 2. The Y intercept is that point with X coordinate equal to 0 which satisfies the equation for the function. In $Y = 0.5X + 2$, Y must equal 2 when $X = 0$.

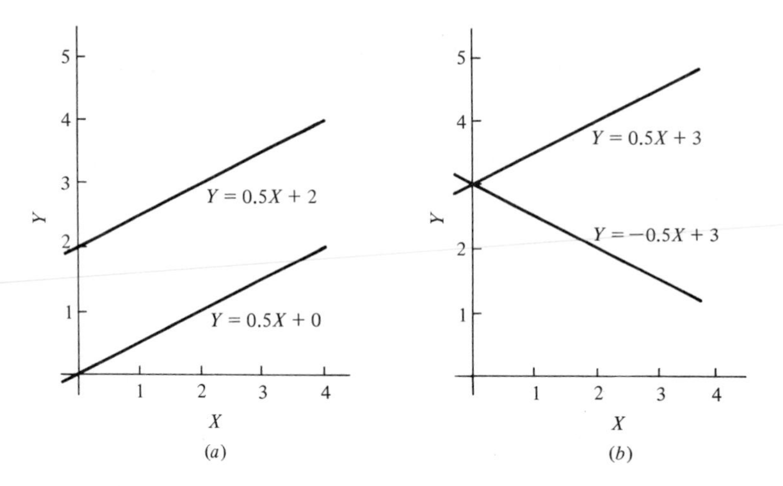

Figure 19-3: (*a*) Functions with different Y intercepts. (*b*) Functions with positive and negative slope.

The Y intercept of any linear equation will always be equal to the constant b. Therefore, the constant b is called the Y intercept. In equations of the form $Y = mX$, the Y intercept is 0; we have already noted that such functions are satisfied by the point with coordinates (0,0).

It is of course possible for a Y intercept to be a negative number, as in the equation $Y = X - 10$. This equation describes a line with slope of 1 and Y intercept at -10, 10 units below the origin. (It will also have an X intercept which is positive; when $Y = 0$, $X = 10$. But the X intercept does not figure as a constant in equations of the form $Y = mX + b$.)

The slope constant m may also be negative. Figure 19-3*b* shows two functions whose equations are the same except for the sign of the slope constant. The line $Y = 0.5X + 3$ slopes upward to the right; its slope constant is positive in sign. The line $Y = -0.5X + 3$ slopes upward to the left; its slope constant is negative. Whenever an increase in X occurs with an increase in Y, $(Y_2 - Y_1)/(X_2 - X_1)$ will be positive in sign and the slope constant is positive. Whenever an increase in X occurs with a *decrease* in Y, $Y_2 - Y_1$ will be negative while $X_2 - X_1$ is positive, and the slope constant will have a negative sign.

Since we are describing Y as a function of X in statements of the form $Y = mX + b$, it is customary to describe functions with positive slope as INCREASING linear functions: as X increases, Y also increases. Functions with negative slope are DECREASING linear functions: as X increases, Y decreases.

19-15 The Y intercept for any linear function is the value of Y when $X =$ _______. *0*

19-16 Take $Y = 2X$. When $X = 0$, Y must equal _______. The Y intercept is _______ for this equation. *0* *0*

19-17 If $Y = 2X + 4$, the Y intercept is _______. If $Y = 2X - 4$, the Y intercept is _______. *4* *−4*

19-18 In the general equation $Y = mX + b$, $Y =$ _______ when $X = 0$. The Y intercept is the same as the constant _______. *b* *b*

19-19 The constant m is the __________ constant. The constant b is the __________. *slope* *Y intercept*

19-20 An increasing linear function differs from a decreasing linear function in the __________ of the slope constant. *sign*

19-21 If the slope constant has a positive sign, the equation describes a(an) __________ linear function. Increases in X are accompanied by __________ in the value of Y. *increasing* *increases*

19-22 A decreasing linear function has a slope constant with a __________ sign. Increases in X are accompanied by __________ in Y. *negative* *decreases*

19-23 If $m = 0$, the equation $Y = mX + b$ becomes simply the equation _______. $Y = b$

19-24 If $Y = 3$, then the value of Y is always _______, no matter what value X assumes. Sketch the graph of $Y = 3$ in the coordinate system of Fig. 19-3*b*. *3*

19-25 A straight line sloping upward to the right indicates a (an) __________ linear function. A straight line sloping downward to the right indicates a(an) __________ linear function. *increasing* *decreasing*

19-26 A horizontal straight line parallel to the X axis indicates a function with slope equal to _______. *0*

19-27 In the function $Y = -15X - 5$, every increase of 1 unit in X is accompanied by a(an) __________ of _______ units in Y. *decrease; 15*

19-28 The equation $Y = -15X - 5$ describes a(an) __________ linear function whose slope constant is _______ and whose Y intercept is at the point with coordinates (_______, _______). *decreasing* *−15* *(0,−5)*

PROBLEMS

1. Find the slope constant for the straight line in the graph below. Find the Y intercept. State the equation for this linear function.

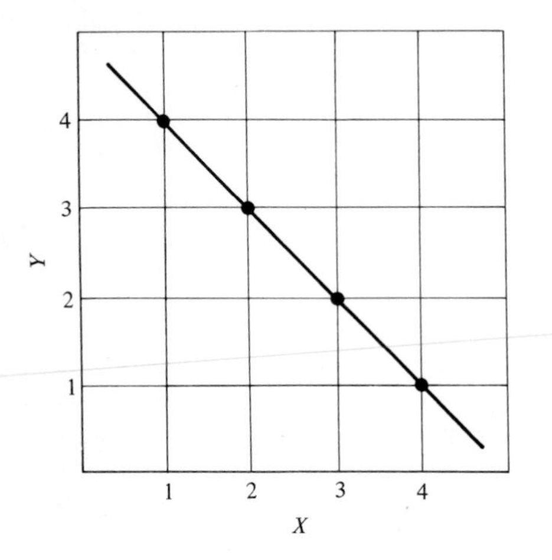

2. What is the equation for each of the following lines?

a. The line which passes through the origin and the point (2,5).

b. The line which passes through the points (0,3) and (4,0).

c. The line which passes through the points (1,2) and (4,5).

3. Which of the following functions pass through the origin?

a. $Y = -10X$.

b. The line passing through the points (2,3) and (5,6).

c. The line with a slope of -0.5 passing through the point (2,1.5).

d. The line with a slope of 2.5 passing through the point (10,25).

4. *a*. What is the slope of the function which passes through the point (5,10) and has a Y intercept of 3.0?

b. What is the slope of the function which passes through the point (4,7) and has a Y intercept of 3?

c. What is the Y intercept of the function which has a slope $= -4.0$ and passes through the point (3,8)?

5. What is the equation for the line passing through the points (2,10) and (10,10)? What is the slope of this line? The Y intercept?

6. A student starts the academic year with $150 in the bank and spends $30 a month. Let Y be his bank balance and X be the number of months that have passed. Assume no further deposits and no bank service charges; write a linear equation showing how Y varies as X increases.

LESSON 20
LINEAR-REGRESSION ANALYSIS IN SINGLE-FACTOR EXPERIMENTS

In a simple analysis of variance we are interested merely in deciding whether any significant effects of factor A are present. In certain experiments it is possible to go farther, once significant effects have been found. Figure 20-1 illustrates such a case; let us examine it briefly.

This figure shows the means of seven subgroups in a one-way analysis of variance; $n = 10$ for each subgroup. Each of the seven levels of factor A is an amount (in milligrams) of a certain drug administered to persons who then took a test in mental arithmetic. The seven dose levels were selected to represent equal intervals within the range from 0 to 30 milligrams. Because these levels are quantitative rather than qualitative, we can represent them in the graph as values of an experimental variable X.

The trend of the subgroup means suggests a systematic relation between Y (performance level) and X (factor A). Whenever a relation between two quantitative variables appears to be systematic, it may be possible to describe it by a mathematical equation or FUNCTION. In this example the observed means lie almost on a straight line, suggesting that at least part of the association between X and Y can be described by a linear function. We therefore undertake to decide whether there is a significant amount of LINEAR ASSOCIATION between X and Y in these data.

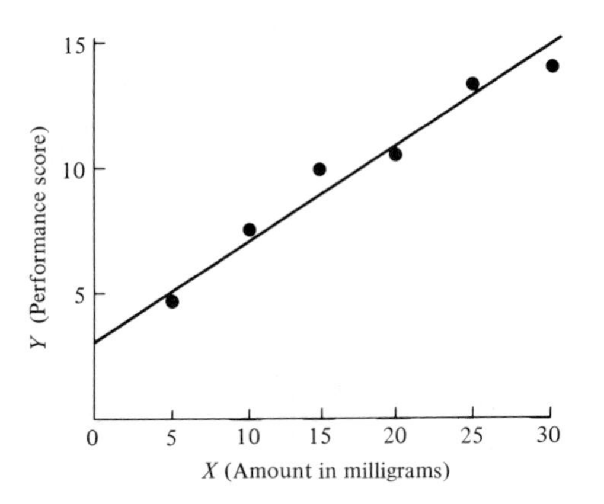

Figure 20-1: Hypothetical means of seven subgroups on a performance test after taking different amounts of a drug.

A. SCORE COMPONENTS IN THE LINEAR-REGRESSION MODEL

We now proceed from our earlier question about the mere existence of an association to questions about the mathematical (or functional) form of that association. In taking this step, we enter upon **REGRESSION ANALYSIS**. We begin with the simplest kind of function, a linear equation.

No straight line can be drawn to pass precisely through all N data points in an experiment like that represented by Fig. 20-1. Even the column means do not lie exactly on one straight line. We cannot expect that a linear function will describe all the variation which is contained in $SS_Y = \sum_1^N (Y_{ij} - \bar{Y})^2$. However, among all the linear functions which might be chosen, there is one which contains as much of SS_Y as can be put into a single equation of linear form. This function is referred to as the **BEST-FITTING STRAIGHT LINE** for these data. We shall write this particular function as $Y'_j = mX_j + b$, where X_j represents the set of seven values (one for each of the C levels) which X can take and Y'_j is the intersection of the best-fitting straight line with its corresponding value of X. We already have seven column means, called $\bar{Y}_j$; we shall now also have seven values of Y'_j.

In Lesson 13 you learned to divide a single deviation score $Y_{ij} - \bar{Y}$ into two components, $Y_{ij} - \bar{Y}_j$ and $\bar{Y}_j - \bar{Y}$. Now that we have introduced a value of Y'_j for each subgroup, we can divide $Y_{ij} - \bar{Y}$ into three distinct parts, $Y_{ij} - \bar{Y}_j$, $\bar{Y}_j - Y'_j$, and $Y'_j - \bar{Y}$. The third of these components, $Y'_j - \bar{Y}$, is the one which contains all of the variation in Y_{ij} which can be described by a linear function. The other two components represent the remainder. Possibly some of this remainder could be described if we used a nonlinear function, but for the moment we regard it simply as error.

In Lesson 14 you learned to associate each component of $Y_{ij} - \bar{Y}$ with a component in the additive model $Y_{ij} = \mu + a_j + e_{ij}$. We shall now do the same thing with these rearranged components, making use of a slightly different **LINEAR-REGRESSION MODEL**. Instead of an effects component a_j, this model contains a linear-regression component r_j; r_j represents only that portion of the effects of factor A which can be described by a linear equation. Quite possibly r_j and a_j will turn out to be equal; there may be no part of a_j which cannot be described by a linear equation. But we do not assume this to be true at first. We write the linear-regression model as $Y_{ij} = \mu + r_j + e'_{ij}$ because the error component e'_{ij} may or may not contain some part of a_j which is left over after r_j has been removed.

Table 20-1 summarizes the score components and sums of squares for the linear-regression model. This model can be applied to the same data to which the simple additive model is applied, provided only that the levels of factor A are quantitative. The linear-regression model provides a test for the existence of a significant amount of linear association in the population from which the sample has been drawn. After analysis of variance has shown that significant effects of factor A are present, the test for linear association is usually the first step taken in a further study of the form of the association between Y and factor A.

Table 20-1 Score Components and Sums of Squares in the Linear-Regression Model

Component	Estimate from Data	Sum of Squares	Title
e'_{ij}	$(Y_{ij} - \bar{Y}_j) +$	$\sum_1^C \sum_1^n (Y_{ij} - \bar{Y}_j)^2$	SS_W (SS within columns)
	$(\bar{Y}_j - Y'_j)$	$\sum_1^C \sum_1^n (\bar{Y}_j - Y'_j)^2$	SS_D (SS deviations from linear regression)
r_j	$Y'_j - \bar{Y}$	$\sum_1^C \sum_1^n (Y'_j - \bar{Y})^2$	SS_L (SS linear regression)
$Y_{ij} - \mu$	$Y_{ij} - \bar{Y}$	$\sum_1^C \sum_1^n (Y_{ij} - \bar{Y})^2$	SS_Y (total SS)

20-1 The total amount of variation present in the Y distribution is measured by the sum of squares called ______.

SS_Y

20-2 When we analyzed SS_Y in Lesson 14, SS_W contained only the variation ______ the columns, and SS_B contained only the variation ______ the columns.

within
between

20-3 Any variation arising from the score component a_j was confined to (SS_B; SS_W).

SS_B

20-4 In linear regression analysis we identify three components of SS_Y. One of these components is unchanged; ______ still represents the variation ______ columns.

SS_W
within

20-5 The other two components result from a partitioning of SS_B into two parts. The part which can be described by a linear function is called SS ______, or SS_L.

linear regression

20-6 SS_L is that part of the variation (*between; within*) columns which can be described by a ______ function.

between
linear

20-7 That part of the variation between columns which *cannot* be described by a linear function is called SS ______ from linear ______, SS_D.

deviations
regression

20-8 In the simple model $Y_{ij} = \mu + a_j + e_{ij}$, the error component e_{ij} gives rise to the SS ______.

within columns

20-9 In the new model $Y_{ij} = \mu + r_j + e'_{ij}$, the error component e'_{ij} gives rise to the two sums of squares, ______ and ______.

$SS_W \leftrightarrow SS_D$

20-10 r_j represents those effects of factor A which can be described by a ______. It gives rise to (SS_D; SS_L).

linear function; SS_L

B. THE PRINCIPLE OF LEAST SQUARES

SS_L contains all the effects of factor A which can be described by a linear function. We want to know whether its size is significant; to find out, we have to compare SS_L with the remainder, $SS_W + SS_D$. How do we calculate SS_L?

We could look for the seven values of Y'_j, then calculate the deviation $Y'_j - \bar{Y}$ for each of the N scores. Adding up the squares of these deviations will give SS_L. To find Y'_j for each column, we need to determine the constants m and b for the best-fitting equation of the form $Y'_j = mX_j + b$. However, determination of the best-fitting equation can be made much simpler by transforming $Y'_j = mX_j + b$ into another linear equation, one in which both Y'_j and X_j are expressed as z scores. The standard deviation of the X distribution ($n = 10$ values of X_1, 10 values of X_2, etc.) is $s_X = \sqrt{SS_X/N}$. Then the equation

$$\frac{Y'_j - \bar{Y}}{s_Y} = r\,\frac{X_j - \bar{X}}{s_X} + k$$

is also a linear equation, with a slope constant r and a Y-intercept constant k. Because we have changed the scale of the variables Y'_j and X_j, we call these constants r and k to remind ourselves that they will not have the same values as the constants m and b in $Y'_j = mX_j + b$. The transformed equation is written

$$z'_Y = rz_X + k$$

where

$$z'_Y = \frac{Y'_j - \bar{Y}}{s_Y} \qquad \text{and} \qquad z_X = \frac{X_j - \bar{X}}{s_X}$$

Now remember that we want $\sum_1^C \sum_1^n (Y'_j - \bar{Y})^2 = SS_L$ to contain as much of the variation in SS_Y as can possibly be described by a linear equation. We want to select r and k so that the remainder, $\sum_1^C \sum_1^n (Y_{ij} - Y'_j)^2$, will be as *small* as possible. Of course, since each Y_{ij} is now z_Y, and since Y'_j is represented by z'_Y, we must really undertake to minimize the sum of squared deviations $\sum_1^C \sum_1^n (z_Y - z'_Y)^2$. By finding the values of r and k which will make this sum as small as possible, we obtain the best-fitting straight line for our set of z scores. This line is called the REGRESSION LINE, and its constants are the REGRESSION CONSTANTS.

The best-fitting straight line for any set of data points is always that line which minimizes the sum of squared deviations of points from the line. This criterion is called the PRINCIPLE OF LEAST SQUARES. Since z_Y represents a data point and z'_Y is that point on the regression line which corresponds to a particular z_Y, minimizing $\sum_1^C \sum_1^n (z_Y - z'_Y)^2$ will produce a straight line which meets the least-squares criterion. Section C will show how the regression constants for this line are chosen.

20-11 If all the effects of factor A can be described by a linear function, the SS ________ will contain all these effects. In this case, a_j (from the simple model) will be completely contained in ______ (from the linear-regression model), and e'_{ij} will be due entirely to ________.

linear regression
r_j
error

20-12 If the effects of factor A cannot be described completely by a linear function, SS_L (*will; will not*) contain all these effects. The component e'_{ij} will then contain both error and ________.

will not
effects

20-13 SS_L must be made to contain all the variation in Y which can be described by a ________ equation. The remainder, $SS_Y - SS_L$, must be made as ________ as possible by our choice of this equation.

linear
small

20-14 Since $SS_L = \sum_1^N (Y'_j - \bar{Y})^2$, $SS_Y - SS_L = \sum_1^N$ ______.

$(Y_{ij} - Y'_j)^2$

20-15 We must find the linear equation that makes $\sum_1^N (Y_{ij} - Y'_j)^2$ as ________ as possible. Such an equation will fulfill the principle of ________.

small
least squares

20-16 The raw-score regression equation will have the form $Y'_j = mX_j + b$. It has two constants: ______ and ______.

$m \leftrightarrow b$

20-17 It is easier to apply the principle of least squares to z-score distributions. Such distributions always have a mean = ______ and a standard deviation = ______.

0; 1

20-18 We can convert each Y'_j to a z score by changing it to a deviation score, ______, and dividing this deviation score by s_Y, the sample ________ for the Y scores.

$|Y'_j - \bar{Y}|$
standard deviation

20-19 We must also convert each X_j to a z score. In Fig. 20-1, $n = 10$ for each subgroup. There are 70 X scores (one for each Y_{ij}), but X_j has only ______ different values.

7

20-20 Since all the subgroups are of equal size, $\bar{X}$ is the mean of these seven values and s_X is their standard deviation. $\bar{X}$ = ______, and $s_X = 10$.

15

20-21 The set of seven different z_X values for Fig. 20-1 will be ______, ______, ______, ______, ______, ______, ______.

1.5; 1.0; 0.5; 0; −0.5; −1.0; −1.5

20-22 The straight line which meets the least-squares criterion is the best-fitting straight line. It is called the ________ line.

regression

C. CHOOSING THE REGRESSION CONSTANTS

The constants r and k are the regression constants for the z-score regression equation $z'_Y = rz_X + k$. We are working with this z-score transformation of $Y'_j = mX_j + b$ for three reasons: (1) it is easier to apply the principle of least squares to z scores; (2) we are seeking $\sum_1^N (Y'_j - \bar{Y})^2$, a sum of squared deviations, rather than Y'_j; and (3) the constant r, which appears in the z-score regression equation, will turn out to be a very important statistic.

This section will show how the principle of least squares is applied in choosing r and k. We must warn you, however, that we are intent upon explaining the source and meaning of the constant r, not upon teaching you how to calculate it from raw data. The computing formula will come later.

We must choose r and k so that $\sum_1^N (z_Y - z'_Y)^2$ will be as small as possible. We can minimize this sum most easily by minimizing its average, $\Sigma(z_Y - z'_Y)^2/N$, and the flow chart shows how we proceed through the first stage.

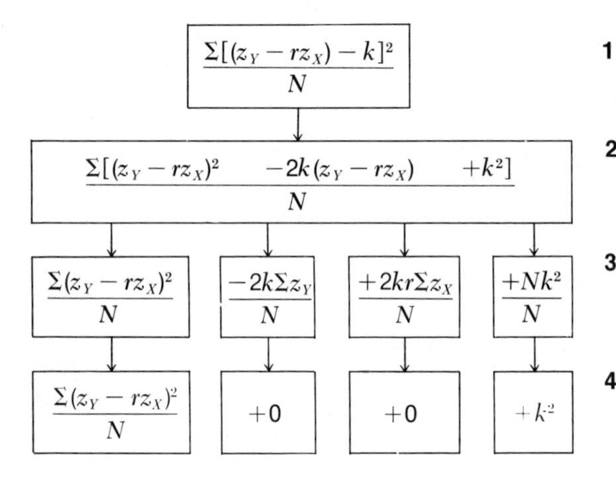

1 $z_Y - z_Y' = z_Y - (rz_X + k)$
$= (z_Y - rz_X) - k$

2 $(A - B)^2 = A^2 - 2AB + B^2$ where $A = z_Y - rz_X$ and $B = k$

3 Summation rules I and II; 2, k, and $1/N$ are constants

4 $\Sigma z/N = 0$; the mean of any distribution of z scores is 0

When we have found that

$$\frac{\Sigma(z_Y - z'_Y)^2}{N} = \frac{\Sigma(z_Y - rz_X)^2}{N} + k^2$$

it is easy to determine the best value for the constant k. There is no value of k which will minimize the whole expression any better than the value $k = 0$. Therefore, we have already found one of the constants for the regression equation $z'_Y = rz_X + k$. Since $k = 0$, we can drop the term k^2 altogether from the result in step 4 above, and we can then do a little more algebra on the remainder, $\Sigma(z_Y - rz_X)^2/N$.

20-23 After we have made a z-score transformation, every value of Y_{ij} is represented as a value of ______, and every value of X_j is represented as a value of ______.

z_Y
z_X

20-24 Since every Y'_j is now a value of z'_Y, the deviation $Y_{ij} - Y'_j$ is written in z-score form as the deviation ______.

$z_Y - z_Y'$

20-25 If we minimize $\Sigma(z_Y - z'_Y)^2/N$, we shall also minimize $\Sigma(z_Y - z'_Y)^2$ itself. $\Sigma(z_Y - z'_Y)^2$ is a sum of squared deviations of z_Y values from the regression line; $\Sigma(z_Y - z'_Y)^2/N$ is the ______ squared deviation of these points from the ______ line.

average
regression

20-26 $z'_Y = rz_X + k$. In $\Sigma(z_Y - z'_Y)^2$ the term $rz_X + k$ can be substituted for ______. When the inner parentheses are removed, $z_Y - (rz_X + k)$ equals ______.

z_Y'
$z_Y - rz_X - k$

20-27 In step 1 of the flow chart, $z_Y - rz_X - k$ is written with the two terms ______ and ______ grouped together.

$z_Y \leftrightarrow rz_X$

20-28 The reason for this grouping is shown in step 2. We treat the combination ______ as term A and the constant k as term B when we square $z_Y - rz_X - k$.

$z_Y \leftrightarrow rz_x$

20-29 We then use our summation rules to break up the expression into a sum of four separate terms. Each of the two middle terms equals ______ because the mean of a distribution of z scores is always ______.

0
0

20-30 $\Sigma z_Y/N$ is the ______ of the z_Y distribution. $\Sigma z_X/N$ is the ______ of the ______ distribution.

mean
mean; z_X

20-31 In step 4 we find that $\Sigma(z_Y - z'_Y)^2/N =$ ______. We can now select a value for the constant ______ which will minimize this expression.

$\dfrac{\Sigma(z_Y - rz_X)^2}{N} + k^2$
k

20-32 The value we select for k is ______. If k is any larger, the expression will have a ______ value.

0
larger

20-33 If k is smaller than 0, its sign will be ______; however, k^2 will have a ______ sign, and the expression $\Sigma(z_Y - rz_X)^2/N + k^2$ (*will; will not*) have its minimum value.

negative
positive
will not

20-34 Since $k = 0$, we can drop k from further consideration. In order for $\Sigma(z_Y - z'_Y)^2/N$ to be minimized, it must equal ______. We still have to find a value for the constant ______.

$\dfrac{\Sigma(z_Y - rz_X)^2}{N}$; r

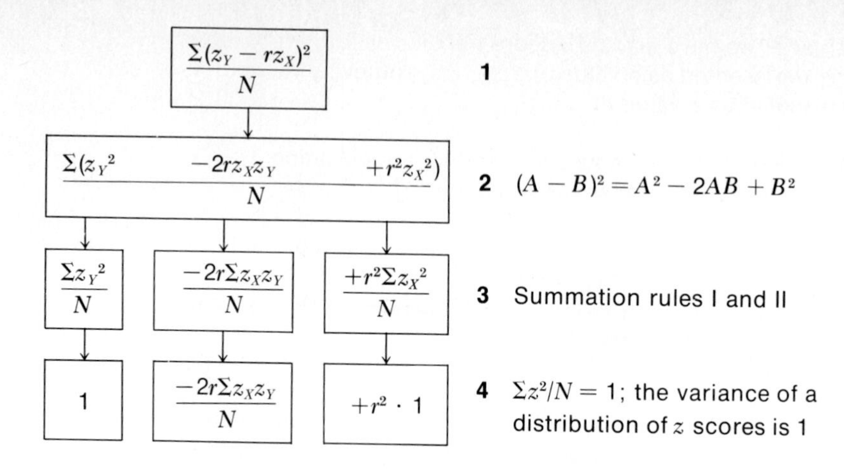

Algebraic manipulation of $\Sigma(z_Y - rz_X)^2/N$ shows that the average squared deviation, $\Sigma(z_Y - z'_Y)^2/N$, has its minimum value when

$$\frac{\Sigma(z_Y - z'_Y)^2}{N} = 1 - 2r\frac{\Sigma z_Xz_Y}{N} + r^2$$

We must now choose a value of r which will minimize this expression. The best choice will be $r = \Sigma z_Xz_Y/N$; we shall show why.

Let us call $\Sigma z_Xz_Y/N$ by the letter v. Suppose we were to choose a value for r which is different from v by a small amount c. Then

$$\begin{aligned} 1 - 2rv + r^2 &= 1 - 2(v + c)v + (v + c)^2 \\ &= 1 - 2v^2 - 2vc + v^2 + 2vc + c^2 \\ &= 1 - v^2 + c^2 \end{aligned}$$

To make this expression as small as possible, we must take $c = 0$. Since $r = v + c$, we find that r must equal $v = \Sigma z_Xz_Y/N$.

The quantity $\Sigma z_Xz_Y/N$ is called the AVERAGE CROSS PRODUCT of the paired scores in our z_X and z_Y distributions. The constant

$$r = \frac{\Sigma z_Xz_Y}{N}$$

is called the PEARSON PRODUCT-MOMENT CORRELATION COEFFICIENT or simply the PEARSON r. It is an index of the degree of correspondence between the paired z scores. With perfect correspondence, $z_X = z_Y$ for every pair, and $\Sigma z_Xz_Y/N$ is the same as $\Sigma z_X{}^2/N$ or $\Sigma z_Y{}^2/N$; r then has its maximum value, +1. With perfect *inverse* correspondence, $z_X = -z_Y$ for every pair, and r will equal −1; this is its minimum value. When there is no correspondence between z_X and z_Y, that is, when a high z_X occurs just as often with low as with high z_Y values, r will equal 0. Thus, the Pearson r is a sample statistic which can vary between +1 and −1. As the degree of correspondence between z-score pairs decreases, r approaches zero.

20-35 If we square each member of a set of z scores and divide the sum of these squares by N, we obtain $\Sigma z^2/N$, the ________ of the set of z scores. *variance*

20-36 The variance of a set of z scores is always equal to ________. In step 3, the term $\Sigma z_Y^2/N$ is the variance of the ________ distribution, and it is equal to ________. *1* / *z_Y; 1*

20-37 The variance of the z_X distribution appears in the term ________ in step 3. Since $\Sigma z_X^2/N = 1$, this term equals ________. *$\frac{r^2 \Sigma z_X^2}{N}$; r^2*

20-38 Step 4 shows that $\Sigma(z_Y - rz_X)^2/N$ is equal to the sum of three terms. If we take $r = \Sigma z_X z_Y/N$, these three terms reduce to a sum of two terms, ________. *$1 - r^2$*

20-39 We choose $r = \Sigma z_X z_Y/N$ in order to give $\Sigma(z_Y - rz_X)^2/N$ its ________ value. *minimum*

20-40 If we choose any other value for r, the average squared deviation $\Sigma(z_Y - z'_Y)^2/N$ will have a value which is ________ than its minimum value. *greater*

20-41 In that case, $\Sigma(Y_{ij} - Y'_j)^2$ will also have a value which is ________ than its minimum value, and $SS_L = \Sigma(Y'_j - \bar{Y})^2$ will have a value which is ________ than its maximum value. *greater* / *smaller*

20-42 Unless SS_L has its maximum value, the regression equation will not describe all the variation in Y which can be described by a ________. *linear function (or linear equation)*

20-43 In order for SS_L to have its maximum value, the constant r must equal the quantity ________, which is called the average ________ of the paired z scores. *$\frac{\Sigma z_X z_Y}{N}$* / *cross product*

20-44 The constant r is the Pearson product-moment ________ coefficient. It can take values from ________ to ________. *correlation; +1 ↔ −1*

20-45 When correspondence between the paired z scores is perfect, so that $z_X = z_Y$ in every case, r has the value ________. *+1*

20-46 When the z scores are paired at random, so that there is no correspondence at all, r has the value ________. *0*

20-47 We speak of perfect "inverse" correspondence when every z_X and z_Y have the same size but have opposite algebraic ________. In such a case, r has the value ________. *signs; −1*

D. OBTAINING THE SUMS OF SQUARES FOR THE LINEAR-REGRESSION MODEL

Once the value of the Pearson r has been found, it is an easy matter to obtain the sums of squares for the linear-regression model. First, we shall show that $SS_L = r^2 SS_Y$. We start again from our z-score regression equation, $z'_Y = rz_X + k$. Since $k = 0$, we can write

$$\frac{Y'_j - \bar{Y}}{s_Y} = r\frac{X_j - \bar{X}}{s_X} \qquad \text{and} \qquad Y'_j - \bar{Y} = r\frac{s_Y}{s_X}(X_j - \bar{X})$$

To obtain a single squared deviation $(Y'_j - \bar{Y})^2$, we square both sides:

$$(Y'_j - \bar{Y})^2 = r^2 \frac{s_Y{}^2}{s_X{}^2}(X_j - \bar{X})^2$$

To obtain SS_L, we sum these squared deviations for all N observations:

$$\sum_1^N (Y'_j - \bar{Y})^2 = r^2 \frac{s_Y{}^2}{s_X{}^2}\sum_1^N (X_j - \bar{X})^2$$

But $\Sigma(X_j - \bar{X})^2$ defines the sum of squares for the X distribution, SS_X. Moreover, $s_Y{}^2 = SS_Y/N$ and $s_X{}^2 = SS_X/N$; thus $s_Y{}^2/s_X{}^2 = SS_Y/SS_X$. Then

$$SS_L = \sum_1^N (Y'_j - \bar{Y})^2 = r^2 \frac{SS_Y}{SS_X} SS_X = r^2 SS_Y$$

Since $SS_L = r^2 SS_Y$, r^2 indicates the *proportion* of SS_Y which is contained in SS_L. SS_Y is a measure of Y variation within the sample; thus r^2 is the proportion of sample variation which can be described by a linear function.

When SS_L has been removed from SS_Y, the remainder is the sum of SS_W (SS within columns) and SS_D (SS deviations from linear regression). $SS_W + SS_D$ forms a new sum of squares which we shall call SS_{W+D}. This sum of squares arises from the score component e'_{ij}, and its mean square will serve to estimate error. Since $SS_{W+D} = SS_Y - SS_L$, we can write

$$SS_{W+D} = SS_Y - r^2 SS_Y = (1 - r^2) SS_Y$$

As in one-way analysis of variance, we use these sums of squares to form mean-square estimates of population variance. Table 20-2 summarizes the computing formulas, degrees of freedom, and mean squares for linear regression analysis. SS_L has only one degree of freedom; only one value of $(Y'_j - \bar{Y})$ can be regarded as independent. Information about just one value of $(Y'_j - \bar{Y})$ will suffice to determine the best-fitting straight line, from which all other values of $(Y'_j - \bar{Y})$ can be determined.

SS_{W+D} has the degrees of freedom of its components, SS_W and SS_D. You will recall that SS_W has df $= (N - C)$. SS_D has $C - 2$ degrees of freedom; it contains the contributions of C columns, but 2 degrees of freedom are lost when the Y'_j values are required to lie along a straight line $Y'_j = mX_j + b$ with two constants, m and b. Therefore, only $C - 2$ of these deviations $Y_j - Y'_j$ are independent; the last two are determined by the others and by the regression equation. SS_{W+D} then has $(N - C) + (C - 2) = N - 2$ degrees of freedom.

Table 20-2 Mean Squares in the Linear-Regression Model

SS	Defining Formula	Computing Formula	df	Mean Square
SS_{W+D}	$\sum_1^N (Y_{ij} - Y'_j)^2$	$(1 - r^2)SS_Y$	$N - 2$	$\frac{SS_{W+D}}{N-2} = MS_{W+D}$ (MS error)
SS_L	$\sum_1^N (Y'_j - \bar{Y})^2$	$r^2 SS_Y$	1	$\frac{SS_L}{1} = MS_L$ (MS linear regression)
SS_Y	$\sum_1^N (Y_{ij} - \bar{Y})^2$	$\sum_1^N Y_{ij}^2 - \frac{G^2}{N}$	$N - 1$	$\frac{SS_Y}{N-1} = MS_Y$

20-48 $SS_L = r^2 SS_Y$. If $r = .1$, SS_L is equal to SS_Y times ______. In percentage terms, SS_L is ______ percent of SS_Y when $r = .1$.

.01
1

20-49 If $r = .5$, SS_L is ______ percent of SS_Y; if $r = .9$, SS_L is ______ percent of SS_Y.

25
81

20-50 The proportion of SS_Y which is contained in SS_L is always given by the square of the Pearson ______.

r

20-51 Therefore, the proportion of sample variation which can be described by a linear function is always equal to ______.

r^2

20-52 The remaining variation is called SS_{W+D}; it is equal to SS_Y times ______.

$1 - r^2$

20-53 When we speak here of variation in Y, we mean variation in the (*population; sample*) which has actually been observed. For this reason, the Pearson r is called a (*parameter; statistic*).

sample
statistic

20-54 SS_L has df = ______. When it is divided by its df, it becomes MS ______.

1
linear regression

20-55 $SS_{W+D} = SS_Y - SS_L$. Since SS_Y has df = ______ and SS_L has df = ______, the df for SS_{W+D} must be ______.

$N - 1$
$1; N - 2$

20-56 If all the effects of factor A can be described by a linear function, ______ will contain all the effects and ______ will contain nothing but error.

SS_L; SS_{W+D}

20-57 MS_{W+D} is then an estimate of ______ variance.

error

20-58 If the effects of factor A cannot be described fully by a linear function, SS_L will not contain all these effects. SS_{W+D} will contain a mixture of ______ and ______ variance, and MS_{W+D} (*will; will not*) be a pure estimate of error variance.

effects ↔ error
will not

E. TESTING THE HYPOTHESIS OF ZERO LINEAR REGRESSION

We applied an F test to mean squares derived from the simple additive model $Y_{ij} = \mu + a_j + e_{ij}$, and this test enabled us to decide whether a significant effect of factor A was present (Lesson 15). Can we apply the same kind of test to the mean squares derived from the linear-regression model?

Recall the reasoning behind application of an F test to the ratio MS_B/MS_W. We reason that if no effects variance is present in the numerator, we are dealing simply with a ratio between two independent estimates of error variance. By turning to a table of the F distribution for the appropriate numerator and denominator degrees of freedom, we can find the probability of getting an F ratio at least as large as the one observed when, in fact, these estimates are two independent estimates of the same variance.

Now consider the ratio $\text{MS}_L/\text{MS}_{W+D}$ from the linear-regression model. If some hypothesis can be found on which both MS_L and MS_{W+D} are expected to be estimates of error variance, $F = \text{MS}_L/\text{MS}_{W+D}$ will provide a test of that hypothesis.

MS_{W+D} will be an estimate of error variance only if we assume that *all* effects of factor A can be described by a linear function. Any other assumption will imply that SS_{W+D} contains a mixture of error and effects. Therefore, we must first assume that no effect of factor A exists which cannot be described by a linear function. This assumption makes $r_j = a_j$ and $e'_{ij} = e_{ij}$.

When we make this assumption, the expectation of MS_{W+D} becomes $E(\text{MS}_{W+D}) = \sigma_e^2$ and the expectation of MS_L becomes $E(\text{MS}_L) = \sigma_e^2 + N\rho^2\sigma_Y^2$, where ρ (small Greek rho) is the population product-moment correlation coefficient (the parameter corresponding to the statistic r) and σ_Y^2 is, of course, the true population variance of Y. MS_L can be an estimate of error variance only if $\rho^2 = 0$. Therefore, H_1 is the hypothesis that $\rho^2 = 0$, and $F = \text{MS}_L/\text{MS}_{W+D}$ provides a test of this hypothesis. The hypothesis $H_1 : \rho^2 = 0$ is called the hypothesis of **ZERO LINEAR REGRESSION IN THE POPULATION**. If there is a clear reason to expect that the population correlation coefficient will be positive (or that it will be negative), H_2 can be directional ($H_2 : \rho > 0$ or $H_2 : \rho < 0$). Since $\sqrt{F}$ for df(1,k) has the same distribution as the t statistic for k degrees of freedom, the square root of this F statistic can be used as a t statistic for a one-tailed test of H_1.

A significant test of $F = \text{MS}_L/\text{MS}_{W+D}$ tells us that effects of factor A exist and that most (if not all) of these effects can be described by a linear function. However, when $F = \text{MS}_L/\text{MS}_{W+D}$ is not significant, we still may not wish to conclude that no linear association is present (i.e., that $\rho^2 = 0$). Use of this F test rests on the assumption that the association is *exclusively* linear; if we have reason to expect a linear association, a nonsignificant F may really mean that this assumption is not justified and that some nonlinear association is also present. Significant nonlinear regression could prevent us from finding evidence of linear regression by this F test. For this reason, the test for zero linear regression must take a different form whenever there is any possibility that some of the association is nonlinear.

20-59 An F ratio is a ratio between two __________ estimates of the same __________. — *independent; variance*

20-60 In one-way analysis of variance, the F ratio MS_B/MS_W is a ratio between two independent estimates of __________ variance when H_1 is true. — *error*

20-61 With factor A fixed, H_1 in one-way analysis of variance is the hypothesis that all the values of a_j are equal to ______. — *0*

20-62 When this H_1 is true, $E(MS_B)$ and $E(MS_W)$ are both equal to ______. — σ_e^2

20-63 MS_L/MS_{W+D} will be an F ratio only when MS_L and MS_{W+D} are independent __________ of the same variance. — *estimates*

20-64 MS_L and MS_{W+D} are independent of each other because MS_L arises from the sum of squared deviations of Y_j' values around ______ while MS_{W+D} arises from the sum of squared deviations of Y_{ij} around ______. — $\bar{Y}$; Y_j'

20-65 $E(MS_{W+D})$ equals σ_e^2 only when all the effects of factor A can be __________ by a __________ function. — *described; linear*

20-66 When all the effects of factor A can be described by a linear function, $E(MS_L) = \sigma_e^2 + N\rho^2\sigma_Y^2$. σ_Y^2 is the true __________ of the population of Y values. — *variance*

20-67 ρ is the true product-moment __________ for the population sampled. If all elements in the population were included in the sample, ρ would be the same as the statistic ______. — *correlation coefficient*; *r*

20-68 s_Y^2 has the same relation to σ_Y^2 as r^2 has to ______. — ρ^2

20-69 $E(MS_L)$ can equal σ_e^2 only when the parameter ρ^2 is equal to ______. — *0*

20-70 When no effects are present which can be described by a linear function, $\rho^2 = 0$. $E(MS_L)$ and $E(MS_{W+D})$ equal ______ when $\rho^2 = 0$. — σ_e^2

20-71 Thus, MS_L/MS_{W+D} is an F ratio when H_1: ______ $= 0$ is true. — ρ^2

20-72 A test of $F = MS_L/MS_{W+D}$ is a test of __________ linear __________. If F is significantly greater than 1, some linear __________ is present in the population. — *zero; regression; regression (or association)*

20-73 F has df (______, ______) in this test. For a one-tailed test, $\sqrt{F}$ is a ______ statistic with df = ______. — *1; N − 2*; *t; N − 2*

REVIEW

20-74 The simple additive model $Y_{ij} = \mu + a_j + e_{ij}$ leads to a partitioning of SS_Y into the two components ______ and ______.

$SS_B \leftrightarrow SS_W$

20-75 The F ratio arising from this partitioning is the ratio ______. In a fixed-factor experiment, where the effects of factor A are represented by a_j, this ratio provides a test of H_1: ______ $= 0$.

$\frac{MS_B}{MS_W}$
all a_j

20-76 The linear-regression model $Y = \mu + r_j + e'_{ij}$ cannot be applied to all single-factor experiments; it applies only to those in which the factor ______ are quantitative.

levels

20-77 When the levels of factor A are ______, they can be regarded as values of a variable X, and we can try to describe some of the Y variation by a linear equation of the form $Y'_j =$ ______.

quantitative
$mX_j + b$

20-78 The model $Y_{ij} = \mu + r_j + e'_{ij}$ leads to a partitioning of SS_Y into the two components ______ and ______.

$SS_L \leftrightarrow SS_{W+D}$

20-79 SS_L is the sum of squares for ______. It contains as much of SS_Y as can be described by a ______.

linear regression
linear function

20-80 SS_{W+D} is the sum of squares arising from the score component ______. It contains one of the sums of squares identified in simple analysis of variance, ______. In addition, it contains the sum of squares for ______ from linear regression, SS_D.

e_{ij}
SS_W
deviations

20-81 $Y'_j = mX_j + b$ is the raw-score ______ equation. It has a slope constant ______ and a Y intercept ______.

regression
m; b

20-82 The constants m and b are the ______ constants for this equation. We have not discussed their calculation yet because we did not need the values of Y'_j in order to calculate SS_L.

regression

20-83 To calculate SS_L, we simply multiply ______ by r^2, the square of the Pearson product-moment ______.

SS_Y
correlation coefficient

20-84 The Pearson r is the ______ constant in the z-score ______ equation, $z'_Y = rz_X + k$.

slope
regression

20-85 We found the Pearson r by applying the principle of ______ to this z-score regression equation.

least squares

20-86 According to the principle of least squares, the best-fitting straight line for any set of data points is the line which makes the sum of the ________ of deviations of points from the line as ________ as possible.

squares
small

20-87 The best-fitting straight line of the form $Y'_j = mX_j + b$ will thus be that line which makes the sum $\sum_1^N (\_\_\_\_\_\_)^2$ as small as possible.

$Y_{ij} - Y'_j$

20-88 This sum is one of the component sums of squares in the linear-regression model. $\sum_1^N (Y_{ij} - Y'_j)^2$ is ________, and when it is as small as possible, the other component, $\sum_1^N (Y'_j - \bar{Y})^2 =$ ________, is as large as possible.

SS_{W+D}

SS_L

20-89 We studied the z-score regression equation and sought to minimize the sum of squared deviations $\sum_1^N (z_Y - z'_Y)^2$. We found that k must equal ________ and that r must equal the quantity ________ in order for this sum to have its minimum value.

0
$\frac{\Sigma z_X z_Y}{N}$

20-90 The quantity $\Sigma z_X z_Y / N$ is called the ________ of the paired z scores.

average cross product

20-91 Since $SS_L = r^2 SS_Y$, r^2 for any set of data tells the amount of Y variation in that sample which can be ________ by a ________ function.

described
linear

20-92 r^2 is a sample statistic. The corresponding population parameter is ________.

ρ^2

20-93 The ratio MS_L/MS_{W+D} is an F ratio with ________ numerator and ________ denominator df. This ratio provides a test of H_1: ________ $= 0$. The test is called a test for zero ________.

1
$N - 2$
ρ^2
linear regression

20-94 If the test is significant, we conclude that the correlation coefficient for the (*population; sample*) is not 0 and that some amount of ________ association between X and Y is present in the (*population; sample*).

population
linear
population

20-95 If the test is *not* significant, we conclude that any effects of factor A which are present in the population (*can; cannot*) be described by a linear function. Such effects (*can; cannot; may or may not*) be described by a nonlinear function.

cannot

may or may not

PROBLEMS

1. Calculate $r = \Sigma z_X z_Y / N$ for the following set of raw scores. Begin by making a z-score transformation of each distribution; $\bar{Y} = 13$, $s_Y = 4$, $\bar{X} = 9$, $s_X = 4$. Since z_X is a constant for each column, find Σz_Y for the column and then multiply Σz_Y by z_X (summation rule II).

After you have found r, write the z-score regression equation for these data. Calculate SS_Y, SS_B, and SS_W as you would in an ordinary one-way analysis of variance (using the raw scores); then determine SS_L, SS_D, and SS_{W+D}.

What is the value of F for the test of zero linear regression on these data? What are its degrees of freedom?

	$X_1 = 4$	$X_2 = 6$	$X_3 = 8$	$X_4 = 12$	$X_5 = 15$
Y_1	11	12	18	20	21
Y_2	10	12	16	18	20
Y_3	10	12	16	17	18
Y_4	9	11	14	17	17
Y_5	8	11	14	16	17
Y_6	8	10	13	16	16
Y_7	7	10	13	15	16
Y_8	7	9	13	14	15
Y_9	6	7	12	14	15
Y_{10}	4	6	11	13	15

2. A single-factor experiment has $N = 62$, $SS_Y = 600$, and $r = .3$. Make a test for zero linear regression. Can H_1 be rejected at $\alpha = .05$? At $\alpha = .01$?

LESSON 21 CORRELATION ANALYSIS

Lesson 20 has introduced the use of regression analysis in single-factor experiments. We now turn our attention to the closely related topic of correlation.

The word "correlation" is sometimes used loosely as if it were a synonym for "association." For example, you will find it said that two variables are "correlated" whenever there is any significant association between them. Without rejecting this very common way of using the word, we shall develop a precise technical meaning for it. Its technical application begins with the *method* by which an association between variables is identified and verified.

There is an operational difference between the experimental and correlational methods of data collection, i.e., a difference in the operations which an investigator carries out in collecting his evidence. In the experimental method he selects his sample randomly and randomly assigns one subgroup to each of C different categories on the independent X variable (factor A). He then makes whatever measurements are necessary to obtain a score on the dependent Y variable for each element and compares the scores found in each of his two or more subgroups.

In the correlational method he treats his sample differently. He himself does no assigning of elements to subgroups on X. Instead, after randomly selecting a sample (usually a set of persons or other organisms), he makes the measurements necessary to obtain two scores from each element in the sample, a score on the X variable and a score on the Y variable. We say that the X and Y values are SAMPLED BY PAIRS. These values may be either discrete or continuous for X as well as for Y. (Remember that the values of the X variable in a single-factor experiment are always levels of factor A and therefore always discrete values, even when factor A is random.)

Once his data are collected, the investigator will use many of the same statistical procedures regardless of the method by which the data were obtained. Correlation analysis requires some steps which are arithmetically identical with those required in regression analysis. However, the conditions assumed in making statistical tests are different. Because of the differences in assumptions and in data-collection methods, interpretation of the results must be made in different terms for correlation and regression problems.

Certain kinds of questions in science and technology are amenable only to the correlational method. This is true, above all, of "naturally occurring" characteristics which cannot be manipulated by an investigator, such as scores on tests, height and other physical characteristics, grades in college, annual family income. Whenever we are interested in the association between two (or more) variables of this sort, we have a potential correlation problem.

A. DIFFERENTIATING BETWEEN CORRELATION AND REGRESSION PROBLEMS

A regression problem always arises from data suitable for analysis of variance, and such data must fall into subgroups on the X variable (factor A). We identify each Y value by two subscripts, j for its subgroup and i for its position within the subgroup. The distribution Y_{ij} is continuous, and its parent population is assumed to be reasonably normal. The distribution of X values (X_j) is always discrete. Whenever the investigator arranges to have subgroups of equal size, the X_j distribution will be rectangular in shape.

A correlation problem arises when X and Y values are sampled by pairs. No restrictions are placed on the values which X may assume, and there is no guarantee that any two measures on X will be exactly the same. It is not customary to talk of subgroups on the X variable, and we therefore do not identify Y values as belonging to any subgroup. The Y distribution is designated simply as Y_i, and the X distribution is designated as X_i. In the correlation problems to be considered in this lesson, both distributions are continuous. Both may also be approximately normal.

Figure 21-1 presents the data from a correlation problem in a form commonly called a SCATTER DIAGRAM. A group of 94 rats, randomly selected from a laboratory population, learned two mazes, X and Y. Each rat has an error score on each maze, and the error scores on maze X form the X_i distribution whose frequencies are summarized in the row f_X. The 94 Y scores are summarized in the last column, f_Y. Although the error scores are discrete whole numbers, both X and Y distributions are treated as continuous variables in the analysis.

The data in this scatter diagram constitute a JOINT X,Y DISTRIBUTION. Each animal has been tallied in just one of the 110 cells representing the joint occurrence of a particular score on X and a particular score on Y. An animal making 9 errors on maze X and 8 on maze Y is tallied in the third row and ninth column.

The tallies are not evenly scattered through the diagram. They group themselves into a roughly oval pattern extending from lower left to upper right, suggesting that much of the variation in Y could be described as an increasing linear function of X. By using the principle of least squares, just as in Lesson 20, the best-fitting linear equation $Y'_i = mX_i + b$ can be found, and the amount of linear association between X and Y can be measured for this sample.

Moreover, the X,Y distribution in Figure 21-1 appears to come from an X,Y population which is a BIVARIATE NORMAL DISTRIBUTION. The two marginal distributions, f_X and f_Y, do not deviate far from the shape of a normal distribution. The distributions within each column and within each row are irregular, but they tend to be unimodal and symmetrical. If the parent population has a normal distribution for each column and for each row, we are dealing with a sample from a bivariate normal population. *For such populations, the only form of association possible is a linear association.* When no linear association is present, X and Y must be independent.

Errors on maze X

Errors on maze Y	1	2	3	4	5	6	7	8	9	10	11	f_Y
10								1	1	1	1	4
9							1		1	2	1	5
8						1	1	2	4	1		9
7						4	5	3	2			14
6					1	6	4	4	1			16
5				1	5	3	5	1				15
4			2	2	4	4	2					14
3		1	3	3	2	1						10
2	1	2	1	1								5
1	1	1										2
f_X	2	4	6	7	12	19	18	11	9	4	2	94

Figure 21-1: Scatter diagram of error scores made by a group of 94 rats in learning each of two mazes, X and Y.

21-1 Column 1 contains tallies for animals which made exactly ______ error on maze X. The number of animals tallied in this column is ______, as shown in the row labeled ______.

1
2
f_X

21-2 f_X stands for the frequencies in the X distribution. The frequencies in the Y distribution are called ______. This scatter diagram shows a ______ X,Y distribution.

f_Y
joint

21-3 Animals with 3 errors on maze X and 4 errors on maze Y are tallied in the cell for $X =$ ______ and $Y =$ ______. There are ______ animals with this X,Y value.

3; 4
2

21-4 When all the row, column, and marginal distributions are normal, an X,Y distribution is said to be a bivariate ______ distribution.

normal

21-5 You (*could; could not*) encounter a sample with a bivariate normal distribution in a regression problem.

could not

21-6 The only form of association possible in a bivariate normal population is a ______ association.

linear

An experimenter wishing to treat the maze problem by experimental method would have proceeded in a different manner. He might take errors on maze X as factor A; if he treats factor A as fixed, he would take a subgroup of animals for each of the 11 levels of this variable. Ideally he would want all 11 subgroups to be of equal size; he would have to observe a very large number of rats on maze X in order to obtain exactly n rats who make 1 error, n who make two errors, and so on for all levels of the factor. Once he has obtained his subgroups of animals, he can run all the animals on maze Y and record their Y scores.

This procedure is formally similar to true experimental method, but there is still an important difference. The investigator has not randomly *assigned* certain animals to achieve particular scores on maze X; he has selected certain animals for study after observing that they did in fact make particular scores on maze X. The data will fit the pattern required for analysis of variance, and such data can also be subjected to linear regression analysis. But the aspect of manipulation and control—so powerful in typical cases of experimental method—is necessarily absent from this application of experimental method to a potential correlation problem.

You may feel certain that no investigator would actually choose to approach the maze problem in this way. The data of Fig. 21-1 are of interest only because they indicate the extent to which the two mazes measure the same thing (ability of rats to learn a particular type of maze). The joint study of error scores on these two mazes would be undertaken in order to find out the *extent of agreement* between the two sets of scores, not to find out whether a certain independent variable X influences scores on maze Y.

Thus, experimental and correlational methods are evidently chosen for answering different sorts of questions. If one is interested in learning the extent of agreement between two continuous variables X and Y neither of which can be manipulated experimentally, the correlational method is chosen. Because values of X and Y are sampled by pairs, the data will be *representative* of the joint X,Y distribution in the population. Correlation analysis focuses attention on the correlation coefficient r, which can be expected to reflect the extent of agreement between values of X and Y in the population.

On the other hand, the experimental method is chosen when one must determine how Y is affected by an X-variable which can be manipulated experimentally. Because variable X is actually manipulated, the experimenter may be able to argue that all his subgroups would have achieved similar scores on Y if he had not subjected them to different treatments on X. Significant effects of X may therefore indicate a *causal* dependence of Y on X. Regression analysis will give information about the form (linear or nonlinear) of the mathematical function best describing this dependence. The statistic r^2 plays a role, but its exact value is of less interest than what it tells about the function relating X and Y.

21-7 We want to determine how intelligence affects college grades. Intelligence (*is; is not*) a variable which can be manipulated experimentally. We should therefore choose the (*correlational; experimental*) method for our study.

is not

correlational

21-8 Because of our choice of this method, our data (*will; will not*) be representative of the joint distribution in the population. They (*can; cannot*) be interpreted as showing a causal dependence of grades on intelligence.

will

cannot

21-9 We have two fourth-grade spelling tests. To find out whether these tests are equivalent, we should choose the ________ method.

correlational

21-10 We give both tests to a random sample of fourth-grade students. The resulting data will be ________ of the joint X,Y distribution of grades on these tests among fourth-grade students.

representative

21-11 Our main interest will be in the value of the ________ between X and Y because it measures the extent of agreement between scores on the two tests.

correlation coefficient

21-12 We wish to determine the effect of "number of homework problems" on grades in algebra. Since we can manipulate the number of problems assigned, we should choose the ________ method for our study.

experimental

21-13 We assign 60 students at random to each of four subgroups: 0, 10, 20, and 40 problems per week. If the 240 students are treated alike in all other ways, we (*can; cannot*) conclude that any differences in subgroup performance are *caused by* differences in homework.

can

21-14 We know that most algebra teachers assign about 10 problems a week and that teachers almost never assign as many as 40 problems. Our data (*will; will not*) be representative of the joint X,Y distribution in the population.

will not

21-15 Our main interest (*will; will not*) be in determining the value of the correlation coefficient between X and Y.

will not

21-16 In regression analysis, we are mainly interested in determining the ________ of the function relating X and Y.

form

21-17 We suspect that grades increase with increases in homework up to a point, then level off or drop with further increases. If so, the form of the function is (*linear; nonlinear*).

nonlinear

B. PREDICTING QUANTITATIVE VARIABLES

Another important difference between correlation and regression problems requires more extended treatment. In this section, we shall study the role played by the concept of prediction in regression and correlation analysis.

When two variables X and Y are related, it may be possible to use knowledge of one variable to assist in predicting the other variable. In regression problems the independent variable X is always the PREDICTOR variable, and Y is always predicted from knowledge of X. When the test of linear regression is significant, use of the linear equation $Y'_j = mX_j + b$ will improve the prediction of Y. It is customary to refer to Y'_j as the PREDICTED VALUE of Y.

For the F test of zero linear regression, we do not need the raw-score regression equation. Therefore, we did not calculate the regression constants m and b for this equation in Lesson 20. These constants are easily found, once the Pearson r has been determined. Since $z'_Y = (Y'_j - \bar{Y})/s_Y$ and $z_X = (X_j - \bar{X})/s_X$, we can write $Y' - \bar{Y} = r(s_Y/s_X)\,(X - \bar{X})$, which leads to a raw-score regression equation

$$Y'_j = r\frac{s_Y}{s_X}X_j + (\bar{Y} - r\frac{s_Y}{s_X}\bar{X})$$

We can identify the regression constants m and b in this equation:

$$m = r\frac{s_Y}{s_X} \qquad \text{and} \qquad b = \bar{Y} - r\frac{s_Y}{s_X}\bar{X}$$

To understand why we might be interested in Y'_j as a prediction let us recall an earlier discussion. Lesson 12 described such predictions for two associated *qualitative* variables, A and B. Knowledge of a person's category on A may reduce the error that would otherwise be made in guessing his category on B. We were able to measure the amount of this error reduction, using the λ index, but our measure applied only to predictions *within the sample,* i.e., predictions in the peculiar sense of guesses about a category which is already known because it is part of the data being analyzed. Measuring error in such predictions has no practical value; it is useful only in measuring the amount of association between A and B present in the sample.

With quantitative variables, predictions can be made about actual scores rather than about mere categories. In any single-factor experiment, knowledge of a person's level on X may reduce the error in guessing his score on Y. If X_j is known for an individual, the best long-run guess at his Y score will be $\bar{Y}_j$, the mean of his subgroup, for all predictions within the sample.

However, we can now begin to think also about predictions *beyond* the observed sample, predictions of future scores to be obtained under the same conditions. For such predictions we need the best possible estimate of the subpopulation mean $\mu + a_j$. Before regression analysis, our best estimate of this mean is $\bar{Y}_j$. But if regression analysis shows a significant amount of linear regression in the population, the best estimate of $\mu + a_j$ becomes Y'_j. Given knowledge of a person's X_j value, our best long-run guess at his *future* Y score is the value of Y'_j which we obtain from the raw-score regression equation.

21-18 When we try to guess the value of one of the Y scores which has actually been observed, we are making a "prediction" within the ________. *sample*

21-19 When we try to guess the value of a Y score which has not yet been observed, i.e., a future Y score, we are making a prediction ________ the sample. *beyond*

21-20 For predictions beyond the sample, we must estimate parameters of the ________ from which the sample is drawn. We can make predictions only about future scores which are expected to belong to that same ________. *population* *population*

21-21 Predictions within the sample have no practical use, but they may provide a measure of the amount of ________ between X and Y present in the sample. *association*

21-22 With qualitative variables, we can only predict a person's ________ on the predicted variable. With quantitative variables, we can try to predict his actual ________. *category* *score*

21-23 We are trying to predict a person's Y score in an experimental study. If we know he is in subgroup X_j, our best prediction *within* the sample is ________, the ________ of his subgroup. $\bar{Y}_j$; *mean*

21-24 If we are making a prediction *beyond* the sample, we need to know the best estimate of the mean of the ________ to which his score will belong. *subpopulation*

21-25 Before regression analysis, the best estimate of $\mu + a_j$ is ________. Therefore, without regression analysis our prediction about his score should be ________. $\bar{Y}_j$ $\bar{Y}_j$

21-26 If regression analysis shows a significant amount of linear regression, the best estimate of $\mu + a_j$ is ________, and this estimate is our prediction. Y'_j

21-27 In order to know the value of Y'_j, we must evaluate the regression constants ________ and ________ in the regression equation for raw scores. $m \leftrightarrow b$

21-28 These constants can be found when r, $\bar{X}$, $\bar{Y}$, s_X, and s_Y are known. The slope constant m is equal to ________. $r\frac{s_Y}{s_X}$

21-29 The Y intercept b is equal to $\bar{Y}$ minus the quantity ________. $r\frac{s_Y}{s_X}\bar{X}$

In correlation problems, neither X nor Y is truly an independent variable, and the predictor may be either X or Y. To predict Y from knowledge of X, we use the equation $Y_i' = mX_i + b$ and call it the REGRESSION EQUATION OF Y ON X. To predict X from knowledge of Y, we use an equation *with different values for the regression constants;* it is $X_i' = mY_i + b$, the REGRESSION EQUATION OF X ON Y. In the absence of subgroups on X or on Y, there are no column means, and Y_i' (or X_i') will be the best guess even for predictions *within* the sample whenever a significant amount of linear regression is present. With no linear regression, the best prediction is always $\bar{Y}$ (or $\bar{X}$).

Correlation analysis is often undertaken for the sake of practical application. Figure 21-2 provides an example. The figure is a scatter diagram of scores on the Scholastic Aptitude Test (X) and freshman grade averages (Y) for the 547 students entering Oberlin College as freshmen in 1968. The descriptive statistics for this sample were calculated by computer from the exact raw scores of each of the 547 students. In the figure, both variables have been divided into a series of discrete classes so that you can examine the row, column, and marginal distributions.

Y (Grade average)	X (SAT) 251–300	301–350	351–400	401–450	451–500	501–550	551–600	601–650	651–700	701–750	751–800	Total
3.76–4.00 (3.88)									2	1		3
3.51–3.75 (3.63)				1				2	8	12	8	31
3.26–3.50 (3.38)					1		4	9	19	15	8	56
3.01–3.25 (3.13)						3	7	29	26	22	15	102
2.76–3.00 (2.88)						5	12	15	31	30	11	104
2.51–2.75 (2.63)			1	1	1	2	12	21	22	17	4	81
2.26–2.50 (2.38)					1	7	5	20	14	12	9	68
2.01–2.25 (2.13)					1	4	7	19	11	7	3	52
1.76–2.00 (1.88)	1			1	2	2	3	8	10			27
1.51–1.75 (1.63)	1				1		2	3	1	2	1	11
1.26–1.50 (1.38)							2		5	2		9
1.01–1.25 (1.13)								1				1
0.76–1.00 (0.88)								1				1
0.51–0.75 (0.63)							1					1
Total	2	0	1	3	7	23	55	128	149	120	59	547

	Means	Standard Deviations	Correlation Coefficient
SAT	$\bar{X} = 663$	$s_X = 74$	
Grade average	$\bar{Y} = 2.74$	$s_Y = 0.53$	$r = +0.3$

Figure 21-2: Scatter diagram of scores on the SAT (Scholastic Aptitude Test) and freshman grade averages for 547 students entering Oberlin College as freshmen in 1968. *(Data and computer analysis supplied by E. L. Van Atta.)*

21-30 In regression analysis, the predictor variable is always _______, and the predicted variable is always _______.

X; Y

21-31 In correlation analysis, the predictor variable may be either X or Y. When it is X, our best prediction of Y is always the value of _______ from the regression equation of _______ on _______, provided linear regression is present.

Y'_i; Y
X

21-32 If no linear regression is present, the best prediction of Y is _______.

$\bar{Y}$

21-33 When the predictor is Y, we use the regression equation of _______ on _______; if no linear regression is present, the best prediction of X is _______.

X; Y
$\bar{X}$

21-34 We are more likely to use such predictions for practical purposes when they result from (*correlation; regression*) analysis.

correlation

21-35 In the example of Fig. 21-2, we might wish to use correlation analysis to aid in predicting __________ from __________ before classes begin.

grade averages; SAT scores

21-36 Since grades are the Y variable, we would need the regression equation whose general form is _______.

$Y'_i = mX_i + b$

21-37 We can find the regression constants for $Y'_i = mX_i + b$ from the statistics in the figure. We need five statistics: _______, _______, _______, _______, _______.

r; $\bar{X}$; $\bar{Y}$; s_X; s_Y

21-38 On page 320, the constant m was shown to equal $r\ s_Y/s_X$. Substitute the values from this example, and calculate m: _______. (Keep three significant figures.)

.00215

21-39 The constant b equals $\bar{Y} - r(s_Y/s_X)\bar{X}$. Calculate b for the data in Fig. 21-2: _______. (Indicate its sign.)

+1.32

21-40 Examine the marginal distributions in Fig. 21-2. These distributions are both (*skewed; symmetrical*).

skewed

21-41 The marginal distributions have tails extending toward (*higher; lower*) values; the direction of the skew is therefore (*negative; positive*).

lower
negative

21-42 The row and column distributions are similarly skewed. This sample (*does; does not*) appear to be drawn from a bivariate normal population.

does not

C. UNDERSTANDING THE REGRESSION LINES

You are able now to find the regression line of Y on X, whether it is in z-score or raw-score form. The regression line of X on Y requires some additional discussion. Figure 21-3 will help, for it compares two graphs of the same z_X, z_Y distribution, whose coordinates are shown in the inset.

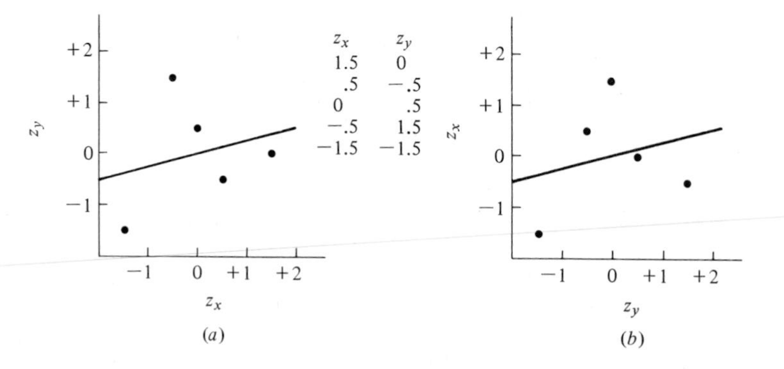

Figure 21-3: Two z-score regression lines for the same set of five data points (*inset*). Regression line of (*a*) z_Y on z_X and (*b*) z_X on z_Y; $r = +.25$ for these data.

Observe that the variable to be predicted is always plotted on the vertical axis, while the predictor variable is plotted on the horizontal axis. Thus, when z_X is the predictor, values of z_X are plotted on the horizontal axis (Fig. 21-3*a*); when z_Y is the predictor, values of z_Y are plotted on the horizontal axis (Fig. 21-3*b*). From this figure you will see at once that the regression lines for these two graphs have identical constants. The slope constant r equals $+.25$ for each graph, and both regression lines pass through the point (0,0). The regression equation of X on Y is therefore $z_X = rz_Y$.

Let us reflect for a moment on Fig. 21-3*b*. If we had started out in the first place to find the regression line of X on Y, we would have sought the constants for $z'_X = rz_Y + k$ by applying the principle of least squares to the sum of squared deviations $\Sigma(z_X - z'_X)^2$. The quantity $z_X - z'_X$ is the deviation of a point from the regression line measured along the z_X axis; it is the deviation of that point's z_X coordinate from the line. The quantity $z_Y - z'_Y$ is the deviation of a point's z_Y coordinate from the line. When we find an equation to minimize $\Sigma(z_X - z'_X)^2$, we are selecting the regression line so as to minimize the distance of points from the line in the vertical direction when z_X is on the vertical axis. Of course, the search for this line will exactly parallel the algebra in Lesson 20C, and we shall find that r must equal $\Sigma z_X z_Y / N$ in order to minimize $\Sigma(z_X - z'_X)^2$. The Pearson r is therefore the slope constant for *both* regression equations when these equations are in z-score form.

Figure 21-4*a* and *b* illustrates cases in which the slope constant r equals $+1$ and 0, respectively.

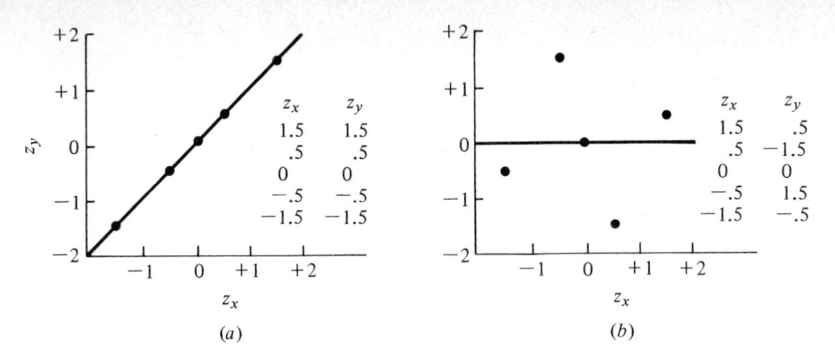

Figure 21-4: Scatter diagrams and regression lines of z_Y on z_X for the two sets of points shown in insets. (*a*) $r = +1.0$; (*b*) $r = 0$.

21-43 The Pearson r is the __________ constant in the z-score regression equation. When $r = +.25$, the regression equation of z_Y on z_X is $z'_Y =$ ________.

slope

$+.25z_X$

21-44 When $r = +.25$, the regression equation of z_X on z_Y is ________.

$z'_X = +.25z_Y$

21-45 In Fig. 21-3, $z'_Y = +.25z_X$ minimizes the sum $\Sigma(z_Y - z'_Y)^2$. The equation $z'_X = +.25z_Y$ minimizes the sum ________.

$\Sigma(z_X - z'_x)^2$

21-46 When $r = +1$, the slope of the regression line is (*downward; upward*) toward the right.

upward

21-47 When $r = +.25$, the slope of the regression line is __________ toward the right. This slope is (*less steep; steeper*) than the slope when $r = +1$.

upward; less steep

21-48 When $r = 0$, the regression line is a perfectly horizontal line at $z_Y =$ ________. This value of z_Y is the __________ of the z_Y distribution.

0; mean

21-49 Calculate $\Sigma(z_Y - z'_Y)^2$ for the five data points in Fig. 21-4*b*, where $r = 0$. Since $z'_Y = 0$ for all five points, $\Sigma(z_Y - z'_Y)^2 = \Sigma z_Y{}^2 =$ ______. (Remember that $\Sigma z_Y{}^2/N = 1$.)

5

21-50 In Fig. 21-3*a*, where $r = +.25$, this sum $\Sigma(z_Y - z'_Y)^2 =$ 4.6875. In Fig. 21-4*a* where $r = +1$, this sum equals ________ because all five data points lie directly on the regression line.

0

21-51 As r *decreases* from +1 to 0, the size of the sum $\Sigma(z_Y - z'_Y)^2$ (*decreases; increases*).

increases

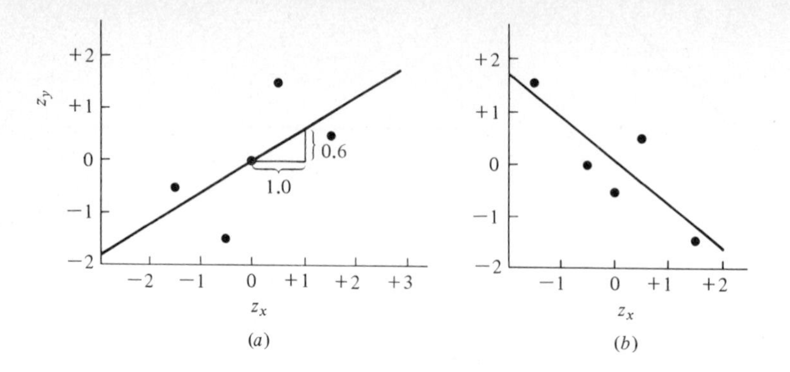

Figure 21-5: Scatter diagrams and regression lines to illustrate correlation coefficients of (*a*) +0.6 and (*b*) −0.85.

Figure 21-5 shows z-score regression lines with $r = .6$ and $r = -.85$. When you compare these graphs with the ones in Fig. 21-4 for $r = +1$ and $r = 0$, you will see that the Pearson r describes both the trend and the scatter of data points. When all the points lie in a straight line, a linear equation can be found which will describe *all* the variation in the data and r will equal +1 or −1. As the amount of scatter increases, r approaches 0; a linear equation is less able to describe the data points, and even the best possible equation will not fit the data points very well.

We can now explain the choice of the word "regression" for these equations. Whenever the absolute value of r is less than 1, the predicted z'_Y will have a smaller absolute value than the z_X score from which it is predicted. Therefore, the predicted Y'_i must always lie closer to its grand mean $\bar{Y}$ than the corresponding X_i lies to its grand mean $\bar{X}$. This fact has been called REGRESSION TOWARD THE MEAN; we start with an X score which deviates a certain amount from $\bar{X}$, and we end by predicting a Y score which deviates by a lesser amount from its mean $\bar{Y}$. Our prediction thus moves back, or regresses, toward the mean value. The name "regression equation" arises from this observation.

Figure 21-6 shows the raw-score regression lines for the example in Fig. 21-2. Since both z-score regression lines pass through the point (0,0) (the means of the z_X and z_Y distributions), you will not be surprised to see that the raw-score regression lines both pass through the point $(\bar{X},\bar{Y})$. You can derive the constants for the regression line of X on Y from the z-score regression equation $z'_X = rz_Y$:

$$X'_i = r\frac{s_X}{s_Y}Y_i + \left(\bar{X} - r\frac{s_X}{s_Y}\bar{Y}\right)$$

Observe that the slope constant for this equation is *not* the same as the slope constant for the regression line of Y on X, which is $r\,s_Y/s_X$. Of course, the constant b is also different. When the statistics from this example are substituted, the equation for predicting SAT from grades is $X'_i = 41.7Y_i + 549$.

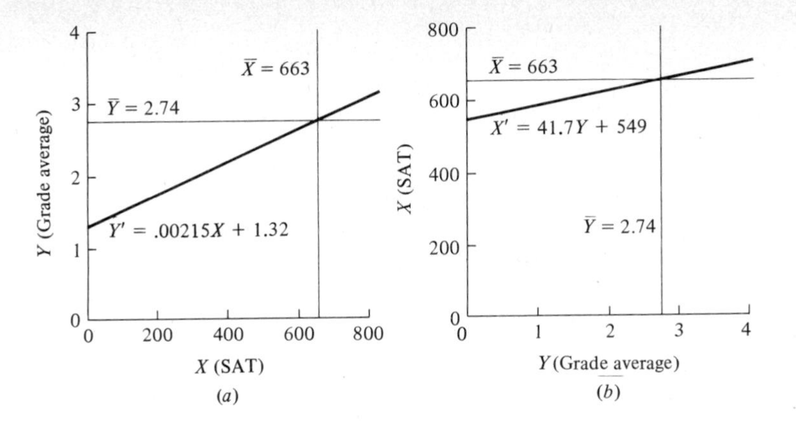

Figure 21-6: Two raw-score regression lines for the data from Fig. 21-2. (*a*) Regression line of Y (grade average) on X (SAT score). (*b*) Regression line of X on Y.

21-52 Figure 21-6*a* shows the regression line of ______ on ______ for the data in Fig. 21-2. When $X = 0$, $Y' =$ ______. When $X = \bar{X}$, $Y' =$ ______.

Y
X; 1.32
2.74

21-53 Figure 21-6*b* shows the regression line of ______ on ______ for the same data. When $Y = 0$, $X' =$ ______; when $Y = \bar{Y}$, $X' =$ ______.

X
Y; 549
663

21-54 Both raw-score regression lines have one point in common; the common point is the ______ of both distributions.

mean

21-55 The slope of the equation for predicting Y is r times the quantity s_Y/s_X. The slope of the equation for predicting X is r times the quantity ______.

$\frac{s_X}{s_Y}$

21-56 Figure 21-5*b* shows a set of points with $r = -.85$. The trend of these points is (*downward; upward*) toward the right; positive values of z_X tend to be paired with (*negative; positive*) values of z_Y.

downward
negative

21-57 In Fig. 21-5*a*, $z'_Y = +.6z_X$. When $z_X = 1.5$, $z'_Y =$ ______. Which is closer to its mean, z_X or z'_Y? ______

.9
z'_Y

21-58 Any z'_Y will be closer to its mean than the z_X from which it is predicted unless $r =$ ______ or ______. This fact is called "______ toward the mean."

$+1 \leftrightarrow -1$
regression

D. CALCULATING THE PEARSON *r* FROM RAW SCORES

You have now seen the importance of the Pearson r in both regression and correlation problems. It is essential to have a convenient means of calculating this statistic. You have already learned to find ΣY and ΣY^2 simultaneously when you are analyzing a distribution of Y scores. When you have a distribution of paired X,Y values, as you will in a correlation problem, most desk calculators will enable you to find ΣX, ΣX^2, and ΣXY at the same time that you are finding the sums for Y. The flow chart begins with $r = \Sigma z_X z_Y / N$ and leads to a computing formula containing *only these five sums and N.*

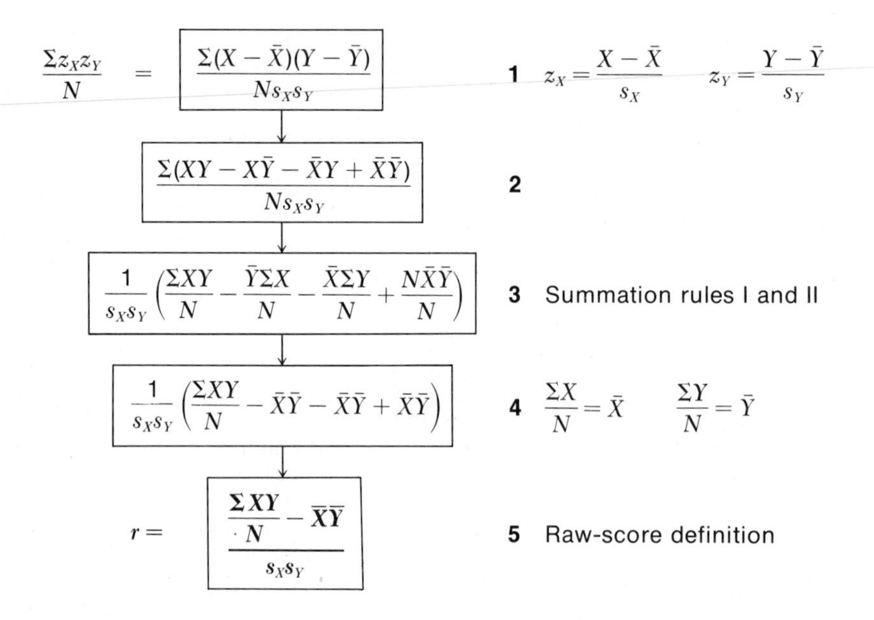

Before continuing the chart, look at the result obtained in step 5. This raw-score definition of the Pearson r contains an important statistic in its numerator; $\Sigma XY/N - XY$ is called the COVARIANCE OF X AND Y, cov XY. It is analogous to the variance of Y, $s_Y{}^2 = SS_Y/N$. Since $SS_Y = \sum_1^N Y^2 - \frac{G^2}{N}$,

$$s_Y{}^2 = \frac{\Sigma Y^2}{N} - \frac{G^2}{N^2} = \frac{-\Sigma Y^2}{N} - \bar{Y}^2$$

This definition of Y variance is visibly parallel to the covariance definition,

$$\text{cov } XY = \frac{\Sigma XY}{N} - \bar{X}\bar{Y}$$

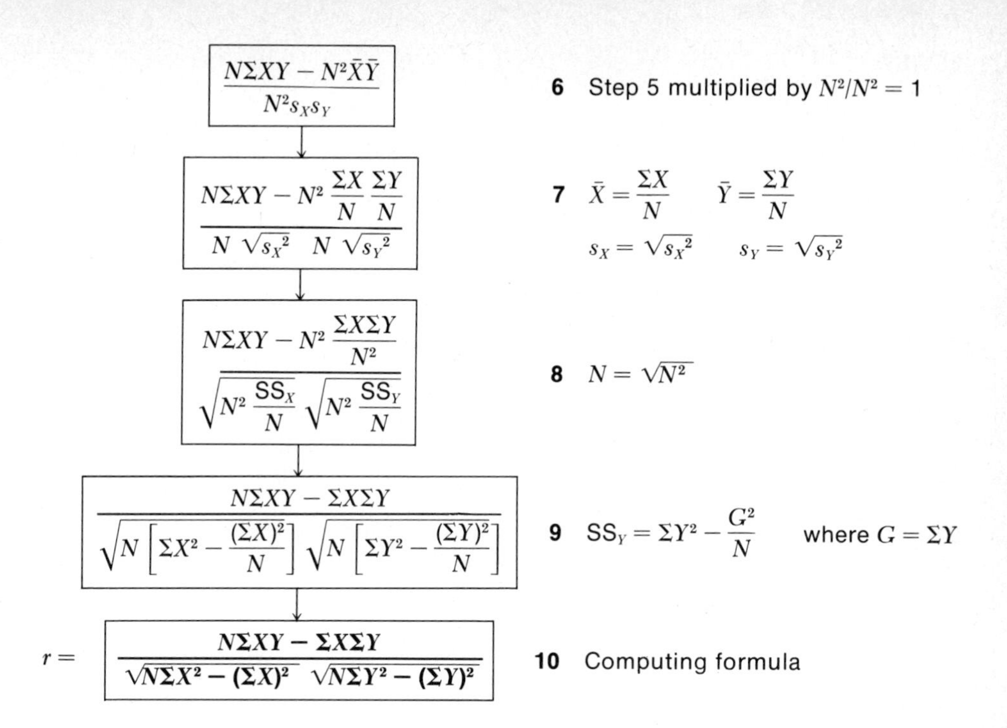

21-59 Step 5 gives the raw-score definition of r. It can be recalled as the ________ of X and Y divided by the product of ______ and ______.

covariance
$s_X \leftrightarrow s_Y$

21-60 The covariance of X and Y is analogous to the variance of Y, s_Y^2. The variance equals $\Sigma Y^2/N - \bar{Y}^2$. The covariance equals ______ minus ______.

$\frac{\Sigma XY}{N}$; $\bar{X}\bar{Y}$

21-61 In order to develop a convenient computing formula, this raw-score definition of r is multiplied in step 6 by ______.

$\frac{N^2}{N^2}$

21-62 The remaining steps manipulate the expression in step 6 to get a formula containing only N and the five sums ______, ______, ______, ______, and ______.

ΣX
ΣX^2; ΣY; ΣY^2; ΣXY

21-63 The z-score definition of r would be inconvenient as a computing formula. To use it in finding the value of r, we would have to convert every raw score to a ________.

z score

21-64 The raw-score definition of r would also be inconvenient. To use it in computing r, we would first have to find the ________ $\bar{X}$ and $\bar{Y}$ and the ________ s_X and s_Y.

means; standard deviations

REVIEW

21-65 Correlation analysis is applied to X and Y values which have been sampled by __________. Both variables are (*continuous; discrete*), and we speak of the data as a __________ X,Y distribution.

pairs
continuous; joint

21-66 A joint X,Y distribution is a bivariate normal distribution when the row, column, and marginal distributions are all __________.

normal

21-67 Regression analysis is applied to data collected by the __________ method. The X variable is always a __________ variable, and the Y values fall into __________ on this variable.

experimental; discrete
subgroups

21-68 The correlational method is designed for studying the relation between two __________ variables neither of which can be __________ experimentally.

continuous
manipulated

21-69 In a correlation problem, attention is focused on the value of the statistic ______. In a regression problem, this statistic is useful mainly because it indicates something about the __________ of the function relating Y and Y.

r

form

21-70 The predictor variable is always X in __________ analysis. It may be either X or Y in __________ analysis.

regression
correlation

21-71 For predictions beyond the sample in regression analysis, the best prediction of Y when X_j is known is the estimated __________ of the __________ at level A_j.

mean; subpopulation

21-72 If $\rho^2 = 0$, the best estimate of $\mu + a_j$ is ______. If $\rho^2 \neq 0$, the best estimate of $\mu + a_j$ is ______.

$\bar{Y}_j$
Y'_j

21-73 For predictions beyond the sample in correlation analysis, the best prediction of Y_i when $\rho = 0$ is ______. When $\rho \neq 0$ and X_i is known, the best prediction of Y_i is ______.

$\bar{Y}$
Y'_i

21-74 Accuracy in prediction may be used to measure the __________ of an association.

strength

21-75 This is the main reason for discussing predictive accuracy in a __________ problem. In a __________ problem, predictions are more often put to practical use.

regression; correlation

21-76 In order to determine the raw-score regression equations, five sample statistics are required. These are ______, ______, ______, ______, ______.

$\bar{X}$
$\bar{Y}$; s_X; s_Y; r

21-77 Both raw-score regression equations pass through that point which represents the __________ of both X and Y distributions. — *mean*

21-78 The slope of the regression equation for predicting Y is ______ times s_Y/s_X. The slope of the regression equation for predicting X is ______. — r; $r\frac{s_X}{s_Y}$

21-79 The statistic r is the average cross product of the paired __________. The z-score definition of r is therefore ______. — *z scores*; $r = \frac{\Sigma z_X z_Y}{N}$

21-80 The raw-score definition of r is $\Sigma XY/N - \bar{X}\bar{Y}$ divided by ______. The quantity in the numerator is called the __________ of X and Y. — $s_X s_Y$; *covariance*

21-81 The covariance of X and Y is analogous to the __________ of Y. — *variance*

21-82 Where the formula for the variance has $\Sigma Y^2/N$, the formula for the covariance has ______. Where the variance has $\bar{Y}^2$, the covariance has ______. — $\frac{\Sigma XY}{N}$; $\bar{X}\bar{Y}$

21-83 The computing formula for r contains N and five sums which can be obtained simultaneously on a desk calculator. These five sums are ______, ______, ______, ______, ______. — ΣY; ΣY^2; ΣX; ΣX^2; ΣXY

21-84 The slope constant for both *z-score* regression equations is ______. Both equations pass through the point with coordinates (______, ______). — r; *(0,0)*

21-85 The word "regression" is used to describe the regression equations because of the phenomenon of regression toward the __________. — *mean*

21-86 Whenever r is not equal to $+1$ or -1, the *predicted* score Y_i' will lie __________ to the mean $\bar{Y}$ than the predictor score X_i lies to its mean $\bar{X}$. Our prediction moves back, or __________, toward its mean. — *closer*; *regresses*

21-87 When the z_X and z_Y points have a trend which is upward to the right, r will have a __________ sign. When their trend is downward to the right, r will have a __________ sign. — *positive*; *negative*

21-88 When the points are clustered close to a straight line, the *absolute* value of r will be close to ______. When the points are widely scattered, the value of r will be close to ______. — *1*; *0*

PROBLEMS

1. To check your understanding of the defining and computing formulas, take the following set of paired X,Y values:

X	65	55	50	45	35
Y	68	52	64	60	76

a. Find, using a desk calculator, the values of ΣX, ΣX^2, ΣY, ΣY^2, and ΣXY.

b. Use these terms in the computing formulas to find $\bar{X}$, $\bar{Y}$, s_X, s_Y, and the covariance of X and Y.

c. Find the value of the Pearson r for these raw scores by each of the following methods: (1) the computing formula on page 329; (2) the defining formula for raw scores on page 328; and (3) the average cross product of z scores. For method 3, you will have to convert all 10 scores into their z score equivalents. Compare the results and efficiency of these methods.

2. For the regression problem given as Prob. 1 on page 314, calculate the Pearson r by the computing formula for raw scores. Find the raw-score regression equation for predicting Y from X.

3. The correlation between two variables X and Y is +.6; $s_X = 4.0$, $s_Y = 3.6$, $\bar{X} = 30$, and $\bar{Y} = 40$. Find the regression equation of Y on X and the regression equation of X on Y.

4. *a.* For the set of five data points in Fig. 21-5*a* (page 326), find the value of $\Sigma z_X z_Y$.

b. For each observed z_Y in this set, find a predicted $z'_Y = +.6z_X$, and then take the sum of the squared deviations $\Sigma(z_Y - z'_Y)^2$ for the set of five points.

c. Recall that $\Sigma(z_Y - z'_Y)^2/N = 1 - r^2$ when r is taken as $\Sigma z_X z_Y/N$. Find $1 - r^2$, and use it to determine $\Sigma(z_Y - z'_Y)^2$. Compare the value obtained with the value of the same statistic found in part *b* above.

d. Now consider another linear function, the one which will pass directly through the three data points (+1.5,+0.5), (0,0), and (−1.5,−0.5); this line has the equation $z''_Y = \frac{1}{3} z_X$. Find the predicted z''_Y values for this equation, and determine the sum of squared deviations of points from this line, $\Sigma(z_Y - z''_Y)^2$. Which of these two linear functions has the lower sum of squared deviations?

5. Calculate the Pearson r for the correlation between error scores on mazes X and Y, using the raw scores given in Fig. 21-1 (page 317). Write the regression equation for the regression of Y on X and the equation for the regression of X on Y.

LESSON 22
INTERPRETING THE PEARSON r

The correlation coefficient r is not an easy number to understand. Its interpretation is relatively simple when it takes the values $+1$, -1, or 0, but these values of r do not occur frequently. For interpreting its more common values, you will need a few additional concepts and a considerable amount of experience.

The first thing to keep in mind is that the Pearson r is always a sample statistic. It is in the same class with a sample mean and sample standard deviation. The Pearson r is subject to sampling variability like all descriptive statistics. It is only an estimate of the population parameter ρ.

When we know the statistic r for a sample, we know the amount of linear association between X and Y *in the sample studied.* However, the relation between r and amount of association is not like the relation between a thermometer reading and temperature. An r of .5 does not usually mean just half as much association as $r = 1.0$, and a change in r from .1 to .2 does not represent as large an increase as a change from .8 to .9. Even to understand what r can tell us about the sample, we must bear in mind its relation to a best-fitting straight line for the sample observations.

Lesson 21 showed how predictions within and beyond the sample depend upon the regression equations. The statistic r is most easily understood in terms of its effect on predictions within the sample. We shall use the relative accuracy—and the relative inaccuracy, or *error*—of such predictions to give the possible values of r some concrete meaning.

A. RELATIVE ERROR IN RELATION TO r^2

Strength of association can be measured in terms of the amount of error which is removed by using the association as a basis for prediction. In Lesson 12 (page 186) we defined total error as the amount of error which is expected when the prediction is made without using the association. We defined residual error as the amount of error which remains even though the association is used. Then we set up a pattern for measuring strength of association:

$$\text{Proportion of error removed} = 1 - \frac{\text{residual error}}{\text{total error}}$$

For qualitative data, the index of predictive association λ measures strength of association according to this pattern. For quantitative data, the *square* of the Pearson r fits this pattern, as we shall now see.

The correlation of .3 between SAT score and Oberlin grades, presented in Fig. 21-2 (page 322), will illustrate how r^2 measures relative predictive error. We can use the regression equation $Y' = .00215X + 1.32$ to find a "predicted" grade average for each one of the 547 students in the sample. Then we can compare the prediction with the actually observed score and calculate the relative error of prediction. Just as in the linear regression analysis of single-factor experiments (Lesson 20), we can determine the deviation of each Y score from the grand mean $\bar{Y}$, and we can then regard this deviation $Y_i - \bar{Y}$ as the sum of the two component deviations $Y_i - Y_i'$ and $Y_i' - \bar{Y}$. The error in our prediction for student i is the deviation $Y_i - Y_i'$, the difference between his actual and his predicted score. Without the association, this error would be $Y_i - \bar{Y}$. Thus, $Y_i - \bar{Y}$ is his contribution to total error, $Y_i - Y_i'$ his contribution to residual error, and $Y_i' - \bar{Y}$ his contribution to error removed by the association.

The computer has made this analysis for each of our 547 cases, but we shall examine only a few cases—just enough to make the point clear. Table 22-1 shows the breakdown of scores for those seven students whose SAT scores fell into the class 451 to 500 in Fig. 21-2.

Table 22-1 Analysis of Scores of Seven Students from Fig. 21-2

Student Number i	SAT Score X	"Predicted" Average Y'	Actual Average Y	$Y - Y'$	$Y' - \bar{Y}$	$Y - \bar{Y}$
54	457	2.30	2.12	−0.18	−0.44	−0.62
131	499	2.39	2.59	+0.20	−0.35	−0.15
213	482	2.35	1.85	−0.50	−0.39	−0.89
229	479	2.35	2.49	+0.14	−0.39	−0.25
370	500	2.39	1.96	−0.43	−0.35	−0.78
460	460	2.31	1.74	−0.57	−0.43	−1.00
468	451	2.29	3.37	+1.08	−0.45	+0.63

The first column in the table shows the index number assigned by computer to the individual students. This index number is our familiar subscript i: when $i = 54$, $X_i = 457$, and $Y_i = 2.12$. By coincidence the student with index number 460 also had an SAT score of 460.

The third column shows the grade average which would be "predicted" for each student on the basis of the regression equation $Y' = .00215X + 1.32$. Of course, these values are not true predictions; the actual Y scores were already known and were used in finding the regression equation. Y' could properly be called a "postdiction," a prediction made after the fact and purely for purposes of measuring the strength of the association between X and Y.

The last three columns in Table 22-1 show the deviation scores and their components. Observe that in each case $(Y - Y') + (Y' - \bar{Y}) = Y - \bar{Y}$.

Figure 22-1 presents the data for these seven students as a graph. All the values of Y' are points on the regression line. The three deviation scores for student number 460 are indicated by arrows in the graph. For all other students, only a single arrow has been drawn to show the deviation $Y - Y'$.

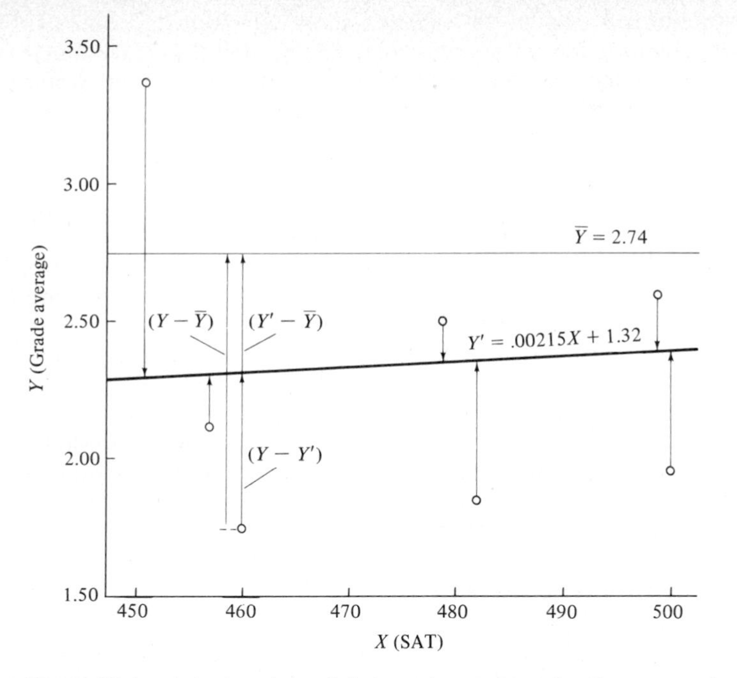

Figure 22-1: Actual and predicted grade averages for the seven students in Table 22-1. Unlabeled arrows show the deviation $Y - Y'$ for each student; for student number 460, labeled arrows show all three deviations $Y - \bar{Y} = (Y - Y') + (Y' - \bar{Y})$.

22-1 The deviation $Y - Y'$ is the distance between the actual grade average and the ________ average. This distance is greatest for student number ______ in Table 22-1.

predicted
468

22-2 The deviation $Y - \bar{Y}$ is the distance between the actual grade average and the ________ of the Y distribution. This distance is greatest for student number ______.

mean
460

22-3 If we did not know the value of X for these students, we would have to predict the same grade average for all 547 students. The best long-run prediction would be ______.

2.74

22-4 For student 468, the distance $Y - \bar{Y}$ is (*larger; smaller*) than the distance $Y - Y'$. In this case $\bar{Y}$ would be closer to the observed Y than the prediction made from knowledge of X.

smaller

22-5 Consider each of the seven students in this group. The distance $Y - \bar{Y}$ is *smaller* than the distance $Y - Y'$ for ______ of the seven students; for the other ______ students, Y' is a closer estimate of the observed Y.

2
5

When we square and sum these deviation-score components for all 547 students, we obtain the sums of squares SS_L, SS_{W+D}, and SS_Y. Without a computer to calculate the 547 predicted scores, such a calculation would be extremely tedious. Fortunately, we know from Lesson 20 that we need only the total sum of squares SS_Y, which can be calculated from $\sum_1^N Y_i^2 - \frac{G^2}{N}$. We have already seen (page 308) that $SS_L = r^2 SS_Y$ and that $SS_{W+D} = (1 - r^2)SS_Y$.

These sums of squares will lead us directly to a measure of strength of *linear association*. Total error is measured in terms of the amount of error which would be made without knowledge or use of the linear association. Without knowledge of X, we would predict $\bar{Y} = 2.74$ for all students; therefore, the deviation $Y - \bar{Y}$ is the amount of total error for any given student, and $\Sigma(Y - \bar{Y})^2 = SS_Y$ is the measure of total error for all 547 students.

Residual error is measured in terms of the amount of error which is still present even when the association is used to aid prediction. With knowledge of X, we predict Y' for each student; the deviation $Y - Y'$ is the amount of residual error for any given student, and $\Sigma(Y - Y')^2 = SS_{W+D}$ is the measure of residual error for all 547 students. Thus,

$$\text{Proportion of error removed} = 1 - \frac{SS_{W+D}}{SS_Y} = \frac{SS_Y - SS_{W+D}}{SS_Y} = \frac{SS_L}{SS_Y}$$

Since $SS_L = r^2 SS_Y$, we find that r^2 itself is the index of strength of linear association. $r^2 SS_Y$ tells the *amount* of predictive error which is removed by the association, and r^2 tells the *proportion* of error removed.

It is customary to state these facts in terms of the sample variance s_Y^2. $s_Y^2 = SS_Y/N$ represents total variance in the sample, and it can serve instead of SS_Y as a measure of total error. Then $r^2 s_Y^2$ can be interpreted as the amount of sample *variance* which no longer produces predictive error when the linear association is used. We say that r^2 represents the proportion of sample variance which is *accounted for* by the linear association between X and Y. Then $1 - r^2$ of course represents the proportion of Y variance which is not accounted for; it is still error variance.

The words "accounted for" may trap an unwary student into believing that r^2 represents the proportion of variance in grades which is *caused by* variations in aptitude. But the variance is "accounted for" only in a statistical sense—in the sense that this variance is no longer a source of predictive error. To avoid the temptation of a causal interpretation, always remind yourself that the proportion r^2 applies equally to both directions of prediction. The proportion of X (SAT) variance accounted for by the linear association with Y (grades) is also equal to r^2. Predicting X from Y may not make as much practical sense in this case, but it is statistically just as meaningful as predicting Y from X. r^2 is the proportion of variance in *each* variable which is accounted for by their linear association.

22-6 If we undertake to "predict" SAT scores within our sample, a measure of (*residual; total*) error in such predictions is given by the variance of the X distribution, s_X^2. — *total*

22-7 We can use the linear association to aid in this prediction of X. When we do, the measure of ______ error will be $(1 - r^2)s_X^2$. — *residual*

22-8 Strength of association = 1 − (residual error)/(total error). For predicting X from Y, the equation will read: strength of association = 1 − ______. — $\dfrac{(1 - r^2)s_X^2}{s_X^2}$

22-9 Since s_X^2 is present in both numerator and denominator, the index becomes $1 - (1 - r^2)$, which equals ______. — r^2

22-10 As a measure of strength of association, ______ represents the proportion of variance in each variable which is accounted for by the ______ association with the other variable. — r^2; *linear*

22-11 Fill in the values of r^2:

r	r^2
+.1	______
+.3	______
+.5	______
−.7	______
+.9	______
−.95	______

.01
.09
.25
.49
.81
.90

22-12 Your table shows that a correlation of plus or minus ______ is required to account for about half the variance and a correlation of plus or minus ______ is required to account for 90 percent of the variance. — *.7*; *.95*

22-13 The correlation between SAT score and grade average in our example is .3. The linear association with SAT scores accounts for ______ percent of the variation in grade average in this sample; ______ percent of this variation remains unaccounted for. — *9*; *91*

22-14 This same linear association accounts for ______ percent of the variation in SAT scores in the sample. — *9*

B. THE STANDARD ERROR OF ESTIMATE

We have just seen that the statistic r^2 divides sample variance into two parts, one which is accounted for by the linear association and another which must still be attributed to error. When we are predicting Y, sample variance is s_Y^2; the proportion accounted for is $r^2 s_Y^2$, and the proportion still unaccounted for is $(1 - r^2)s_Y^2$. These facts apply equally to regression and correlation problems.

Because variance is expressed in different units, it cannot be visualized in relation to a scatter diagram of the X,Y distribution. For this reason, it may be more meaningful to state these facts in relation to the square root of sample variance, the sample standard deviation. $\sqrt{s_Y^2} = s_Y$ is the standard deviation of a set of Y values from their mean $\bar{Y}$. After regression or correlation analysis, we can describe the standard deviation of these same Y values from the regression line of Y on X; this standard deviation is $\sqrt{(1 - r^2)s_Y^2} = s_Y\sqrt{1 - r^2}$, and it is commonly called the **STANDARD ERROR OF ESTIMATE**. Since this statistic is the standard deviation which remains, although knowledge of the linear association with X is given, we designate it as the standard error of Y *given* X, $s_{Y|X}$.

In correlation problems we may also wish to know the standard error of X given Y, $s_{X|Y}$. The standard deviation of the X values without information about the linear association is s_X, the standard deviation of the X distribution around its mean $\bar{X}$. The standard deviation of these same X values from the regression line of X *on* Y is the standard error of estimate $s_{X|Y} = s_X\sqrt{1 - r^2}$.

Table 22-2 illustrates how these statistics vary with the value of r. Calculations have been made for five different X,Y distributions. Each distribution has $\bar{X} = \bar{Y} = 50$, $s_Y = 5$, and $s_X = 10$. Each of the five distributions has a different r, resulting from a different pairing of the X and Y values.

Table 22-2 Sample Statistics Used in Interpreting the Pearson *r*

r	Y on X	$r^2 s_Y^2$	$s_Y\sqrt{1-r^2}$	X on Y	$r^2 s_X^2$	$s_X\sqrt{1-r^2}$
.1	$Y' = .05X + 47.5$	.25	4.97	$X' = .2Y + 40$	1	9.95
.3	$Y' = .15X + 42.5$	2.25	4.77	$X' = .6Y + 20$	9	9.54
.5	$Y' = .25X + 37.5$	6.25	4.33	$X' = 1.0Y$	25	8.66
.7	$Y' = .35X + 32.5$	12.25	3.57	$X' = 1.4Y - 20$	49	7.14
.9	$Y' = .45X + 27.5$	20.25	2.18	$X' = 1.8Y - 40$	81	4.36

Never forget that $s_{Y|X}$ and $s_{X|Y}$ are descriptive sample statistics, like r itself. The two standard errors of estimate are the standard deviations of actually observed sample points from the two regression lines when those lines have been chosen according to the least-squares principle. These sample statistics do *not* indicate the amount of error to be expected when predictions beyond the sample are attempted.

22-15 The first distribution in Table 22-2 has a Pearson $r =$ ______. Since $s_Y^2 = 25$, the variance accounted for by the linear association with X is $r^2 s_Y^2 =$ ______.

.1
.25

22-16 Total Y variance is 25; .25 is accounted for. The unaccounted-for variance is $(1 - r^2)s_Y^2 = s_Y^2 - r^2 s_Y^2 =$ ______.

24.75

22-17 The square root of $(1 - r^2)s_Y^2$ is called the standard ______ of ______, $s_{Y|X}$. Its value is ______ for this distribution.

error; estimate; 4.97

22-18 $s_{Y|X}$ can be compared directly with the sample standard deviation s_Y. In this case, $s_Y =$ ______ and $s_{Y|X} =$ ______.

5; 4.97

22-19 s_Y is the standard deviation of this Y distribution around ______. $s_{Y|X}$ is the standard deviation of the same Y values around the regression line $Y' =$ ______.

$\bar{Y}$
$.05X + 47.5$

22-20 The second distribution in Table 22-2 has $r =$ ______. For this distribution, $s_{Y|X} =$ ______. This is the standard deviation of these Y values from the line ______.

.3
4.77
$Y' = .15X + 42.5$

22-21 For this second distribution, the standard deviation of the X values around $\bar{X}$ is ______; the standard deviation of these same values around $X' = .6Y + 20$ is ______.

10
9.54

22-22 As r increases from .1 to .5, the slope of the regression line of Y on X (*decreases; increases*). The proportion of Y variance accounted for (*decreases; increases*). The standard deviation of Y values around their regression line (*decreases; increases*).

increases
increases
decreases

22-23 For all five distributions, $\bar{Y} =$ ______ and $s_Y =$ ______. For all five distributions, $\bar{X} =$ ______ and $s_X =$ ______. The only difference among the five X,Y distributions is in the way the X values are ______ with the Y values.

50; 5
50; 10
paired

22-24 In the first distribution, X and Y values are paired almost randomly. There is only a slight tendency for high X values to go with (*high; low*) Y values. The regression lines are not steep, and the amount of scatter of points around these lines is (*great; small*), as indicated by the standard errors of ______.

high
great; estimate

22-25 In the last distribution, there is a strong tendency for high X values to go with ______ Y values. The regression lines are steep, and the amount of scatter around these lines is ______.

high
small

C. USING r TO MAKE INFERENCES ABOUT THE POPULATION

The sample statistic r is only an estimate of the population correlation coefficient ρ. Since we do not know the exact value of ρ, it follows that we also do not know the population parameters which correspond to r^2, $s_{Y|X}$, $s_{X|Y}$, and the regression constants m and b for either regression line. Consequently, we do not know how much the linear association can reduce $\sigma_Y{}^2$ or $\sigma_X{}^2$, nor do we know the population regression equations which would be desirable for making estimates of *future* Y scores (or future X scores).

Inferences about the population correlation coefficient and related parameters can generally be made with confidence only when the X,Y distribution is approximately a bivariate normal distribution. Such distributions are rarer than one might suppose. In most of the cases for which you will calculate a Pearson r, you should be content to take it only as a useful index of the amount of linear association you have actually observed.

When the X,Y distribution is approximately a bivariate normal distribution, a statistical test can be used to determine whether the true value of the population correlation coefficient is different from 0. The ratio

$$t = \frac{r\sqrt{N-2}}{\sqrt{1-r^2}}$$

is a t statistic with $N - 2$ degrees of freedom. It provides either one- or two-tailed tests of H_1: $\rho = 0$. The test should be one-tailed when there is a clear advance expectation about the sign of ρ, that is, when H_2 asserts either $\rho > 0$ or $\rho < 0$. Since, for such a bivariate normal population, only linear association is possible, accepting H_1 implies that X and Y are independent in this population.

This t ratio is precisely the same statistic used in Lesson 20 for tests of zero linear regression. We wrote the statistic at that time as an F ratio:

$$F = \frac{MS_L}{MS_{W+D}} = \frac{SS_L/1}{SS_{W+D}/(N-2)} = \frac{r^2 SS_Y(N-2)}{(1-r^2)SS_Y}$$

This F statistic is the square of the t ratio given above; $F = r^2(N-2)/(1-r^2)$ with df$(1, N-2)$.

The test for zero linear regression arose from further analysis of a single-factor experiment. It therefore assumes that the usual requirements for analysis of variance have been met: independence of observations, approximate population normality for the Y distribution, and approximately equal subgroups. But since the investigator controls the distribution of X values used in a single-factor experiment, no assumption is made about the shape of the population distribution from which these X values are drawn. That distribution could occasionally be normal, but usually it is not.

The conditions required for inferences about ρ are thus more restrictive in correlation problems than in regression problems. Accordingly, the conclusions which can be drawn are more sweeping when the conditions are met. In regression analysis, accepting H_1: $\rho^2 = 0$ does not automatically mean that X and Y are independent; one may still look for nonlinear forms of association.

22-26 To test whether $\rho = 0$ in correlation analysis, we must have an X,Y distribution which is approximately a ________ distribution. *bivariate normal*

22-27 To test whether $\rho^2 = 0$ in regression analysis, we must have data which meet the three requirements for ________ of ________. *analysis variance*

22-28 These three requirements do not include any statement about the population distribution of the $(X;Y)$ variable. *X*

22-29 When we have a bivariate normal distribution, we test $H_1: \rho = 0$ by using a ______ statistic with ______ degrees of freedom. *t; $N - 2$*

22-30 The only variables in this t ratio are ______ and ______. Therefore, for any given value of r, the outcome of the test will depend upon the ________ of the sample. *$N \leftrightarrow$ r size*

22-31 For $r = .3$, $\sqrt{1 - r^2} = .95$. If we obtain $r = .3$ for a bivariate normal distribution with $N = 123$, the t statistic is $t = r\sqrt{N-2}/\sqrt{1-r^2} =$ ______, with df = ______. *3.47; 121*

22-32 When df = 100, the critical value of t at $\alpha = .002$ (two-tailed) is ______. With $N = 123$ and $r = .3$, $H_1: \rho = 0$ can be ________ by a two-tailed test at $\alpha = .002$. *3.160 rejected*

22-33 If we obtain $r = .3$ for a bivariate normal distribution with $N = 38$, the t statistic is $t =$ ______ with df = ______. *1.89; 36*

22-34 At df = 30, the critical value of t at $\alpha = .05$ is ______ for a two-tailed test and ______ for a one-tailed test. *2.042 1.697*

22-35 With $N = 38$ and $r = .3$, we could reject H_1 at $\alpha = .05$ by a ________-tailed test, but we could not reject H_1 at this level by a ________-tailed test. *one two*

22-36 If we obtain $r = .3$ for a bivariate normal distribution with $N = 27$, $t =$ ______ with df = ______. *1.58; 25*

22-37 At df = 25 the critical value of t at $\alpha = .05$ (one-tailed) is ______. With $N = 27$ and $r = .3$, we must (*accept; reject*) H_1 at $\alpha = .05$. *1.708; accept*

22-38 The only form of association possible in a bivariate normal distribution is a ________ association. If $H_1: \rho = 0$ must be accepted, it can be concluded that X and Y are ________. *linear independent*

D. INTERPRETING *r* IN REGRESSION AND CORRELATION PROBLEMS

Regression and correlation analysis require some of the same calculations. The Pearson r is fundamental to both forms of analysis. To decide whether a significant amount of linear association is present, we use a single statistic based on r, sometimes in the form of an F ratio and sometimes as $\sqrt{F} = t$. In correlation analysis, the t statistic is more commonly used; in regression problems, the F statistic is ordinarily used unless H_2 specifies the sign of the correlation coefficient.

The similarity in procedures does not mean that the results of regression and correlation analysis can be interpreted in the same way. There are important differences. Some of these differences arise from the different conditions which must be met before the significance test can be performed; we have just discussed these conditions. In both forms of analysis, a significant Pearson r means that some amount of linear association exists between X and Y. But correlation analysis requires a bivariate normal population in order for such a test to be valid. Thus when r is not significant, correlation analysis permits the conclusion that X and Y are independent; from regression analysis we conclude only that any relation present between X and Y must be nonlinear.

Some of the differences in interpretation stem from differences between the correlational and experimental methods. Regression analysis is based on experimental method; individuals have been randomly assigned to subgroups on X, and differences among the subgroups may have been literally brought about by differences in X. The proportion of Y variance accounted for by X may often be interpreted, therefore, as the proportion of Y variation caused by variation in X.

On the other hand, correlational method permits no such causal interpretation. We can say that r^2 is the proportion of variance in *either* variable which is accounted for by the linear association; we cannot say that either variable has brought about changes in the other variable.

The difference in method also affects the usefulness of r as a descriptive sample statistic. Correlational method produces data which are representative of the natural relation between X and Y. When the bivariate normal condition is met, the Pearson r in correlation analysis is more likely to describe the way X and Y covary in nature. Even when this condition is obviously not met, as in our distribution of grades and aptitude scores, the sample statistics from a correlation analysis with large N will provide a reliable indication of the shape of the X,Y distribution in the population. Experimental method of course precludes any such interpretation of the sample statistics in regression analysis.

22-39 In both correlation and regression analysis, a significant Pearson r means that some ________ association is present. *linear*

22-40 If r is not significant in correlation analysis, it can be concluded that X and Y are ________. If r is not significant in regression analysis, it can be concluded that any association which exists between X and Y is ________. *independent* *nonlinear*

22-41 This difference in meaning cannot arise from a difference in the statistical procedures used; these procedures are the same. The difference arises from a difference in the ________ which must be met before the significance of r can be tested. *conditions*

22-42 When r has been calculated, we can say that the linear association ________ for a certain proportion of the Y variance; this proportion is equal to ________. *accounts* r^2

22-43 Because regression analysis is based on ________ method, we may sometimes be able to say that this proportion of Y variance is actually ________ by the variation in X. *experimental* *caused*

22-44 If correlation analysis showed that $\rho = .5$ between college grades and aptitude scores, we (*could; could not*) conclude that 25 percent of the variation in grades is caused by differences in aptitude. *could not*

22-45 The Pearson r is more likely to indicate how X and Y covary in nature when it is obtained through ________ analysis. *correlation*

22-46 In Lesson 21 we discussed how the error scores of 94 rats on two mazes might be studied by correlational and experimental method. If we want to know the population correlation coefficient between scores on these two mazes, we shall get a better estimate by using the ________ method. *correlational*

22-47 If the experimental method is used, we (*can; cannot*) calculate a value of r for this association, and we (*can; cannot*) test its statistical significance. *can* *can*

22-48 But the r statistic may differ from the population coefficient ρ because the X distribution in the sample is not ________ of the X distribution in the population. *representative*

22-49 The value of r is of more interest as a sample statistic in ________ analysis. It is of more interest for what it tells about the form of the association in ________ analysis. *correlation* *regression*

E. THE SPEARMAN RANK CORRELATION COEFFICIENT

Interpretation of the Pearson r can be further illuminated by contrasting it with the product-moment correlation coefficient calculated for *ordinal* data. With such data, the Y distribution is the set of consecutive integers from 1 to N, the X distribution is this same set of numbers, and the joint X,Y distribution is some particular pairing of the X and Y ranks. A set of consecutive integers has peculiar properties; it is meaningless to speak of the shape of the distribution or of its variance. Therefore, the product-moment correlation coefficient for a joint distribution of ranks is given a special name. It is called the SPEARMAN RANK CORRELATION COEFFICIENT, and we shall symbolize it as r_S.

When there are no ties in the ranks on X or Y, r_S should be calculated by a simplified formula

$$r_S = 1 - \frac{6 \sum_{1}^{N} d^2}{N(N^2 - 1)}$$

where d represents the rank difference $X_i - Y_i$ for each of the N individuals. If ties are present in the rank orders, it is more accurate to use the standard procedure for calculating a Pearson r, treating the rank numbers as if they were raw scores.

The meaning of r_S can be suggested by an example. Fifty girls are drawn at random from a college population; each is given a rank on beauty (X) and a rank on college grades (Y). The coefficient r_S indicates the extent of agreement between the two rank orders. r_S may also be interpreted to give information about the association of X and Y in the parent population. If beauty and grades are not associated in this population, then all the $N!$ possible pairings of X and Y ranks are equally likely to occur and the expected value of r_S is 0. However, r_S may also be approximately 0 when the association between beauty and grades is NONMONOTONIC. Therefore, a test of $H_1: r_S = 0$ is a test for a MONOTONIC ASSOCIATION between X and Y.

Monotonic functions are of two kinds. If Y always increases as X increases, Y is an INCREASING MONOTONIC FUNCTION of X. If Y always decreases as X increases, Y is a DECREASING MONOTONIC FUNCTION of X. But if the best grades tend to be made by girls of middling beauty, the function is nonmonotonic; Y first increases, then decreases with increasing X. Other nonmonotonic forms are also possible.

We can test H_1: $r_S = 0$ by examining the probability distribution of r_S when H_1 is true. For $N < 10$, exact probability tables are available for this distribution. For $N = 10$ or more, the distribution of $r_S\sqrt{N-2}/\sqrt{1-r_S^2}$ is approximately a t distribution with $N - 2$ degrees of freedom. Since the shape of a population distribution cannot be reflected in ordinal data, this test does not require a bivariate normal X,Y population.

Thus the conditions for a test of H_1: $r_S = 0$ are less restrictive than those for a test of H_1: $\rho = 0$. The conclusions which can be drawn are also less specific. A significant r_S cannot tell us whether the association is linear; it can only tell us that it is monotonic.

22-50 The Spearman correlation coefficient r_S is obtained by calculating a Pearson r for ________ data. *ordinal*

22-51 There is a simplified formula for calculating r_S which can be used when there are no ________ in the ranks on X or Y. *ties*

22-52 Interpreted as a sample statistic, r_S shows the extent of ________ between the rank orders on X and Y. *agreement*

22-53 If X and Y are not associated in the population, the value of r_S is expected to be ________. *0*

22-54 When N is at least ________, the test of $H_1: r_S = 0$ is based on a statistic parallel to the statistic $t = r\sqrt{N-2}/\sqrt{1-r^2}$. The statistic for testing r_S is ________ with ________ df. *10*; $\frac{r_S\sqrt{N-2}}{\sqrt{1-r_S^2}}$; $N-2$

22-55 The test of H_1: $r_S = 0$ (*does; does not*) depend on the shape of the parent population. *does not*

22-56 The population distribution of grades (in the example) may be skewed; this skewness (*will; will not*) be reflected in the distribution of ranks on Y. *will not*

22-57 The population distribution of beauty may be normal; it is certainly unknown. Its distribution (*can; cannot*) affect the distribution of ranks on X. *cannot*

22-58 When r_S is significantly different from 0, we conclude that a ________ association between X and Y is present in the (*population; sample*). *monotonic* *population*

22-59 When the Pearson r is significantly different from 0, we conclude that a ________ association between X and Y is present in the population. *linear*

22-60 All linear functions are monotonic. If the Pearson r is significant, we know that r_S for ranks taken from the same data (*will; will not*) be significant. *will*

22-61 Some monotonic functions are nonlinear. $Y = X^3$ is a (*linear; nonlinear*) function, but Y must increase as X increases; this function is therefore (*monotonic; nonmonotonic*). *nonlinear* *monotonic*

22-62 When r_S is not significant, there are two possibilities. Either X and Y are ________, or they are related by a ________ function. *independent* *nonmonotonic*

22-63 When the Pearson r is not significant, X and Y must be ________. *independent*

REVIEW

22-64 Each of two judges ranks a group of 10 girls in a beauty contest. The statistic r_S could be used to measure the extent of __________ between the two rank orders. *agreement*

22-65 r_S is the Spearman __________ correlation coefficient. It is obtained by calculating a Pearson r for a set of N pairs of __________ on X and Y. *rank* *ranks*

22-66 We could convert the 547 Oberlin grade averages into a set of ranks. The distribution of grades is negatively skewed; the set of ranks (*would; would not*) be affected by this skewness. *would not*

22-67 The corresponding set of 547 SAT scores is also negatively skewed. With such a large sample, we can be sure that the parent X,Y population (*does; does not*) have a bivariate normal distribution. *does not*

22-68 This sample does not meet the conditions required for a statistical test of H_1: $\rho =$ ________. We can apply such a test with confidence only when the row, __________, and __________ distributions for the sample are all approximately __________. *0* *column* *marginal* *normal*

22-69 If we convert both the grade and the SAT distributions into sets of ranks, we can calculate the coefficient r_S. We (*can; cannot*) test H_1: $r_S = 0$ without a bivariate normal population. *can*

22-70 If r_S is significantly different from 0, we can conclude that grades and aptitude scores are related by a __________ function in this population. *monotonic*

22-71 If r_S is *not* significantly different from 0, we (*can; cannot*) conclude that grades and aptitude scores are independent in this population. *cannot*

22-72 When $r_S = 0$, X and Y may be independent, but it is also possible that they are related by a __________ function. *nonmonotonic*

22-73 *If the sample had met the bivariate normal condition,* we could have applied a statistical test to the Pearson r. If we found r to be significantly different from 0, we could conclude that grades and aptitude scores are related by a __________ function in this population. *linear*

22-74 The bivariate normal distribution of the population would have excluded any form of association except a ________ one. *linear*

22-75 In that case, if we found r *not* significantly different from 0, we could conclude that grades and aptitude scores are ________ in this population. *independent*

22-76 Because the conditions for this test were not met, we (*can; cannot*) be sure that the *only* form of association between grades and aptitude scores is a linear one. *cannot*

22-77 However, with $r = .3$, we can state that a linear association between grades and aptitude scores accounts for ________ percent of the variance in grades in this (*population; sample*). *9* *sample*

22-78 In order to account for 25 percent of the sample variance, we would have to obtain an r of at least ________. *.5*

22-79 We could also use the standard error of estimate to describe the scatter of the 547 grade averages around their ________. *regression line*

22-80 This statistic is the standard ________ of estimate for Y, given X. It is designated ________, and it is equal to ________. *error* $s_{Y|X}$; $s_Y \sqrt{1 - r^2}$

22-81 The statistic $s_{X|Y}$ describes the scatter of the (X,Y) values around the regression line of ________ on ________. *X* *X; Y*

22-82 When we are interested in obtaining a good estimate of ρ for the population, we would usually choose to do a study using the ________ method. *correlational*

22-83 When we are interested in showing that variation in X has a causal influence upon Y, we would usually choose to use the ________ method. *experimental*

22-84 In regression analysis, if ρ^2 is significantly different from 0, we can conclude that (*all; some*) of the effects of X can be described by a linear function. *some*

22-85 In regression analysis, if ρ^2 is *not* significantly different from 0, we (*can; cannot*) conclude that X and Y are independent. *cannot*

22-86 When $\rho^2 = 0$, X and Y may be independent, but it is also possible that they are related by a ________ function. *nonlinear*

PROBLEMS

1. The standard deviation of a certain Y distribution is 10. What is the standard error of estimating Y from X when the correlation coefficient is $+.714$?

2. In order to obtain a value of $s_{Y|X}$ which is only 40 percent of s_Y, how large must r be? Could this r be negative?

3. With a correlation coefficient of $+.37$, what proportion of variance in Y is accounted for by variation in X? What proportion remains unaccounted for? What is the size of $s_{Y|X}$, relative to s_Y?

4. The correlation coefficient between X and Y is $+.4$ in the maze example used in Lesson 21. What is the proportion of total variance which is contributed by variance due to linear regression? What proportion of total variance is contributed by variance due to error? How much of the variation in Y is accounted for by variation in X?

5. Calculate r_S by the simplified formula for each of the following sets of paired ranks:

a.

X	1	2	3	4
Y	1	3	4	2

b.

X	1	2	3	4
Y	3	2	1	4

c.

X	1	2	3	4
Y	2	4	1	3

d.

X	1	2	3	4
Y	4	1	2	3

6. Calculate r_S to determine the extent of agreement between the rankings given by judges X and Y in the example on page 90. Use the simplified formula for r_S, then use the standard computing formula for the Pearson r. Which method is more accurate, and how much difference does it make?

7. Find r_S, and write a regression equation for predicting *rank on* Y from *rank on* X for the following set of paired ranks. Since a set of ranks has no variance, assume that the two sets of nine consecutive integers have equal variance, and consider s_Y/s_X as equal to 1.

X	1	2	3	4	5	6	7	8	9
Y	1	3	5	8	2	7	9	4	6

APPENDIX

Tables
Answers to Problems
References

TABLES

Table A-1 Areas under the Normal Curve†

$z = \frac{x}{\sigma}$	Area (Mean to z)	$z = \frac{x}{\sigma}$	Area (Mean to z)	$z = \frac{x}{\sigma}$	Area (Mean to z)	$z = \frac{x}{\sigma}$	Area (Mean to z)
.05	.0199	.90	.3159	1.35	.4115	1.80	.4641
.10	.0398	.91	.3186	1.36	.4131	1.81	.4649
.15	.0596	.92	.3212	1.37	.4147	1.82	.4656
.20	.0793	.93	.3238	1.38	.4162	1.83	.4664
.25	.0987	.94	.3264	1.39	.4177	1.84	.4671
.26	.1026	.95	.3289	1.40	.4192	1.85	.4678
.28	.1103	.96	.3315	1.41	.4207	1.86	.4686
.30	.1179	.97	.3340	1.42	.4222	1.87	.4693
.32	.1255	.98	.3365	1.43	.4236	1.88	.4699
.34	.1331	.99	.3389	1.44	.4251	1.89	.4706
.36	.1406	1.00	.3413	1.45	.4265	1.90	.4713
.37	.1443	1.01	.3438	1.46	.4279	1.91	.4719
.38	.1480	1.02	.3461	1.47	.4292	1.92	.4726
.40	.1554	1.03	.3485	1,48	.4306	1.93	.4732
.42	.1628	1.04	.3508	1.49	.4319	1.94	.4738
.44	.1700	1.05	.3531	1.50	.4332	1.95	.4744
.46	.1772	1.06	.3554	1.51	.4345	1.96	.4750
.48	.1844	1.07	.3577	1.52	.4357	1.97	.4756
.50	.1915	1.08	.3599	1.53	.4370	1.98	.4761
.52	.1985	1.09	.3621	1.54	.4382	1.99	.4767
.54	.2054	1.10	.3643	1.55	.4394	2.00	.4772
.56	.2123	1.11	.3665	1.56	.4406	2.05	.4798
.58	.2190	1.12	.3686	1.57	.4418	2.08	.4812
.60	.2257	1.13	.3708	1.58	.4429	2.12	.4830
.62	.2324	1.14	.3729	1.59	.4441	2.17	.4850
.64	.2389	1.15	.3749	1.60	.4452	2.20	.4861
.66	.2454	1.16	.3770	1.61	.4463	2.24	.4875
.68	.2517	1.17	.3790	1.62	.4474	2.29	.4890
.70	.2580	1.18	.3810	1.63	.4484	2.33	.4901
.72	.2642	1.19	.3830	1.64	.4495	2.37	.4911
.74	.2703	1.20	.3849	1.65	.4505	2.41	.4920
.76	.2764	1.21	.3869	1.66	.4515	2.46	.4931
.77	.2794	1.22	.3888	1.67	.4525	2.50	.4938
.78	.2823	1.23	.3907	1.68	.4535	2.54	.4945
.79	.2852	1.24	.3925	1.69	.4545	2.58	.4950
.80	.2881	1.25	.3944	1.70	.4554	2.61	.4955
.81	.2910	1.26	.3962	1.71	.4564	2.65	.4960
.82	.2939	1.27	.3980	1.72	.4573	2.70	.4965
.83	.2967	1.28	.3997	1.73	.4582	2.75	.4970
.84	.2995	1.29	.4015	1.74	.4591	2.81	.4975
.85	.3023	1.30	.4032	1.75	.4599	2.88	.4980
.86	.3051	1.31	.4049	1.76	.4608	2.97	.4985
.87	.3078	1.32	.4066	1.77	.4616	3.09	.4990
.88	.3106	1.33	.4082	1.78	.4625	3.29	.4995
.89	.3133	1.34	.4099	1.79	.4633	3.72	.4999

† Adapted and condensed from E. S. Pearson and H. O. Hartley (eds.), *Biometrika Tables for Statisticians*, vol. 1, 3d ed., 1966, table 1. Reproduced with the permission of E. S. Pearson and the trustees of *Biometrika*.

Table A-2 Critical Values of Chi Square†

df	Probability under H_1 of Obtaining Chi Square Greater than or Equal to the Value Entered in the Table											
	0.990	0.975	0.900	0.750	0.500	0.250	0.100	0.050	0.025	0.010	0.005	0.001
1	.000157	.000982	.015791	.10153	.45494	1.3233	2.7055	3.8415	5.0239	6.6349	7.8794	10.828
2	.020101	.050636	.21072	.57536	1.3863	2.7726	4.6052	5.9915	7.3778	9.2103	10.597	13.816
3	.11483	.21580	.58437	1.2125	2.3660	4.1083	6.2514	7.8147	9.3484	11.345	12.838	16.266
4	.29711	.48442	1.0636	1.9226	3.3567	5.3853	7.7794	9.4877	11.143	13.277	14.860	18.467
5	.55430	.83121	1.6103	2.6746	4.3515	6.6257	9.2364	11.070	12.832	15.086	16.750	20.515
6	.87209	1.2373	2.2041	3.4546	5.3481	7.8408	10.645	12.592	14.449	16.812	18.548	22.458
7	1.2390	1.6899	2.8331	4.2548	6.3458	9.0372	12.017	14.067	16.013	18.475	20.278	24.322
8	1.6465	2.1797	3.4895	5.0706	7.3441	10.219	13.362	15.507	17.534	20.090	21.955	26.125
9	2.0879	2.7004	4.1682	5.8988	8.3428	11.389	14.684	16.919	19.023	21.666	23.589	27.877
10	2.5582	3.2470	4.8652	6.7372	9.3418	12.549	15.987	18.307	20.483	23.209	25.188	29.588
11	3.0535	3.8158	5.5778	7.5841	10.341	13.701	17.275	19.675	21.920	24.725	26.757	31.264
12	3.5706	4.4038	6.3038	8.4384	11.340	14.845	18.549	21.026	23.337	26.217	28.300	32.909
13	4.1069	5.0088	7.0415	9.2991	12.340	15.984	19.812	22.362	24.736	27.688	29.820	34.528
14	4.6604	5.6287	7.7895	10.165	13.339	17.117	21.064	23.685	26.119	29.141	31.319	36.123
15	5.2294	6.2621	8.5468	11.036	14.339	18.245	22.307	24.996	27.488	30.578	32.801	37.697
16	5.8122	6.9077	9.3122	11.912	15.338	19.369	23.542	26.296	28.845	32.000	34.267	39.252
17	6.4078	7.5642	10.085	12.792	16.338	20.489	24.769	27.587	30.191	33.409	35.718	40.790
18	7.0149	8.2308	10.865	13.675	17.338	21.605	25.989	28.869	31.526	34.805	37.156	42.312
19	7.6327	8.9065	11.651	14.562	18.338	22.718	27.204	30.144	32.852	36.191	38.582	43.820
20	8.2604	9.5908	12.443	15.452	19.337	23.828	28.412	31.410	34.170	37.566	39.997	45.315
21	8.8972	10.283	13.240	16.344	20.337	24.935	29.615	32.671	35.479	38.932	41.401	46.797
22	9.5425	10.982	14.042	17.240	21.337	26.039	30.813	33.924	36.781	40.289	42.796	48.268
23	10.196	11.689	14.848	18.137	22.337	27.141	32.007	35.172	38.076	41.638	44.181	49.728
24	10.856	12.401	15.659	19.037	23.337	28.241	33.196	36.415	39.364	42.980	45.558	51.179
25	11.524	13.120	16.473	19.939	24.337	29.339	34.382	37.652	40.646	44.314	46.928	52.618
26	12.198	13.844	17.292	20.843	25.336	30.435	35.563	38.885	41.923	45.642	48.290	54.052
27	12.878	14.573	18.114	21.749	26.336	31.528	36.741	40.113	43.194	46.963	49.645	55.476
28	13.565	15.308	18.939	22.657	27.336	32.620	37.916	41.337	44.461	48.278	50.993	56.892
29	14.256	16.047	19.768	23.567	28.336	33.711	39.088	42.557	45.722	49.588	52.336	58.301
30	14.954	16.791	20.599	24.478	29.336	34.800	40.256	43.773	46.979	50.892	53.672	59.703
40	22.164	24.433	29.050	33.660	39.335	45.616	51.805	55.758	59.342	63.691	66.766	73.402
50	29.707	32.357	37.689	42.942	49.335	56.334	63.167	67.505	71.420	76.154	79.490	86.661
60	37.485	40.482	46.459	52.294	59.335	66.982	74.397	79.082	83.298	88.379	91.952	99.607
70	45.442	48.758	55.329	61.698	69.334	77.577	85.527	90.531	95.023	100.42	104.22	112.32
80	53.540	57.153	64.278	71.144	79.334	88.130	96.578	101.88	106.63	112.33	116.32	124.84
90	61.754	65.647	73.291	80.625	89.334	98.650	107.56	113.14	118.14	124.12	128.30	137.21
100	70.065	74.222	82.358	90.133	99.334	109.14	118.50	124.34	129.56	135.81	140.17	149.45

† Adapted and condensed from E. S. Pearson and H. O. Hartley (eds.), *Biometrika Tables for Statisticians*, vol. 1, 3d ed., 1966, table 8. Reproduced with the permission of E. S. Pearson and the trustees of *Biometrika*.

Table A-3 Critical Values of F ($\alpha = .05$)†

Denominator Degrees of Freedom	Numerator Degrees of Freedom																	
	2	3	4	5	6	7	8	9	10	12	15	20	24	30	40	60	120	∞
1	199.5	215.7	224.6	230.2	234.0	236.8	238.9	240.5	241.9	243.9	245.9	248.0	249.1	250.1	251.1	252.2	253.3	254.3
2	19.00	19.16	19.25	19.30	19.33	19.35	19.37	19.38	19.40	19.41	19.43	19.45	19.45	19.46	19.47	19.48	19.49	19.50
3	9.55	9.28	9.12	9.01	8.94	8.89	8.85	8.81	8.79	8.74	8.70	8.66	8.64	8.62	8.59	8.57	8.55	8.53
4	6.94	6.59	6.39	6.26	6.16	6.09	6.04	6.00	5.96	5.91	5.86	5.80	5.77	5.75	5.72	5.69	5.66	5.63
5	5.79	5.41	5.19	5.05	4.95	4.88	4.82	4.77	4.74	4.68	4.62	4.56	4.53	4.50	4.46	4.43	4.40	4.36
6	5.14	4.76	4.53	4.39	4.28	4.21	4.15	4.10	4.06	4.00	3.94	3.87	3.84	3.81	3.77	3.74	3.70	3.67
7	4.74	4.35	4.12	3.97	3.87	3.79	3.73	3.68	3.64	3.57	3.51	3.44	3.41	3.38	3.34	3.30	3.27	3.23
8	4.46	4.07	3.84	3.69	3.58	3.50	3.44	3.39	3.35	3.28	3.22	3.15	3.12	3.08	3.04	3.01	2.97	2.93
9	4.26	3.86	3.63	3.48	3.37	3.29	3.23	3.18	3.14	3.07	3.01	2.94	2.90	2.86	2.83	2.79	2.75	2.71
10	4.10	3.71	3.48	3.33	3.22	3.14	3.07	3.02	2.98	2.91	2.85	2.77	2.74	2.70	2.66	2.62	2.58	2.54
11	3.98	3.59	3.36	3.20	3.09	3.01	2.95	2.90	2.85	2.79	2.72	2.65	2.61	2.57	2.53	2.49	2.45	2.40
12	3.89	3.49	3.26	3.11	3.00	2.91	2.85	2.80	2.75	2.69	2.62	2.54	2.51	2.47	2.43	2.38	2.34	2.30
13	3.81	3.41	3.18	3.03	2.92	2.83	2.77	2.71	2.67	2.60	2.53	2.46	2.42	2.38	2.34	2.30	2.25	2.21
14	3.74	3.34	3.11	2.96	2.85	2.76	2.70	2.65	2.60	2.53	2.46	2.39	2.35	2.31	2.27	2.22	2.18	2.13
15	3.68	3.29	3.06	2.90	2.79	2.71	2.64	2.59	2.54	2.48	2.40	2.33	2.29	2.25	2.20	2.16	2.11	2.07
16	3.63	3.24	3.01	2.85	2.74	2.66	2.59	2.54	2.49	2.42	2.35	2.28	2.24	2.19	2.15	2.11	2.06	2.01
17	3.59	3.20	2.96	2.81	2.70	2.61	2.55	2.49	2.45	2.38	2.31	2.23	2.19	2.15	2.10	2.06	2.01	1.96
18	3.55	3.16	2.93	2.77	2.66	2.58	2.51	2.46	2.41	2.34	2.27	2.19	2.15	2.11	2.06	2.02	1.97	1.92
19	3.52	3.13	2.90	2.74	2.63	2.54	2.48	2.42	2.38	2.31	2.23	2.16	2.11	2.07	2.03	1.98	1.93	1.88
20	3.49	3.10	2.87	2.71	2.60	2.51	2.45	2.39	2.35	2.28	2.20	2.12	2.08	2.04	1.99	1.95	1.90	1.84
21	3.47	3.07	2.84	2.68	2.57	2.49	2.42	2.37	2.32	2.25	2.18	2.10	2.05	2.01	1.96	1.92	1.87	1.81
22	3.44	3.05	2.82	2.66	2.55	2.46	2.40	2.34	2.30	2.23	2.15	2.07	2.03	1.98	1.94	1.89	1.84	1.78
23	3.42	3.03	2.80	2.64	2.53	2.44	2.37	2.32	2.27	2.20	2.13	2.05	2.01	1.96	1.91	1.86	1.81	1.76
24	3.40	3.01	2.78	2.62	2.51	2.42	2.36	2.30	2.25	2.18	2.11	2.03	1.98	1.94	1.89	1.84	1.79	1.73
25	3.39	2.99	2.76	2.60	2.49	2.40	2.34	2.28	2.24	2.16	2.09	2.01	1.96	1.92	1.87	1.82	1.77	1.71
26	3.37	2.98	2.74	2.59	2.47	2.39	2.32	2.27	2.22	2.15	2.07	1.99	1.95	1.90	1.85	1.80	1.75	1.69
27	3.35	2.96	2.73	2.57	2.46	2.37	2.31	2.25	2.20	2.13	2.06	1.97	1.93	1,88	1.84	1.79	1.73	1.67
28	3.34	2.95	2.71	2.56	2.45	2.36	2.29	2.24	2.19	2.12	2.04	1.96	1.91	1.87	1.82	1.77	1.71	1.65
29	3.33	2.93	2.70	2.55	2.43	2.35	2.28	2.22	2.18	2.10	2.03	1.94	1.90	1.85	1.81	1.75	1.70	1.64
30	3.32	2.92	2.69	2.53	2.42	2.33	2.27	2.21	2.16	2.09	2.01	1.93	1.89	1.84	1.79	1.74	1.68	1.62
40	3.23	2.84	2.61	2.45	2.34	2.25	2.18	2.12	2.08	2.00	1.92	1.84	1.79	1.74	1.69	1.64	1.58	1.51
60	3.15	2.76	2.53	2.37	2.25	2.17	2.10	2.04	1.99	1.92	1.84	1.75	1.70	1.65	1.59	1.53	1.47	1.39
120	3.07	2.68	2.45	2.29	2.17	2.09	2.02	1.96	1.91	1.83	1.75	1.66	1.61	1.55	1.50	1.43	1.35	1.25
∞	3.00	2.60	2.37	2.21	2.10	2.01	1.94	1.88	1.83	1.75	1.67	1.57	1.52	1.46	1.39	1.32	1.22	1.00

† Condensed from E. S. Pearson and H. O. Hartley (eds.), *Biometrika Tables for Statisticians*, vol. 1, 3d ed., 1966, table 18. Reproduced with the permission of E. S. Pearson and the trustees of *Biometrika*.

Table A-4 Critical Values of F ($\alpha = .01$)†

Denominator Degrees of Freedom	Numerator Degrees of Freedom																	
	2	3	4	5	6	7	8	9	10	12	15	20	24	30	40	60	120	∞
1	4,999.5	5,403	5,625	5,764	5,859	5,928	5,981	6,022	6,056	6,106	6,157	6,029	6,235	6,261	6,287	6,313	6,339	6,366
2	99.00	99.17	99.25	99.30	99.33	99.36	99.37	99.39	99.40	99.42	99.43	99.45	99.46	99.47	99.47	99.48	99.49	99.50
3	30.82	29.46	28.71	28.24	27.91	27.67	27.49	27.35	27.23	27.05	26.87	26.69	26.60	26.50	26.41	26.32	26.22	26.13
4	18.00	16.69	15.98	15.52	15.21	14.98	14.80	14.66	14.55	14.37	14.20	14.02	13.93	13.84	13.75	13.65	13.56	13.46
5	13.27	12.06	11.39	10.97	10.67	10.46	10.29	10.16	10.05	9.89	9.72	9.55	9.47	9.38	9.29	9.20	9.11	9.02
6	10.92	9.78	9.15	8.75	8.47	8.26	8.10	7.98	7.87	7.72	7.56	7.40	7.31	7.23	7.14	7.06	6.97	6.88
7	9.55	8.45	7.85	7.46	7.19	6.99	6.84	6.72	6.62	6.47	6.31	6.16	6.07	5.99	5.91	5.82	5.74	5.65
8	8.65	7.59	7.01	6.63	6.37	6.18	6.03	5.91	5.81	5.67	5.52	5.36	5.28	5.20	5.12	5.03	4.95	4.86
9	8.02	6.99	6.42	6.06	5.80	5.61	5.47	5.35	5.26	5.11	4.96	4.81	4.73	4.65	4.57	4.48	4.40	4.31
10	7.56	6.55	5.99	5.64	5.39	5.20	5.06	4.94	4.85	4.71	4.56	4.41	4.33	4.25	4.17	4.08	4.00	3.91
11	7.21	6.22	5.67	5.32	5.07	4.89	4.74	4.63	4.54	4.40	4.25	4.10	4.02	3.94	3.86	3.78	3.69	3.60
12	6.93	5.95	5.41	5.06	4.82	4.64	4.50	4.39	4.30	4.16	4.01	3.86	3.78	3.70	3.62	3.54	3.45	3.36
13	6.70	5.74	5.21	4.86	4.62	4.44	4.30	4.19	4.10	3.96	3.82	3.66	3.59	3.51	3.43	3.34	3.25	3.17
14	6.51	5.56	5.04	4.69	4.46	4.28	4.14	4.03	3.94	3.80	3.66	3.51	3.43	3.35	3.27	3.18	3.09	3.00
15	6.36	5.42	4.89	4.56	4.32	4.14	4.00	3.89	3.80	3.67	3.52	3.37	3.29	3.21	3.13	3.05	2.96	2.87
16	6.23	5.29	4.77	4.44	4.20	4.03	3.89	3.78	3.69	3.55	3.41	3.26	3.18	3.10	3.02	2.93	2.84	2.75
17	6.11	5.18	4.67	4.34	4.10	3.93	3.79	3.68	3.59	3.46	3.31	3.16	3.08	3.00	2.92	2.83	2.75	2.65
18	6.01	5.09	4.58	4.25	4.01	3.84	3.71	3.60	3.51	3.37	3.23	3.08	3.00	2.92	2.84	2.75	2.66	2.57
19	5.93	5.01	4.50	4.17	3.94	3.77	3.63	3.52	3.43	3.30	3.15	3.00	2.92	2.84	2.76	2.67	2.58	2.49
20	5.85	4.94	4.43	4.10	3.87	3.70	3.56	3.46	3.37	3.23	3.09	2.94	2.86	2.78	2.69	2.61	2.52	2.42
21	5.78	4.87	4.37	4.04	3.81	3.64	3.51	3.40	3.31	3.17	3.03	2.88	2.80	2.72	2.64	2.55	2.46	2.36
22	5.72	4.82	4.31	3.99	3.76	3.59	3.45	3.35	3.26	3.12	2.98	2.83	2.75	2.67	2.58	2.50	2.40	2.31
23	5.66	4.76	4.26	3.94	3.71	3.54	3.41	3.30	3.21	3.07	2.93	2.78	2.70	2.62	2.54	2.45	2.35	2.26
24	5.61	4.72	4.22	3.90	3.67	3.50	3.36	3.26	3.17	3.03	2.89	2.74	2.66	2.58	2.49	2.40	2.31	2.21
25	5.57	4.68	4.18	3.85	3.63	3.46	3.32	3.22	3.13	2.99	2.85	2.70	2.62	2.54	2.45	2.36	2.27	2.17
26	5.53	4.64	4.14	3.82	3.59	3.42	3.29	3.18	3.09	2.96	2.81	2.66	2.58	2.50	2.42	2.33	2.23	2.13
27	5.49	4.60	4.11	3.78	3.56	3.39	3.26	3.15	3.06	2.93	2.78	2.63	2.55	2.47	2.38	2.29	2.20	2.10
28	5.45	4.57	4.07	3.75	3.53	3.36	3.23	3.12	3.03	2.90	2.75	2.60	2.52	2.44	2.35	2.26	2.17	2.06
29	5.42	4.54	4.04	3.73	3.50	3.33	3.20	3.09	3.00	2.87	2.73	2.57	2.49	2.41	2.33	2.23	2.14	2.03
30	5.39	4.51	4.02	3.70	3.47	3.30	3.17	3.07	2.98	2.84	2.70	2.55	2.47	2.39	2.30	2.21	2.11	2.01
40	5.18	4.31	3.83	3.51	3.29	3.12	2.99	2.89	2.80	2.66	2.52	2.37	2.29	2.20	2.11	2.02	1.92	1.80
60	4.98	4.13	3.65	3.34	3.12	2.95	2.82	2.72	2.63	2.50	2.35	2.20	2.12	2.03	1.94	1.84	1.73	1.60
120	4.79	3.95	3.48	3.17	2.96	2.79	2.66	2.56	2.47	2.34	2.19	2.03	1.95	1.86	1.76	1.66	1.53	1.38
∞	4.61	3.78	3.32	3.02	2.80	2.64	2.51	2.41	2.32	2.18	2.04	1.88	1.79	1.70	1.59	1.47	1.32	1.00

† Condensed from E. S. Pearson and H. O. Hartley (eds.), *Biometrika Tables for Statisticians*, vol. 1, 3d ed., 1966, table 18. Reproduced with the permission of E. S. Pearson and the trustees of *Biometrika*.

Table A-5 Critical Values of t†

	Probability under H_1 of Obtaining t Greater than or Equal to the Critical Value						
One-Tailed Test:	0.1	0.05	0.025	0.01	0.005	0.001	0.0005
Two-Tailed Test:	0.2	0.1	0.05	0.02	0.01	0.002	0.001
df							
1	3.078	6.314	12.706	31.821	63.657	318.31	636.62
2	1.886	2.920	4.303	6.965	9.925	22.327	31.598
3	1.638	2.353	3.182	4.541	5.841	10.214	12.924
4	1.533	2.132	2.776	3.747	4.604	7.173	8.610
5	1.476	2.015	2.571	3.365	4.032	5.893	6.869
6	1.440	1.943	2.447	3.143	3.707	5.208	5.959
7	1.415	1.895	2.365	2.998	3.499	4.785	5.408
8	1.397	1.860	2.306	2.896	3.355	4.501	5.041
9	1.383	1.833	2.262	2.821	3.250	4.297	4.781
10	1.372	1.812	2.228	2.764	3.169	4.144	4.587
11	1.363	1.796	2.201	2.718	3.106	4.025	4.437
12	1.356	1.782	2.179	2.681	3.055	3.930	4.318
13	1.350	1.771	2.160	2.650	3.012	3.852	4.221
14	1.345	1.761	2.145	2.624	2.977	3.787	4.140
15	1.341	1.753	2.131	2.602	2.947	3.733	4.073
16	1.337	1.746	2.120	2.583	2.921	3.686	4.015
17	1.333	1.740	2.110	2.567	2.898	3.646	3.965
18	1.330	1.734	2.101	2.552	2.878	3.610	3.922
19	1.328	1.729	2.093	2.539	2.861	3.579	3.883
20	1.325	1.725	2.086	2.528	2.845	3.552	3.850
21	1.323	1.721	2.080	2.518	2.831	3.527	3.819
22	1.321	1.717	2.074	2.508	2.819	3.505	3.792
23	1.319	1.714	2.069	2.500	2.807	3.485	3.767
24	1.318	1.711	2.064	2.492	2.797	3.467	3.745
25	1.316	1.708	2.060	2.485	2.787	3.450	3.725
26	1.315	1.706	2.056	2.479	2.779	3.435	3.707
27	1.314	1.703	2.052	2.473	2.771	3.421	3.690
28	1.313	1.701	2.048	2.467	2.763	3.408	3.674
29	1.311	1.699	2.045	2.462	2.756	3.396	3.659
30	1.310	1.697	2.042	2.457	2.750	3.385	3.646
40	1.303	1.684	2.021	2.423	2.704	3.307	3.551
60	1.296	1.671	2.000	2.390	2.660	3.232	3.460
100	1.289	1.658	1.980	2.358	2.617	3.160	3.373
∞	1.282	1.645	1.960	2.326	2.576	3.090	3.291

† Adapted and condensed from E. S. Pearson and H. O. Hartley (eds.), *Biometrika Tables for Statisticians*, vol. 1, 3d ed., 1966, table 12. Reproduced with the permission of E. S. Pearson and the trustees of *Biometrika*.

Table A-6 Critical Values of T in the Wilcoxon Matched-Pairs Signed Ranks Test†

	Probability under H_1 of Obtaining T Greater than or Equal to the Critical Value				Probability under H_1 of Obtaining T Greater than or Equal to the Critical Value		
One-Tailed Test:	.025	.01	.005	One-Tailed Test:	.025	.01	.005
Two-Tailed Test:	.05	.02	.01	Two-Tailed Test:	.05	.02	.01
N				N			
6	0	–	–	16	30	24	20
7	2	0	–	17	35	28	23
8	4	2	0	18	40	33	28
9	6	3	2	19	46	38	32
10	8	5	3	20	52	43	38
11	11	7	5	21	59	49	43
12	14	10	7	22	66	56	49
13	17	13	10	23	73	62	55
14	21	16	13	24	81	69	61
15	25	20	16	25	89	77	68

† Adapted from F. Wilcoxon, *Some Rapid Approximate Statistical Procedures*, American Cyanamid Company, New York, 1949, table I. Reproduced with the permission of author and publisher.

Table A-7 Probabilities Associated with Values as Small as Observed Values of U in the Mann-Whitney Test†

$n_2 = 3$

U	n_1 = 1	2	3
0	.250	.100	.050
1	.500	.200	.100
2	.750	.400	.200
3		.600	.350
4			.500
5			.650

$n_2 = 4$

U	n_1 = 1	2	3	4
0	.200	.067	.028	.014
1	.400	.133	.057	.029
2	.600	.267	.114	.057
3		.400	.200	.100
4		.600	.314	.171
5			.429	.243
6			.571	.343
7				.443
8				.557

$n_2 = 5$

U	n_1 = 1	2	3	4	5
0	.167	.047	.018	.008	.004
1	.333	.095	.036	.016	.008
2	.500	.190	.071	.032	.016
3	.667	.286	.125	.056	.028
4		.429	.196	.095	.048
5		.571	.286	.143	.075
6			.393	.206	.111
7			.500	.278	.155
8			.607	.365	.210
9				.452	.274
10				.548	.345
11					.421
12					.500
13					.579

$n_2 = 6$

U	n_1 = 1	2	3	4	5	6
0	.143	.036	.012	.005	.002	.001
1	.286	.071	.024	.010	.004	.002
2	.428	.143	.048	.019	.009	.004
3	.571	.214	.083	.033	.015	.008
4		.321	.131	.057	.026	.013
5		.429	.190	.086	.041	.021
6		.571	.274	.129	.063	.032
7			.357	.176	.089	.047
8			.452	.238	.123	.066
9			.548	.305	.165	.090
10				.381	.214	.120
11				.457	.268	.155
12				.545	.331	.197
13					.396	.242
14					.465	.294
15					.535	.350
16						.409
17						.469
18						.531

$n_2 = 7$

U	n_1 = 1	2	3	4	5	6	
0	.125	.028	.008	.003	.001	.001	.000
1	.250	.056	.017	.006	.003	.001	.001
2	.375	.111	.033	.012	.005	.002	.001
3	.500	.167	.058	.021	.009	.004	.002
4	.625	.250	.092	.036	.015	.007	.003
5		.333	.133	.055	.024	.011	.006
6		.444	.192	.082	.037	.017	.009
7		.556	.258	.115	.053	.026	.013
8			.333	.158	.074	.037	.019
9			.417	.206	.101	.051	.027
10			.500	.264	.134	.069	.036
11			.583	.324	.172	.090	.049
12				.394	.216	.117	.064
13				.464	.265	.147	.082
14				.538	.319	.183	.104
15					.378	.223	.130
16					.438	.267	.159
17					.500	.314	.191
18					.562	.365	.228
19						.418	.267
20						.473	.310
21						.527	.355
22							.402
23							.451
24							.500
25							.549

† Reproduced with the permission of authors and publisher from H. B. Mann and D. R. Whitney, On a Test of Whether One of Two Random Variables is Stochastically Larger than the Other, *Ann. Math. Stat.*, **18:** 52–54 (1947).

Table A-7 (Cont.)

$n_2 = 8$

U	n_1 = 1	2	3	4	5	6	7	8	t	Normal
0	.111	.022	.006	.002	.001	.000	.000	.000	3.308	.001
1	.222	.044	.012	.004	.002	.001	.000	.000	3.203	.001
2	.333	.089	.024	.008	.003	.001	.001	.000	3.098	.001
3	.444	.133	.042	.014	.005	.002	.001	.001	2.993	.001
4	.556	.200	.067	.024	.009	.004	.002	.001	2.888	.002
5		.267	.097	.036	.015	.006	.003	.001	2.783	.003
6		.356	.139	.055	.023	.010	.005	.002	2.678	.004
7		.444	.188	.077	.033	.015	.007	.003	2.573	.005
8		.556	.248	.107	.047	.021	.010	.005	2.468	.007
9			.315	.141	.064	.030	.014	.007	2.363	.009
10			.387	.184	.085	.041	.020	.010	2.258	.012
11			.461	.230	.111	.054	.027	.014	2.153	.016
12			.539	.285	.142	.071	.036	.019	2.048	.020
13				.341	.177	.091	.047	.025	1.943	.026
14				.404	.217	.114	.060	.032	1.838	.033
15				.467	.262	.141	.076	.041	1.733	.041
16				.533	.311	.172	.095	.052	1.628	.052
17					.362	.207	.116	.065	1.523	.064
18					.416	.245	.140	.080	1.418	.078
19					.472	.286	.168	.097	1.313	.094
20					.528	.331	.198	.117	1.208	.113
21						.377	.232	.139	1.102	.135
22						.426	.268	.164	.998	.159
23						.475	.306	.191	.893	.185
24						.525	.347	.221	.788	.215
25							.389	.253	.683	.247
26							.433	.287	.578	.282
27							.478	.323	.473	.318
28							.522	.360	.368	.356
29								.399	.263	.396
30								.439	.158	.437
31								.480	.052	.481
32								.520		

Table A-8 Critical Values of U for a One-Tailed Test at $\alpha = .05$ or for a Two-Tailed Test at $\alpha = .10$†

n_1	n_2: 9	10	11	12	13	14	15	16	17	18	19	20
1											0	0
2	1	1	1	2	2	2	3	3	3	4	4	4
3	3	4	5	5	6	7	7	8	9	9	10	11
4	6	7	8	9	10	11	12	14	15	16	17	18
5	9	11	12	13	15	16	18	19	20	22	23	25
6	12	14	16	17	19	21	23	25	26	28	30	32
7	15	17	19	21	24	26	28	30	33	35	37	39
8	18	20	23	26	28	31	33	36	39	41	44	47
9	21	24	27	30	33	36	39	42	45	48	51	54
10	24	27	31	34	37	41	44	48	51	55	58	62
11	27	31	34	38	42	46	50	54	57	61	65	69
12	30	34	38	42	47	51	55	60	64	68	72	77
13	33	37	42	47	51	56	61	65	70	75	80	84
14	36	41	46	51	56	61	66	71	77	82	87	92
15	39	44	50	55	61	66	72	77	83	88	94	100
16	42	48	54	60	65	71	77	83	89	95	101	107
17	45	51	57	64	70	77	83	89	96	102	109	115
18	48	55	61	68	75	82	88	95	102	109	116	123
19	51	58	65	72	80	87	94	101	109	116	123	130
20	54	62	69	77	84	92	100	107	115	123	130	138

† Adapted and condensed from D. Auble, Extended Tables for the Mann-Whitney Statistic, *Bull. Inst. Educ. Res. Ind. Univ.*, **1** (2): (1953), table 1. Reproduced with the permission of author and publisher.

Table A-9 Critical Values of U for a One-Tailed Test at $\alpha = .025$ or for a Two-Tailed Test at $\alpha = .05$†

n_1	n_2: 9	10	11	12	13	14	15	16	17	18	19	20
1												
2	0	0	0	1	1	1	1	1	2	2	2	2
3	2	3	3	4	4	5	5	6	6	7	7	8
4	4	5	6	7	8	9	10	11	11	12	13	13
5	7	8	9	11	12	13	14	15	17	18	19	20
6	10	11	13	14	16	17	19	21	22	24	25	27
7	12	14	16	18	20	22	24	26	28	30	32	34
8	15	17	19	22	24	26	29	31	34	36	38	41
9	17	20	23	26	28	31	34	37	39	42	45	48
10	20	23	26	29	33	36	39	42	45	48	52	55
11	23	26	30	33	37	40	44	47	51	55	58	62
12	26	29	33	37	41	45	49	53	57	61	65	69
13	28	33	37	41	45	50	54	59	63	67	72	76
14	31	36	40	45	50	55	59	64	67	74	78	83
15	34	39	44	49	54	59	64	70	75	80	85	90
16	37	42	47	53	59	64	70	75	81	86	92	98
17	39	45	51	57	63	67	75	81	87	93	99	105
18	42	48	55	61	67	74	80	86	93	99	106	112
19	45	52	58	65	72	78	85	92	99	106	113	119
20	48	55	62	69	76	83	90	98	105	112	119	127

† Adapted and condensed from D. Auble, Extended Tables for the Mann-Whitney Statistic, *Bull. Inst. Educ. Res. Ind. Univ.*, **1** (2): (1953), table 3. Reproduced with the permission of author and publisher.

Table A-10 Critical Values of U for a One-Tailed Test at $\alpha = .01$ or for a Two-Tailed Test at $\alpha = .02$†

	n_2											
n_1	9	10	11	12	13	14	15	16	17	18	19	20
1												
2					0	0	0	0	0	0	1	1
3	1	1	1	2	2	2	3	3	4	4	4	5
4	3	3	4	5	5	6	7	7	8	9	9	10
5	5	6	7	8	9	10	11	12	13	14	15	16
6	7	8	9	11	12	13	15	16	18	19	20	22
7	9	11	12	14	16	17	19	21	23	24	26	28
8	11	13	15	17	20	22	24	26	28	30	32	34
9	14	16	18	21	23	26	28	31	33	36	38	40
10	16	19	22	24	27	30	33	36	38	41	44	47
11	18	22	25	28	31	34	37	41	44	47	50	53
12	21	24	28	31	35	38	42	46	49	53	56	60
13	23	27	31	35	39	43	47	51	55	59	63	67
14	26	30	34	38	43	47	51	56	60	65	69	73
15	28	33	37	42	47	51	56	61	66	70	75	80
16	31	36	41	46	51	56	61	66	71	76	82	87
17	33	38	44	49	55	60	66	71	77	82	88	93
18	36	41	47	53	59	65	70	76	82	88	94	100
19	38	44	50	56	63	69	75	82	88	94	101	107
20	40	47	53	60	67	73	80	87	93	100	107	114

† Adapted and condensed from D. Auble, Extended Tables for the Mann-Whitney Statistic, *Bull. Inst. Educ. Res. Ind. Univ.*, **1** (2): (1953), table 5. Reproduced with the permission of author and publisher.

Table A-11 Critical Values of U for a One-Tailed Test at $\alpha = .001$ or for a Two-Tailed Test at $\alpha = .002$†

	n_2											
n_1	9	10	11	12	13	14	15	16	17	18	19	20
1												
2												
3									0	0	0	0
4		0	0	0	1	1	1	2	2	3	3	3
5	1	1	2	2	3	3	4	5	5	6	7	7
6	2	3	4	4	5	6	7	8	9	10	11	12
7	3	5	6	7	8	9	10	11	13	14	15	16
8	5	6	8	9	11	12	14	15	17	18	20	21
9	7	8	10	12	14	15	17	19	21	23	25	26
10	8	10	12	14	17	19	21	23	25	27	29	32
11	10	12	15	17	20	22	24	27	29	32	34	37
12	12	14	17	20	23	25	28	31	34	37	40	42
13	14	17	20	23	26	29	32	35	38	42	45	48
14	15	19	22	25	29	32	36	39	43	46	50	54
15	17	21	24	28	32	36	40	43	47	51	55	59
16	19	23	27	31	35	39	43	48	52	56	60	65
17	21	25	29	34	38	43	47	52	57	61	66	70
18	23	27	32	37	42	46	51	56	61	66	71	76
19	25	29	34	40	45	50	55	60	66	71	77	82
20	26	32	37	42	48	54	59	65	70	76	82	88

† Adapted and condensed from D. Auble, Extended Tables for the Mann-Whitney Statistic, *Bull. Inst. Educ. Res. Ind. Univ.*, **1** (2): (1953), table 7. Reproduced with the permission of author and publisher.

ANSWERS TO PROBLEMS

Lesson 1

1. Yes. For example, two very unusual samples of heights, one from heights of women and one from heights of men, could happen to be so much alike that we would mistakenly take them to be from the same population, even though the population of men's heights definitely has a higher average than the population of women's heights.

2. The association, though weak, is so *unexpected* in terms of our current scientific concepts that to establish any association at all would be of great importance.

3. Yes, because the difference is between samples. If we compared the average of all 35 scores in one class with the average of all 35 scores in the other class, we would not have to question the statistical significance of the difference in test scores; we would know that in fact the classes do differ. However, without further analysis we could not conclude that these test scores signify a difference in mathematical knowledge.

4. No, because each test score is based on a sample from the infinite population of "ability-indicating behavior" of that individual, and a small amount of difference between the two scores would be expected to appear through sampling variability alone.

5. Since the batting average is an exact quantity, there can be no question of significance of difference as long as only the average itself is the object of attention. But if the batting average is to be taken as an indicator of batting ability, the question of significance does arise, for each average is the result of a set of observations which make up a sample of the population of possible batting behaviors of each player. Many chance factors entered into the drawing of each sample.

6. Yes, whenever there is appreciable variability among repeated measurements of the same length. If the distances between cities are measured in thousandths of miles, there will be appreciable variability even under the best conditions of measurement. The situation will then be described in exactly the same terms as any other comparison between two samples from infinite populations: does the measured distance between New York and Boston differ *more* from the measured distance between New York and Washington than two successive measurements of the same distance, e.g., New York to Boston, would differ from each other?

7. A score of 79 equals a percent correct of 67.5; a score of 78 equals 66.7. It is not possible, in this case, for any percentage to fall between 66.7 and 67.5 since achievement score is a discrete variable. Strictly speaking, then, there are gaps in the percentage values also, and "percent correct" is not strictly a continuous variable. However, percentages and proportions are conventionally treated as continuous variables.

8. Definitely not. "Statistical association" must never be understood to imply causality; it is a relation between the values which the two variables assume, and it has nothing direct to say about the processes by which the events being observed may be influenced. Even if "college" and "height" are associated, it would be quite incorrect to claim that one's college determines one's height or that one's height determines one's choice of college. Neither does the association between aptitude scores and college grades mean that grades are caused by aptitude; the statistical association does not imply this kind of influence any more than it implies an influence of grades upon aptitude.

9. A statistically trained physician will ask, "Is the difference statistically significant?" before concluding that the same difference will apply to other surgical patients. The two variables are "pain reliever," with two levels ("new drug" and "aspirin"); and "number of requests for additional medication," which is a discrete variable whose values may be zero or any positive number.

Lesson 2

1. No, because it violates the condition of independence. This procedure is like sequential sampling in that it precludes the selection of the same element more than once. Unless the population is infinite, the grabbag procedure has the same faults as sequential sampling without replacement.

2. *a.* No; the condition of independence is violated, since each member of a set of five scores from a particular person is not independent of the selection of the other four scores. This example is important, because it is very common in the social sciences to find the use of repeated measures on the same individual. A set of N observations on k persons, where each person contributes N/k of the N observations, is *not* a random sample from the population of such observations.

b. Yes; the 100 elements are now independent of each other. When repeated measures can be converted to a single score, so that each individual observed contributes just one summary observation (such as an average), the independence condition is met. Use of an average often helps to reduce the effects of chance variation *within* an individual's performance.

c. No, because selection of any one score is not independent of selection of the other scores from the same class. This example is another situation common in social science; it is more convenient to get measures from existing groups, but such a procedure does not provide independent observations.

d. The list is random, but the completed-test sample is probably biased. It is impossible to know whether the factors determining "response to the requests" were random, and since it is quite possible that both equal chance and independence were violated, it is best to assume that the sample is not random.

3. *a.* That the attitudes following the film do not differ from attitudes before seeing the film.

b. The .02 level indicates the greater degree of significance.

c. Yes; no; risk of type II error is increased at the .02 level.

4. Although this is a question of judgment, most scientists would advise the young man to do another experiment before publishing a result significant at the .10 level; the two experiments together may reach a higher level of significance. If he particularly wants to draw attention to his hypothesis, he should meet the conventional minimum significance level; otherwise, colleagues are likely to dismiss the study as unconvincing.

Lesson 3

1. Yes; it is a hypothesis of no difference between the population with $P = 1/2$ and the population from which the sample is drawn.

2. *a.* That he is only guessing and the proportion of correct choices $P = 1/2$.

b. $P \neq 1/2$, $P > 1/2$, $P < 1/2$; he could conceivably make so few correct choices that you would be convinced he was not merely guessing but discriminating between the samples and giving them the wrong names.

3. *a.* $1/3$, $2/3$. *b.* That the zoology class is a random sample from a population containing one-third women and two-thirds men; that it is a random sample from a population containing more than one-third women. *c.* One-sample. *d.* Sex and election of zoology.

4. *a.* $1/4$. *b.* $1/4$. *c.* $12/51$, $13/51$. *d.* $1/2$, $1/2$. *e.* $1/52$, $1/13$, $2/13$. *f.* $3/13$.

5. $1/6$.

6. $1/8$, $3/8$.

Lesson 4

1. 9; 27; 3^N.

2. $\frac{1}{100}$, $\frac{81}{100}$; $\frac{18}{100}$. No; with $P = \frac{1}{2}$, there is no probability smaller than $\frac{1}{4}$; therefore there is no way to get a result whose probability is not greater than .01. Yes, with $P = \frac{1}{10}$, the probability of $K = 2$ is .01; if $K = 2$ occurs, this hypothesis can be rejected at the .01 level.

3. $2^9 = 512$; 10 subsets; $\frac{1}{512}$; $9!/6!3! = 84$; $\frac{84}{512}$.

4. $\frac{1}{6}$; $\frac{1}{3}$.

5. $\frac{1}{36}$; $\frac{1}{36}$; $\frac{1}{18}$; $\frac{1}{6}$.

6. Because the two events are not necessarily independent. His call on the first toss may affect his choice of a call on the second toss.

7. $\frac{15}{64}$; $\frac{1}{64}$.

8. $p = \frac{1}{4}$, $q = \frac{3}{4}$, $\frac{2{,}187}{16{,}384}$. $\frac{945}{16{,}384}$. The probability is $\frac{54}{256}$ times $\frac{9}{64}$, or $\frac{486}{16{,}384}$. Less, because it restricts the number of ways the total of 4 hearts can be arranged; the first probability allows such arrangements as 3 in the first 4 and 1 in the second 3, and so on.

Lesson 5

1. $\dfrac{12!}{11!1!}\left(\dfrac{1}{4}\right)^{11}\left(\dfrac{3}{4}\right)^{1} = 12\dfrac{3}{16{,}777{,}216} = \dfrac{9}{4{,}194{,}304}$

2. With $N = 5$, all 5 ($p = .0003$); with $N = 10$, at least 7 ($p = .0007$); with $N = 20$, at least 11 ($p = .0006$).

3. Since the alternate hypotheses specify *higher* values of K, H_1: $p = \frac{1}{3}$ will be rejected only when K is too high; the critical value of K will therefore be found at the upper end of the binomial distribution. With $N = 10$, K must be at least 7; the probability of K at least 6 is .0764. On H_2: $p = \frac{1}{2}$, the probability of K at least 7 is $\frac{176}{1{,}024} = .1718$; this is the probability $1 - \beta$, or the power of the test to detect H_2: $p = \frac{1}{2}$. On H_2: $p = \frac{2}{3}$, the power is .5591. When $N = 20$, H_1 will be rejected only if K is at least 11; the probability of K at least 11 is .4119 on H_2: $p = \frac{1}{2}$ and .9068 on H_2: $p = \frac{2}{3}$.

4. His α level, the closest H_2 which he wishes to be sure of detecting, and the minimum acceptable probability $1 - \beta$ that he will actually detect this H_2.

5. The distributions for $p = \frac{1}{8}$ and $p = \frac{7}{8}$ are mirror images; they have peaks at $K = 1$ and $K = 9$, respectively, and $p = .3727$ at these peaks. The distributions for $p = \frac{1}{4}$ and $p = \frac{3}{4}$ are mirror images with peaks at $K = 3$ and $K = 7$ ($p = .2502$). The distributions for $p = \frac{3}{8}$ and $p = \frac{5}{8}$ are mirror images with peaks at $K = 4$ and $K = 6$ ($p = .2475$). The most symmetrical distributions are $p = \frac{3}{8}$ and $p = \frac{5}{8}$, and these distributions have the lowest peaks and the greatest dispersion.

6. Critical $K = 12$; $1 - \beta = .2812$. Power increases from .2440 ($N = 10$) to .2812 ($N = 14$); it would increase to .7361 for $N = 20$.

Lesson 6

1. *a*. The modal class only. *b*. All three. *c*. Modal class (modal rank) and median rank assigned by the 20 judges.

2. *a*. 2 students will receive A's, 13 B's, 7 C's, 3 D's, and 5 F's.

b. The mean is 41.9 points. By this procedure, there will be 19 students with A or B and 11 students with C, D, or F.

c. The median score is 44.5. By this procedure, there will of course be 15 students with A or B and 15 with C, D, or F.

d. The top 3 should get A's, but there are 3 students tied for third place; probably the teacher will give A's only to the top 2. The cutoff score between B and C will be between 42 and 43, and that between C and D is just below 38 or 39. The failing mark might be anything below 33 or below 32, depending on the teacher's inclination.

3. No, because the sample of 96 persons who finished the task is not a random sample; it excludes automatically any persons who cannot complete the task within 60 minutes.

Lesson 7

1. For this distribution, $s = 7.4$ and $\bar{X} = 41.9$. The cutoff point between A and B will be 52; only 1 student gets an A. Between B and C, 42; 18 get B's. Between C and D, 34.5; 5 get C's. Between D and F, 27.1; 4 get D's and 2 get F's, if the teacher is strict in adhering to the failing cutoff point.

2. The raw score 115 corresponds to a z score of -0.27. The raw score of 134 corresponds to a z score of $+1.45$. The raw score of 99 corresponds to a z score of -1.73.

3. Range = 10; SS = 390; $s^2 = 5.20$, and $s = 2.28$.

4. Since these are extremely skewed data (a J curve), the mean and standard deviation should not be used; they are too sensitive to the few extremely high values. The median and AAD taken from the median should be used.

Lesson 8

1. *a*. 51.0. *b*. 71.2. *c*. 10.2.

2. *a*. 34 percent, approximately. *b*. 84 percent, approximately. *c*. 2 percent, approximately.

3. Mean = 65; $s = 8$; only 2 percent will be above 81.

4. In a normal distribution, the median and mean are equal; the z score for the median is therefore 0. Only 16 percent of the distribution lies above $z = +1.00$, and 98 percent of the scores lies below +2.00.

5.

H_2	z	$1 - \beta$
$p = .55$	+0.70	.242
$p = .60$	−0.31	.622
$p = .65$	−1.36	.913
$p = .70$	−2.51	.994
$p = .75$	−3.81	1.000

6.

N	Critical K	β
36	25	>.5
64	42	~.5
100	62	~.145
144	86	.039
196	115	

To have β as low as .05, N must be greater than 100.

Lesson 9

1. $16/12 = 1.33$.

2. SS = 121; estimated $\sigma^2 = 1.22$, estimated $\sigma_{\bar{X}}^2 = .0122$, estimated $\sigma_{\bar{X}} = .110$.

3. Estimated $\sigma_{\bar{X}}^2 = 100$, estimated $\sigma_{\bar{X}} = 10$. The probability that $\bar{X} = 365$ lies within 1 standard error of X_T is .68. If $\bar{X}$ is too low by 1 standard error, $X_T = 375$; if $\bar{X}$ is too high by 1 standard error, $X_T = 355$. The probability is then .68 that X_T lies between 355 and 375.

Lesson 10

1. The first hypothesis requires a one-tailed test; $p = .075$. The second hypothesis requires a two-tailed test; $p = 0.15$.

2. The hypothesis can be rejected at the .05 level if *either* 480 or more heads occur *or* 420 or fewer heads occur, since the direction of the bias is not specified.

3. $z = 1.67$ for the obtained result. Since no direction is specified, a two-tailed test is required, and the probability of such a large z in either direction is 2(.0475) = .0950. The hypothesis of no difference cannot be rejected at the .05 level.

4. Mean = 18, $\sigma = 3.46$; $z = 2.02$; $H_1: p = 1/3$ can be rejected at the .02 level in favor of $H_2: p > 1/3$.

Lesson 11

1. $\chi^2 = 62.19$, df = 2, probability under H_1 is less than .001.

2. $\chi^2 = 14.084$; df = 1. No, because all the *expected* frequencies are at least 10.

3. $\chi^2 = 0.6$; df = 3; $p = .90$, H_1 must be accepted.

4. $\chi^2 = 413.4$, df = 6; $p < .01$, H_1 is rejected.

Lesson 12

1. *a.* $\chi^2 = 5.25$ with df = 2. The critical value of χ^2 is 5.99; the relationship between family position and letter frequency is not quite significant at the .05 level.

b. φ' is approximately .2, indicating that the variables "family position" and "letter frequency" are not entirely independent in this sample of 60 students. However, the degree of departure from independence is only .2 on a scale where 1.0 represents complete dependence.

c. $\lambda_L = .22$; error in predicting letter frequency is reduced 22 percent when information about family position is used. $\lambda_P = .175$; error in guessing family position is reduced 17.5 percent when information about letter frequency is used. These statistics apply only to this particular sample.

2. *a.* χ^2 with correction for continuity is 3.675, df = 1, $p > .05$. *b.* $\varphi' = \sqrt{.08} = .28$. *c.* $\lambda_C = 0$; $\lambda_R = .266$.

3. *a.* $\chi^2 = 2.27$, df = 4, $p > .05$. *b.* $\varphi' = \sqrt{.01575} = .125$.

4. *a.* .36. *b.* .196. *c.* .45.

Lesson 13

1. Random: choose the six dose levels randomly from the set of numbers 1 through 100. Fixed: study a systematically selected set of levels covering the range 0 to 100 (such as 0, 20, 40, 60, 80, and 100).

2. $T_1 = 1{,}168$; $T_2 = 1{,}113$; $T_3 = 1{,}043$; $G = 3{,}324$; $\sum_1^C \sum_1^n Y_{ij}^2 = 249{,}634$; $SS_W = 3{,}578$; $SS_B = 657$; and $SS_W = 4{,}235$.

Lesson 14

1. SS_W has 42 df, SS_B has 2, and SS_Y has 44; $MS_W = 85.2$, $MS_B = 328.5$, and $MS_Y = 96.2$. The mean squares are not additive because they are ratios (SS/df); division by the degrees of freedom removes the additivity present in the sums of squares.

2. $SS_W = 86{,}431$, $MS_W = 543.6$; $SS_B = 7{,}484$, $MS_B = 1{,}871$; $SS_Y = 93{,}916$, $MS_Y = 706.1$.

3. Estimates are: for μ, 26.34; for a_1, −0.04; for a_2, −5.84; for a_3, −2.84; for a_4, 12.66; for a_5, −3.94. $\sum_1^C a_j = 0$.

Lesson 15

1.

Source	SS	df	MS	F
Between groups	657	2	328.5	3.85 (p < 0.5)
Within groups	3,578	42	85.2	
Total	4,235	44		

2. $F = 3.44$, df 4 and 159; p approximately .01.

Lesson 16

1. Estimated population variance is 1.332; estimated standard deviation of the sampling distribution of the differences between means of paired samples is 0.32. The value of t is -0.719; with df $= 50$, this value does not approach significance even at the .10 level.

2. Estimated σ^2 is $99.2250/9 = 11.025$. Estimated $\sigma_{\bar{Y}}^2 = 1.1025$; estimated $\sigma_{\bar{Y}}$ for samples of size 10 from this population is 1.05. $t = 2.86$; with a two-tailed test, this value of t has a probability between .01 and .02, and H_1 can be rejected at $\alpha = .05$. The limits for rejection at the .01 level are determined by the critical value (for a two-tailed test) with 9 df at that level; the critical t is 3.25. Since estimated $\sigma_{\bar{Y}} = 1.05$, the maximum deviation giving t less than or equal to 3.25 is 3.41. Therefore, the value of $\bar{Y}$ could be as low as 43.59 or as high as 50.41 before we would have to reject H_1 at $\alpha = .01$.

Lesson 17

1. *a.* By a sign test, these 16 positive and 5 negative differences just fail to reach the .05 level in a directional test.

b. By the Wilcoxon test, $T = 23$; this T is significant beyond .005 in a directional test.

c. By the t test, $V = 3.76$, estimated $\sigma_V^2 = 1.53$, estimated $\sigma_V = 1.24$, and $t = 3.0$; with df $= 20$, the difference is significant beyond the .01 level in a two-tailed test and beyond .005 in a directional test.

2. *a.* $p = .005$. *b.* For $T = 60$ (two-tailed), $p = .003$; for $T = 70$ (two-tailed), $p = .007$; for $T = 100$, $p = .027$ (one-tailed) or .054 (two-tailed).

3. $\mu_T = 22.5$, $\sigma_T = 8.44$. At .025, 5.95; at .01, 2.83; at .005, .73. The table gives critical values of 6, 3, and 2, respectively; the normal approximation is poor only at the .005 level.

4. Since $T = 39$ and $E(T)$ also equals 39, H_1 is accepted.

Lesson 18

1. $U = 45$, $U' = 99$. Critical value for a two-tailed test at the .05 level is 37; the hypothesis cannot be rejected.

2. U' (for class I) is 364, U is 164. Mean of the normalized U distribution is 264, variance (without correction for ties) is 2,068, giving a standard deviation of 45.5. The deviation from the mean is 100; in z-score units, it is 2.198. The probability on a one-tailed test is given by the normal-curve table as .0139 for a z score of 2.20; the hypothesis of no difference can be rejected at the .05 level. With the correction for ties, $\sigma_U^2 = 2{,}061$, and $\sigma_U = 45.4$. The z score becomes 2.203, but the probability to two decimal places is still .014. Thus, the correction for ties has no effect on the statistical decision in this case.

3. Both distributions are U-shaped. $U = U_W = 242.5$, with probability (one-tailed) approximately .02.

4. $SS_B = 69{,}270$; $SS_T = 164{,}565$; $MS_T = 3{,}740$; $H = 18.5$. The chi-square table for df $= 2$ shows this H to be significant beyond the .01 level. In the F test, $p = .01$.

Lesson 19

1. Slope constant is -1, Y intercept is 5. $Y = 5 - X$.

2. *a.* $Y = 2.5X$. *b.* $Y = -0.75X + 3$. *c.* $Y = X + 1$.

3. *a* and *d* pass through the origin.

4. *a.* +1.4. *b.* +1.0. *c.* 20.

5. Equation: $Y = 10$. Slope $= 0$, Y intercept $= 10$.

6. $Y = 150 - 30X$.

Lesson 20

1. $r = .825$; $z'_Y = .825z_X$; $SS_Y = 800$; $SS_B = 600$; $SS_W = 200$; $SS_L = 544.5$; $SS_D = 55.5$; $SS_{W+D} = 255.5$; $F = 544.5/5.32 = 102$, with df(1,48).

2. $F = 6$, with df(1,60); H_1 can be rejected at the .05 level but not at the .01 level.

Lesson 21

1. *a.* $\Sigma X = 250$; $\Sigma X^2 = 13{,}000$; $\Sigma Y = 320$; $\Sigma Y^2 = 20{,}800$; $\Sigma XY = 15{,}840$.

b. $\bar{X} = 50$, $\bar{Y} = 64$, $s_X = 10$, $s_Y = 8$, covariance $= -32$.

c. All three methods give $r = -0.4$. The z-score method is the most laborious. Method 2 is most direct provided that you have already calculated the terms required in part *b* of this problem. Method 1 is most convenient when only the terms required in part *a* of this problem have been calculated, i.e., when there is no need to obtain information on means, standard deviations, and covariance.

2. $r = .825$; $Y' = .825X + 5.575$.

3. $Y' = .54X + 23.8$; $X' = .67Y + 3.2$.

4. *a.* 3.0. *b.* 3.2. *c.* $1 - r^2 = .64$; $\Sigma(z_Y - z'_Y)^2/N = 3.2/5 = .64$. *d.* $\Sigma(z_Y - z''_Y)^2 = 3.55$, a value greater than $\Sigma(z_Y - z'_Y)^2$.

5. $r = +.4$; $Y' = 0.83X + 0.41$; $X' = 0.19Y + 1.07$.

Lesson 22

1. 7.

2. Either +.8 or −.8.

3. 0.1369, or about 14 percent. About 86 percent. $s_{Y|X} = .93s_Y$.

4. 16 percent; 84 percent; 16 percent.

5. *a.* +.6. *b.* +.2. *c.* 0. *d.* −.2.

6. By short method, .667; by long method, .665. Since there are ties, the long method is nominally more accurate; however, the difference is very small in this case.

7. $r_s = +.4$. $Y' = .4X + 3$.

REFERENCES

Hays, William L.: *Statistics for Psychologists,* Holt, Rinehart and Winston, Inc., New York, 1963.

Kemeny, John G., J. Laurie Snell, and Gerald L. Thompson: *Introduction to Finite Mathematics,* Prentice-Hall, Inc., Englewood Cliffs, N.J., 1966.

Siegel, Sidney: *Nonparametric Statistics for the Behavioral Sciences,* McGraw-Hill Book Company, New York, 1956.

Winer, B. J.: *Statistical Principles in Experimental Design,* McGraw-Hill Book Company, New York, 1962.

Lessons 1 and 2 Winer, pp. 4–11; Hays, chap. 9, Hypothesis Testing and Interval Estimation, pp. 245–300.

Lessons 3 and 4 *Probability:* Kemeny, pp. 127–139; Hays, chap. 2, Elementary Probability Theory, pp. 47–67. *Conditional probability:* Kemeny, pp. 144–147, 158–163; Hays, chap. 4, Joint Events and Independence, pp. 108–130. *Independent trials with two outcomes:* Kemeny, pp. 165–182.

Lesson 5 *Partitions and counting:* Kemeny, pp. 84–106. *Binomial distribution:* Kemeny, pp. 109–111; Hays, chap. 5, A Theoretical Distribution: the Binomial, pp. 131–156. *Power:* Winer, pp. 12–13; Hays, pp. 269–275.

Lessons 6 and 7 Hays, chap. 3, Frequency and Probability Distributions, pp. 68–107; chap. 6, Central Tendency and Variability, pp. 157–191. *Summation operations:* Hays, appendix A, pp. 659–666. *Expected value:* Kemeny, pp. 189–192. *Categorical, ordinal, and interval data:* Siegel, pp. 21–30.

Lesson 8 Hays, chap. 8, Normal Population and Sampling Distributions, pp. 218–244.

Lesson 9 Hays, chap. 7, Sampling Distributions and Point Estimation, pp. 192–217. *Central limit theorem:* Kemeny, pp. 201–207.

Lesson 10 *Binomial test:* Siegel, pp. 36–42. *Sign test:* Siegel, pp. 68–75; Hays, pp. 625–628.

Lesson 11 *Chi-square distributions:* Hays, pp. 336–348. *χ^2 tests:* Hays, pp. 578–597, 613–614; Siegel, pp. 42–47, 104–111, 175–179.

Lesson 12 Hays, pp. 603–610.

Lessons 13 to 15 Winer, pp. 46–65; Hays, pp. 348–352 (*F distributions*), pp. 356–385 (*fixed-factor experiments*), pp. 413–429 (*random-factor experiments*). *Algebra of expectations:* Hays, appendix B, pp. 666–671. *Planned and post hoc comparisons:* Hays, pp. 459–489.

Lessons 16 and 17 Winer, pp. 14–24; Hays, chap. 10, Inferences About Population Means, pp. 301–335. *Relations among t, F, χ^2, and normal distributions:* Hays, pp. 352–355. *Nonparametric tests:* Hays, pp. 615–620; Siegel, pp. 20–21, 30–34. *Wilcoxon test:* Hays, pp. 635–637; Siegel, pp. 75–83. *Mann-Whitney U test:* Hays, pp. 633–635; Siegel, pp. 116–127. *Kruskal-Wallis H test:* Hays, pp. 637–639; Siegel, pp. 184–193; Winer, pp. 622–623.

Lessons 19 to 22 Hays, chap. 14, Problems in Linear Regression and Correlation, pp. 490–538. *Spearman rank correlation coefficient:* Hays, pp. 641–647; Siegel, pp. 202–213.

INDEX